Computational Intelligence in Systems and Control Design and Applications

edited by

SPYROS G. TZAFESTAS
Department of Electrical and Computer Engineering,
National Technical University of Athens,
Greece

KLUWER ACADEMIC PUBLISHERS
DORDRECHT / BOSTON / LONDON

A C.I.P. Catalogue record for this book is available from the Library of Congress.

ISBN-13: 978-1-4020-0394-3 e-ISBN-13: 978-94-010-9040-7
DOI: 10.1007/ 978-94-010-9040-7
Transferred to Digital Print 2001

Published by Kluwer Academic Publishers,
P.O. Box 17, 3300 AA Dordrecht, The Netherlands.

Sold and distributed in North, Central and South America
by Kluwer Academic Publishers,
101 Philip Drive, Norwell, MA 02061, U.S.A.

In all other countries, sold and distributed
by Kluwer Academic Publishers,
P.O. Box 322, 3300 AH Dordrecht, The Netherlands.

Printed on acid-free paper

Computational Intelligence in Systems and Control Design and Applications

International Series on

MICROPROCESSOR-BASED AND INTELLIGENT SYSTEMS ENGINEERING

VOLUME 22

The titles published in this series are listed at the end of this volume.

CONTENTS

PART II
FUZZY LOGIC-BASED INTELLIGENT ESTIMATION AND CONTROL

PART III
GENETIC ALGORITHM-BASED SYSTEM ANALYSIS AND DESIGN

19 ROUTE SELECTION IN A NEURO-FUZZY VEHICLE NAVIGATION SYSTEM
G.K.H. Pang

20 HIERARCHICAL POSITION CONTROL USING FUZZY LOGIC : OPPORTUNITIES AND LIMITATIONS
A. Zilouchian , T. Jordanides and D.W. Howard

21 IMPEDANCE CONTROL FOR MULTI-DOF MANIPULATOR BY LEARNING ENVIRONMENT MODELS
F. Nagata, K. Watanabe , K. Sato and K. Izumi

PART V
COMPUTATIONAL INTELLIGENCE IN ENGINEERING APPLICATIONS

22 ENGINEERING APPLICATIONS OF N-STEP AHEAD NEUROCONTROL
A. Hountras , I . Antoniadis and A. Kanarachos

23 NEURAL CLASSIFICATION OF DANGEROUS SHIPS IN COLLISION SITUATIONS
J. Lisowski, A. Rak and A. Grzenia

24 NEURAL COMPUTING IN FABRICS INSPECTION SYSTEM
H. Liu, C.-H. Chan, T. Kwan and G. Pang

PREFACE

This book contains thirty timely contributions in the emerging field of **Computational Intelligence** (CI) with reference to system control design and applications.

The three basic constituents of CI are neural networks (NNs), fuzzy logic (FL) / fuzzy reasoning (FR), and genetic algorithms (GAs). NNs mimic the distributed functioning of the human brain and consist of many, rather simple, building elements (called artificial neurons) which are controlled by adaptive parameters and are able to incorporate via learning the knowledge provided by the environment, and thus respond intelligently to new stimuli. Fuzzy logic (FL) provides the means to build systems that can reason linguistically under uncertainty like the human experts (common sense reasoning). Both NNs and FL / FR are among the most widely used tools for modeling unknown systems with nonlinear behavior. FL suits better when there is some kind of knowledge about the system, such as, for example, the linguistic information of a human expert. On the other hand, NNs possess unique learning and generalization capabilities that allow the user to construct very accurate models of nonlinear systems simply using input-output data. GAs offer an interesting set of generic tools for systematic random search optimization following the mechanisms of natural genetics. In hybrid Computational Intelligence - based systems these three tools (NNs, FL, GAs) are combined in several synergetic ways producing integrated tools with enhanced learning, generalization, universal approximation, reasoning and optimization abilities. For example, neuro-fuzzy systems are able to incorporate prior knowledge to the learning mechanisms, and provide a suitable symbolic expression of the learning results in some form of fuzzy rules.

The volume is organized in five parts. Part I deals with the NN-based intelligent solution of estimation and control problems. Part II is concerned with the solution of estimation and control problems via fuzzy logic and fuzzy reasoning. Part III is devoted to the application of GAs for solving system analysis and design problems. Part IV provides a number of hybrid CI techniques applied to system analysis and control, and, finally, Part V presents a set of computational intelligence engineering applications. The particular themes covered by these parts are as follows.

PART I contains five contributions that present new and/or recent results concerning:

- intelligent forecasting and fault diagnosis using neural networks
- parameter identification of linear dynamic systems using single-layer neural networks
- automated PID tuning using three neural networks (MLPs) each one adjusting one of the PID parameters

- NN-based self-tuning control that combines the self-tuning principle with reinforcement learning.
- NN-based PID controller tuning for single-input multi-output systems, with application to the inverted pendulum stabilization.

PART II contains six contributions dealing with new aspects of:
- fuzzy model-based short-term load forecasting (STLF) using orthogonal least squares
- unsupervised FL-based classification of non elliptic classes using fuzzy C-means and the principle of ambiguity rejects
- stability of dynamic fuzzy systems (new theorems are provided for the stability of the Takagi Sugeno model)
- fuzzy-behavior-based control with local learning that combines the subsumption architecture with fuzzy reasoning
- stable-robust fuzzy controllers for robot position control
- stability of fuzzy self-tuning control of nonlinear systems with limit points.

PART III is dedicated to the application of GAs to several practical problems, namely:
- cellular dynamic channel allocation (DCA) for uniform traffic distribution
- workshop cell decomposition following the concept of group technology (GT)
- video content representation (feature extraction) including a scene cut detection mechanism and a color / motion segmentation procedure
- short-term economic dispatch of hydrothermal power systems (SEDHPS)
- inverse kinematics of parallel platforms aiming at determining the leg lengths of the platform actuators and achieving the desired positioning and orientation of the moving platform.

PART IV presents a variety of novel hybrid CI techniques for system analysis and control problems, namely:
- modeling of a biological batch process (consisting of an animal cell culture) via classical mass balance combined with a NN
- estimation of state variables in power systems by combining a mathematical model with a radial basis function network δ-estimator
- dynamic route selection / guidance (DRG) based on a neuro-fuzzy expert system that models the driver's preference
- hierarchical robot position control using an adaptive neuro-fuzzy inference system (ANFIS) for both the inverse kinematics solution and the controller design
- multi-DOF robot impedance control using fuzzy environment models learned through a genetic algorithm.

PART V contains nine contributions dealing with engineering applications of CI, namely:
- application of "chain-back propagation" (CBP) neurocontrol on a 2-axes articulated manipulator and a 3-axes cartesian heavy duty hydraulic manipulator
- application of NN classification to avoid collision of dangerous ships on the basis of ARPA (Automatic Radar Plotting Aids) information

- application of a high performance hybrid neuro-fuzzy technique to automated visual inspection (AVI)
- application of NN-based observers to the control of a fed-batch bioprocess, namely the production of Saccharomyces Cerevisiae (baker's yeast).
- application of a NN-based technique to the model identification of a naval turbocharger
- application of fuzzy reasoning to the optimal design of a semi-active suspension system for ground vehicles modeled by linear differential equations subject to excitation from road surfaces
- application of a fuzzy nonhomogeneous Markov system (F-NHMS) to model and control the dynamics of ecological populations which involve state parameters such as temperature and pollution, and three population parameters, namely: fecundity (F), mortality (M) and longevity (L)
- application of discrete and continuous Hopfield networks to system identification and control
- application of fuzzy estimation to the cooperative control of multi robotic systems.

Taken together, the thirty contributions of this volume provide a good and well-balanced picture of the work currently conducted in the area of applying computational intelligence techniques to system analysis, identification, optimization, and control problems.

The editor is indebted to the contributors for their high-level contributions, their patience through the lengthy editorial process, and for their work in revising and refining the manuscripts.

It is believed that the volume will be useful to the researcher and practitioner in the field, inspiring both senior and young workers towards new achievements and new applications of computational intelligence.

May 1999

Spyros G. Tzafestas
NTUA, Athens, Greece

CONTRIBUTORS

AL-HADITHI B.	DISAM ,U.P. Madrid, Spain
ANTONIADIS I.	Natl. Tech. Univ. of Athens , Greece
ATANASOVA T.	Bulgarian Academy of Sciences, Bulgaria
AVRITHIS Y.S.	Natl. Tech. Univ. of Athens, Greece
BAFAS G.V.	Natl. Tech. Univ. of Athens, Greece
BAKIRTZIS A.G.	University of Thessaloniki , Greece
BILLAUDEL P.	LAM, Univ. of Reims, France
BISKAS D.	University of Thessaloniki, Greece
BOGAERTS PH.	Univ. Libre de Bruxelles; Belgium
BOUTALIS Y.S.	University of Thrace, Xanthi, Greece
CHAN C.-H..	Univ. of Hong Kong, Hong Kong
CHERLET M.	Univ. Libre de Bruxelles, Belgium
DEEKSHATULU B.L.	CSSTEAP, Dehradum, India
DELCHAMBRE A.	Univ. Libre de Bruxelles, Belgium
DeLIT P.	Univ. Libre de Bruxelles, Belgium
DEVILLEZ A.	LAM, Univ. of Reims, France
DOULAMIS A.D.	Natl. Tech. Univ. of Athens , Greece
DOULAMIS N.D.	Natl. Tech. Univ. of Athens , Greece
DOULGERI Z.	University of Thessaloniki , Greece
FALKENAUER E.	Univ. Libre de Bruxelles, Belgium
GION TH.	Natl. Tech. Univ. of Athens , Greece
GOTZIAS N.	Natl. Tech. Univ. of Athens , Greece
GRABEC I.	Univ. of Ljubljana, Slovenia
GRAEFE J.	Univ. Libre de Bruxelles, Belgium
GRZENIA A.	Gdynia Maritime Acad., Poland
HANOMOLO A.	Univ. Libre de Bruxelles, Belgium
HANUS R.	Univ. Libre de Bruxelles, Belgium
HIRAO T.	Saga University, Japan
HOUNTRAS A.	Natl. Tech. Univ. of Athens , Greece
HOWARD D.W.	Florida Atlantic Univ., U.S.A.
IZUMI K.	Saga University, Japan
JIMÉNEZ A.	DISAM, U.P. Madrid, Spain
JORDANIDES T.	California State Univ. , U.S.A.
KANARACHOS A.E.	Natl. Tech. Univ. of Athens , Greece
KIARTZIS S.J.	Univ. of Thessaloniki, Greece
KOLLIAS S.D.	Natl. Tech. Univ. of Athens, Greece
KOSMIDOU O.I.	Univ. of Thrace, Xanthi, Greece
KOULOHERIS D.	Natl. Tech. Univ. of Athens, Greece
KWAN T.	University of Hong Kong, Hong Kong

LISOWSKI J.	Gdynia Maritime Acad., Poland
LIU H.	University of Hong Kong, Hong Kong
LUBENOVA V.	Bulgarian Academy of Sciences, Bulgaria
MASTOROKOSTAS P.A.	Univ. of Thessaloniki, Greece
MATÍA F.	DISAM, U.P.Madrid, Spain
MENG X.	Northeastern Univ. , P.R. China
MISUMI M.	Saga University, Japan
NAGATA F.	Fukuoka Ind. Tech. Center, Japan
NOUR ELDIN H.A.	Univ. of Wuppertal, Germany
OMATU S.	Osaka Prefecture Univ., Japan
PANG G.K.H.	Univ. of Hong Kong , Hong Kong
PANTELELIS N.G.	Natl. Tech. Univ. of Athens , Greece
POTOČNIK P.	Univ. of Ljubljana, Slovenia
RAK A.	Gdynia Maritime Acad., Poland
RODRIGUEZ-TELLEZ J.	Univ. of Bradford, U.K.
ROUSSIS K.S.	Natl. Tech. Univ. of Athens , Greece
RUANO A.E.B.	Univ. of Algarve, Portugal
SAGIAS D.I.	Natl. Tech. Univ. of Athens, Greece
SANDALIDIS H.G.	T.S.I., Tech. Univ. of Crete, Greece
SARAFIS E.N.	Natl. Tech. Univ. of Athens, Greece
SATO K.	Saga University, Japan
SIETTOS C.I.	Natl. Tech. Univ. of Athens, Greece
SIVAKUMAR L.	BHEL R&D, Vikas Nagar, India
STAMOU G.B.	Natl. Tech. Univ. of Athens , Greece
STAVROULAKIS P.P.	T.S.I., Tech. Univ. of Crete, Greece
SYMEONAKI M.A.	Natl. Tech. Univ. of Athens, Greece
TAN W.	Changchun W.C. & E.P. Inst. P.R. China
THEOCHARIS J.B.	Univ. of Thessaloniki, Greece
TZAFESTAS E.S.	Natl. Tech. Univ. of Athens, Greece
TZAFESTAS S.G.	Natl. Tech. Univ. of Athens, Greece
VILLERMAIN LECOLIER G.	Univ. of Reims, France
VOGIATZIS D.	Natl. Tech. Univ. of Athens, Greece
YADAIAH N.	SCDE JNT Univ. , India
YANG X.	Changchun W.C. & E.P. Inst. P.R. China
YOSHIOKA M.	Osaka Prefecture Univ. , Japan
WATANABE K.	Saga University, Japan
WÉRENNE J.	Univ. Libre de Bruxelles , Belgium
ZAPRIANOV J.	Bulgarian Academy of Sciences , Bulgaria
ZHANG H.	Northeastern Univ., P.R. China
ZILOUCHIAN A.	Florida Atlantic Univ. , U.S.A.

PART I

NEURAL NETWORK-BASED INTELLIGENT ESTIMATION AND CONTROL

1

INTELLIGENT FORECASTING AND FAULT DIAGNOSIS USING NEURAL ESTIMATORS

E.S. TZAFESTAS and S.G.TZAFESTAS
Intelligent Robotics and Automation Laboratory
Department of Electrical and Computer Engineering
National Technical University of Athens (NTUA)
Zografou Campus, Zografou, 15773, Athens, Greece

1. Introduction

Neural Networks (NNs) possess important learning capabilities that enable them to provide results with small forecasting/diagnostic error. To generate the forecast and fault diagnosis result the NN is trained with historical data concerning the temporal/operational behavior of the system, i.e. the dynamic relationships between the measured input and output quantities. The NN training is accomplished through the minimization of an index function of the forecasting or the fault detection / classification error.

Among the various NN models available, the *multilayer perceptron* (MLP) using the *back-propagation* (BP) algorithm and the *radial basis function* (RBF) network are the most popular and appropriate (Tzafestas, 1997a).

Among the RBF networks the most popular one for several reasons, is the Gaussian RBF (GRBF) which is capable to approximate accurately mild nonlinear functions for estimation and forecasting purposes. Both types of NNs have been applied.

Two problems encountered when training NNs is the *overtraining problem* in MLPs (Tzafestas et.al.,1996) and the *data reduction* so as to ensure higher efficiency especially in short-term forecasting. Data reduction can be achieved through the *principal component analysis* (PCA) techniques (Baldi and Hornik, 1989; Chen and Chang, 1995) or through heuristic rules embodied in a *knowledge based system* (KBS) (Tzafestas, 1997). Both methods were used and tested. Through error monitoring the KBS can also select new data for retraining the NN.

This paper describes some recent work carried out by the authors on intelligent forecasting and fault diagnosis. Specifically, it is devoted to the neural-network approach combined with knowledge-based techniques. The two NN models used are briefly presented along with the techniques employed for reducing/selecting the training data from among the available data sets. A discussion on intelligent fault diagnosis via NNs is

S.G. Tzafestas (ed.), Advances in Intelligent Systems, 3–16.

provided. The particular implementation architectures employed are outlined and representative experimental results are included.

2. The two neural models

Here the two neural network models used in our work, namely, the *multilayer perceptron* (MLP) / *Back Propagation* (BP) and the *radial-basis function* (RBF) models are briefly presented.

2.1 MULTILAYER PERCEPTRON

Multilayer perceptrons (Figure1) are networks in which the processing units (neurons) are arranged in layers. The nodes of each layer take as input the outputs of the nodes of the previous layer and perform a non-linear (usually sigmoidal) transformation of them in order to produce their outputs. The units belonging to one layer are connected only to the nodes of the previous and the next layer and not to each other. The network learns by adjusting its weights according to the back-propagation algorithm (hence MLPs are often called back-propagation networks), which computes the error between the actual and the desired output (supervised learning) and "backpropagates" it to the units which adjust their associated weight values so that the actual response of the network moves closer to the desired response.

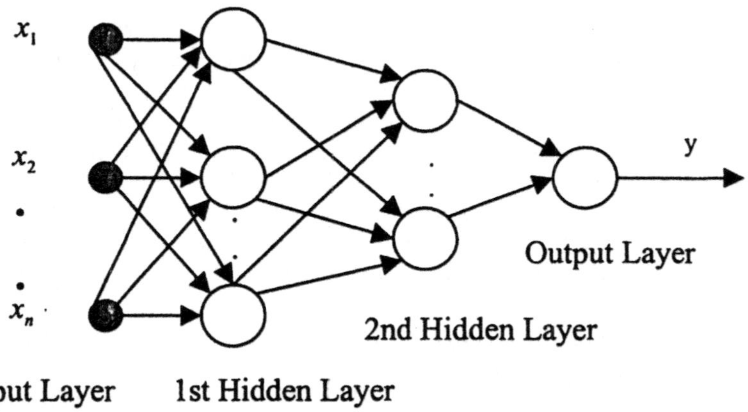

Figure 1 : Structure of a MLP

The output of the p-th node of the network corresponding to the i-th input pattern is equal to: $o_{pi} = \sigma(\sum w_{ij} o_{pj} + \theta_i)$, where σ is usually the sigmoidal function $\sigma = \dfrac{1}{(1+e^{-x})}$.

It has been proved that an MLP trained with the back-propagation algorithm can approximate arbitrarily well every function belonging to L_2 class (Kolgomorov's theorem).

A very important characteristic of the MLP networks is their ability to generalize, i.e. to give valid (meaningful) output for input patterns to which they have not been trained.

2.2 RADIAL BASIS FUNCTION (RBF) NETWORK

An RBF network approximates an input-output mapping using a linear combination of radially symmetric functions (see Figure 2).

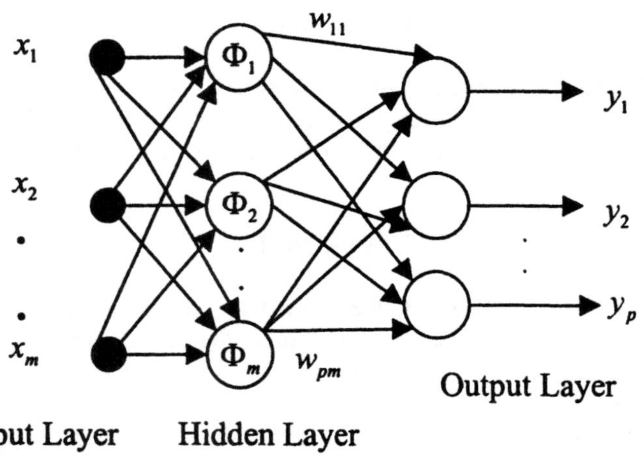

Figure 2 : Structure of the RF network

The kth output y_k is given by ($r_i \geq 0, \sigma_i \geq 0$) :

$$y_k(\underline{x}) = \sum_{i=1}^{m} w_{ki}\Phi_i(\underline{x}),\ \Phi_i(\underline{x}) = \Phi(\|\underline{x} - \underline{c}_i\|) = \Phi(r_i) = \exp(-\frac{r_i^2}{2\sigma_i^2}) \tag{1}$$

The RBF networks have always one hidden layer of computational nodes with nonmonotonic transfer functions $\Phi(\cdot)$. It was shown (Powell,1987) that the choice of $\Phi(\cdot)$ is not very crucial for the effectiveness of the network. In most cases the Gaussian RBF (shown in (1)) is used, where $\underline{c}_i$ and σ_i (i =1,2,....,m) are selected centers and widths respectively. The RBF training procedure is as follows :

Step 1 : Group the training patterns in M clusters (using e.g. the k-means algorithm) and select their centers.

Step 2 : Select the widths σ_i (i=1,2,...,m) using some heuristic technique (e.g. the p-nearest algorithm).

Step 3 : Compute the RBF activation functions $\Phi_i(\underline{x})$ for the training inputs using (1).

Step 4 : Compute the weight vectors $\underline{w}_k = [w_{k1},\ldots, w_{km}]$ by least squares :

$$\underline{w}_k = \underline{A}^+ \underline{b}_k,\ \underline{A}^+ = (\underline{A}^T \underline{A})^{-1} \underline{A}^T$$

where $\underline{A}^+$ is the generalized inverse of the matrix $\underline{A}$ in the matrix form of y_k in equation (1).

3. Neural network based forecasting

Before the application of any forecasting technique (including the NN-based one) the data should be suitably reduced in order to increase the efficiency of the forecasting. Two techniques used in our work are the PCA and the heuristic (KBS) techniques.

3.1 DATA REDUCTION VIA PRINCIPAL COMPONENT ANALYSIS (PCA)

The PCA problem is the following. Given an n-D data vector, find a transformation that allows the representation of the data elements with a reduced number m of principal components (features). If $\underline{x}$ is an n-D random data vector with zero mean value ($E[\underline{x}]$=0), then its projection $\underline{\alpha}$ along the direction of the n-D unit vector $\underline{u}$ is given by $\underline{\alpha} = \underline{x}^T \underline{u} = \underline{u}^T \underline{x}$ with $\|u\| = 1$. The projection $\underline{\alpha}$ is a random variable with zero mean and variance $\sigma^2 = E[\underline{\alpha}^2] = \underline{u}^T \underline{Ru}$ where $\underline{R} = \underline{R}^T = E[\underline{xx}^T]$. The variance σ^2 is a function of u, namely :

$$\Sigma(\underline{u}) = \sigma^2 = \underline{u}^T \underline{Ru}$$

The PCA problem is then reduced to that of finding the directions (unit vectors u) along which $\Sigma(\underline{u})$ has extrema. These unit vectors are found to be the eigenvectors of R, i.e. $\underline{Ru} = \lambda \underline{u}$ where λ is the eigenvalue associated to $\underline{u}$. It can be easily shown that $\Sigma(\underline{u}) = \lambda_j$, j=0,1,2,....,n-1, i.e. the eigenvalue λ_j gives the extreme value of the variance $\Sigma(\underline{u})$. The data vector can be reconstructed using the principal projections $\alpha_j = \underline{u}_j^T \underline{x} = \underline{x}^T \underline{u}_j$ (j=0,1,2,...,n-1) by the formula $\underline{x} = \sum_{j=o}^{n-1} \alpha_j \underline{u}_j$.

The number of characteristics (principal projections) α_j needed for effectively reconstructing $\underline{x}$ can be reduced by omitting the terms that have small variances, i.e. $\underline{x}' = \sum_{j=0}^{m-1} \alpha_j \underline{u}_j$, m<n, where $\underline{x}'$ is the approximation of $\underline{x}$. The variance of the error vector $\underline{e} = \underline{x} - \underline{x}'$ is equal to $\sum_{j=m}^{n-1} \sigma_j^2 = \sum_{j=m}^{n-1} \lambda_j$ where $\lambda_m, \ldots\ldots, \lambda_{n-1}$ are the n-m smallest eigenvalues of $\underline{R}$. It is easy to verify that a simple linear neural network with Hebbian learning can perform the PCA function on the input data. This single neuron model can then be used to produce a multi-neuron single-layer model of the feedforward type with m<n outputs. The first neuron converges to the first principal component, the second to the second principal component, and so on. The above NN-based PCA technique, in particular the so called adaptive learning-based PCA technique (Chen and Chang, 1995), has been employed in our experiments.

3.2 DATA REDUCTION VIA THE KB TECHNIQUE

The values used for data selection refer to the following : the size of the training data set, the data periodicity / seasonality, the training data trend (increasing or decreasing), the forecasting error evolution, the dimension of the NN input array in relation to the forecasting error, and the NN size (the number of hidden layers and neurons) versus the size of training data set and the training epochs required. The rules used for data classification involve rules of creating classes of data sets (limits, trends, input array size etc.), frames and subframes for representing classes and subclasses of data sets with generic and more specific characteristics, and rules for extracting the data set classes out of the frame knowledge structures. The KBS uses the above knowledge to determine a training set (out of the entire data set available) that produces a minimum average forecasting error. This means that a training set with small size is selected which gives the best results for the NN's inputs concerned. This data set, the trained NN and the corresponding forecasting inputs are stored and used again when similar input patterns appear. Through continuous monitoring of the forecasting error the KBS can decide (using appropriate rules) whether it is necessary to retrain the NN with a new updated data set. Of course at each training epoch one must ensure that the BP algorithm renders the forecasting error at the global minimum, otherwise the size of the data set obtained may not be the correct one.

4. Intelligent fault diagnosis

4.1 GENERAL ISSUES

When a fault occurs in a technological system (which may due to many reasons) it is imperative to limit the impact of outages to the minimum and to restore the faulted facilities as quickly as possible. This calls for techniques of identifying the location and type of the fault. This identification process is known as 'fault diagnosis' (FD) (Patton et. al., 1989;Tzafestas,1989). FD is the most fundamental fault handling function of power plant supervisory and control systems such as EMS (energy management systems) and SCADA (supervisory control and data acquisition) systems. The conventional approach to fault diagnosis is the use of a table of possible faults connecting the fault types and the fault locations with the observed symptoms. This approach is suitable for locating and diagnosing simple faults, not covering the cases of multiple faults or undesired operational performance of sensor and control devices. Therefore intelligent fault diagnosis techniques based on the AI and expert system methodologies are required. In this approach the knowledge acquisition process from experts in order to build the diagnostic knowledge base and inference engine plays a primary role (Tzafestas and Tzafestas, 1997). Within IFS the NTUA group is working in the area of developing hybrid expert-system/knowledge-based, neural network-based and fuzzy logic – based FD techniques. An expert system makes inferences to estimate the faulted sections on the basis of conceptual and simplified knowledge and information in almost the same manner as human experts (e.g. expert dispatchers of power systems) do. An FD expert system can answer questions of the type :

- Which is the most likely location where the fault causing the present situation has occurred ?
- How did the fault cause the present situation ?
- What is the best way to face (accommodate) the error ?
- On the basis of current and past data is any fault anticipated to occur?

4.2 NEURAL NETWORKS IN FD

When applied to process fault diagnosis, NN's use sensor measurements and process alarms as their inputs, and they represent particular fault types with their outputs. After determining the association among process and NN input-output characteristics, the learning process takes place. Input vectors of process measurements are introduced to the network, and its response is compared with the associated target vector. The learning phase stops when the selected stop-criterion is satisfied. Once the weights on the network connections are finally adjusted in this phase, new vectors of measurements can be given to the nodes of the network. The degree of the accomplished misclassification depends on the ability of the network to learn and generalize on the given data. The training samples selected must contain signals from every possible fault situation of the plant. But in real environments, this is very difficult to be achieved. Hence the need to develop more stable approaches with increasing generalization capabilities.

In our work the RBF network and an extended dynamic back-propagation (EDBP) network (Tzafestas et. al., 1996) have been used. This EDBP network has one hidden layer of nodes and the sigmoid transition function $\frac{1}{(1+e^{-x})}$ is used. A new energy function is employed namely :

$$E = c_1E_c + c_2E_0 + c_3E_w = c_1\sum_p E_p + c_2\sum \frac{n_0\sum_i O_{ip}^2}{\sum O_{ip}^2 + n_o k_1} + c_3\frac{n_w\sum w_{ij}^2}{\sum w_{ij}^2 + n_w k_2}$$

where O_{ip} is the output of the node i for the pattern p, n_o is greater than or equal to the number of nodes in the hidden layers, and n_w is greater than or equal to the number of weights. In this case, the minimization of E is equivalent to the minimization of $E_c = \sum E_p$ under the constraints $E_o = 0$ and $E_w = 0$, which reduces the network dimension creating dependencies among O_{ip} and w_{ij}. The new algorithm reduces the number of weights and nodes by rendering some of them to zero. For appropriate k_1, k_2, n_w, n_o, the E_w, E_o are equal to the number of weights and nodes. Therefore, by minimizing E_w and E_o one obtains the network with the optimal number of parameters. This implies reduced complexity, reduced computational effort and reduced probability to be trapped at a local minimum. Among the many advantages of this algorithm, is the reduction of the dependence of the training process from the initial weight vector. This property reduces the time needed to achieve convergence.

5. Implementations and representative results

5.1 IMPLEMENTATIONS

5.1.1. Forecasting

A first KB architecture employed for forecasting is as shown in Figure 3 (Tzafestas and Mekras, 1998).

The role of the KB is as described in Section 3.2. The NN is a 2-5-1 network with the tangsigmoid function $\tan sig(x) = (e^x - e^{-x}) / (e^x + e^{-x})$ for the hidden layers and the linear function for the output layer. This architecture was enhanced by closing optimization loops from the NN's size and the number of training epochs. That is , different NNs with less or more hidden networks are tested. In all cases one data set was found that gave the best results for each specific ANN.

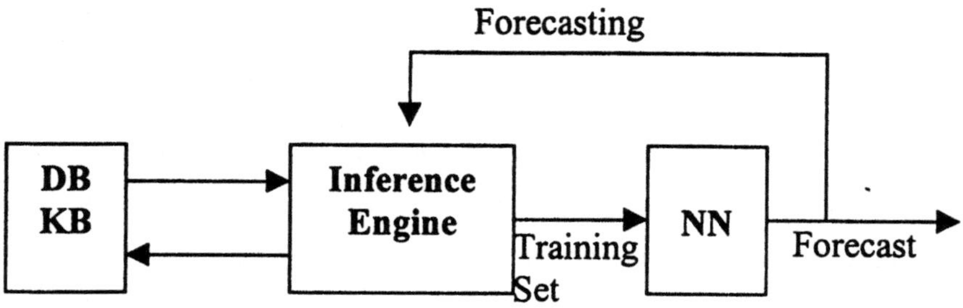

Figure 3 : Implemented neuro-expert forecasting architecture.

5.1.2. Fault Diagnosis

A hybrid knowledge-based and NN architecture has also been tried for fault diagnosis. In this architecture (see Figure 4), the NN is embedded to an expert system as a module that interprets the data received by it and acts as a candidate generator. The expert system interacts with the user, retrieves the sensor measurements, makes the proper data preprocessing, interprets the NNs results, and performs a post-processing diagnostic inference before passing the diagnostic result to the user or to the automatic controllers.

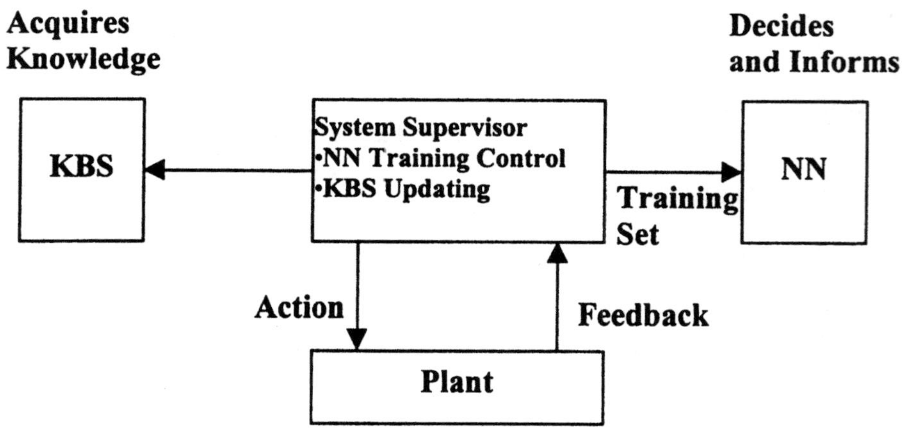

Figure 4 : Hybrid KBS/NN FD architecture.

The system supervisor helps to automate the preprocessing of the data fed to the NN, to control the NN dynamics, to interface the system to the user, and to provide necessary information for the diagnostic system's operation and its performance assessment under various conditions. Rules for scaling the input measurements and associating the output values with the proper responses have to be stored in its knowledge base.

5.1.3. Fault Detection and Accommodation

The NN-based method has also been applied for fault detection and accommodation of interconnected power systems. The hybrid model-based and neural-based architecture used has the form shown in Figure 5.

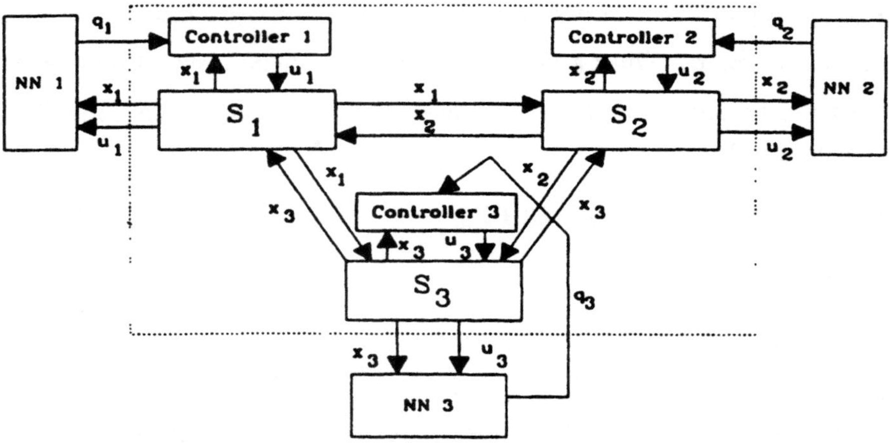

Figure 5 : Hybrid model-based / NN-based architecture for FD and fault accommodation.

The system in general involves N interconnected power subsystems $S_1, S_2, \ldots, S_N$ with their own decentralized optimal controllers:

$$\underline{u}_i = -\underline{K}_i \underline{x}_i, \quad i = 1,2,\ldots..,N$$

which minimize local cost functions $J_i = \frac{1}{2}\int_0^\infty \left(\underline{x}_i^T \underline{Q}_i \underline{x}_i + \underline{u}_i^T \underline{R}\underline{u}_i\right)dt$ and guarantee stability of the overall system $S = \{S_1, S_2, \ldots, S_N\}$ which is described by $\dot{\underline{x}} = \left[\hat{\underline{A}} + \tilde{\underline{A}}\right]\underline{x} + \underline{B}\underline{u}$ where :

$$\underline{x} = \left[\underline{x}_1^T, \underline{x}_2^T, \ldots \underline{x}_N^T\right],\ \underline{x}_i \in R^n,\ \underline{u} = \left[\underline{u}_1^T, \underline{u}_2^T, \ldots \underline{u}_N^T\right],\ \underline{u}_i \in R^m$$

$$\hat{\underline{A}} = block - diag\left[\underline{A}_1, \underline{A}_2, \ldots, \underline{A}_N\right],\ \underline{\mathrm{B}} = block - diag\left[\underline{B}_1, \underline{B}_2, \ldots, \underline{B}_N\right]$$

and $\tilde{\underline{A}}$ is the interconnection matrix :

$$\tilde{\underline{A}} = \begin{bmatrix} \underline{0} & \underline{A}_{12} & \cdots & \underline{A}_{1N} \\ \underline{A}_{21} & \underline{0} & \cdots & \underline{A}_{2N} \\ \vdots & \vdots & \ddots & \\ \underline{A}_{N1} & \underline{A}_{N2} & \cdots & \underline{0} \end{bmatrix}$$

Each subsystem S_i has an associated neural identifier which accepts data $\{\underline{u}_i, \underline{x}_i\}$ and provides estimates of the subsystem's state space matrices A_i, B_i in the form of a vector q_i. When a fault is detected, the nonlinear model provided by the estimator is linearized and the new parameters q_i are passed to the controller (fault accommodation).

5.2 EXPERIMENTAL RESULTS

5.2.1 Forecasting

The neuro-expert technique of Figure 3 was applied to several cases. Here we give the results obtained for the case of a gas consumption DB (Gonzales, 1992). The NN was trained with a 4-year data set and tested for a 6-month forecasting interval. The input data array for each period involves : (i) month at period t, (ii) monthly average of minimum temperatures, (iii) monthly gas consumption at period t, (iv) monthly gas consumption at period t-1, and monthly gas consumption at period t-2. For the 2-5-1 NN used, the mean absolute error (MAE) obtained was the best for a relatively small data set and with data excluded (rejected) by the KB rules. The results obtained with the entire data set or with no data rejection were worst as shown in Figure 6. The best data set found is N=18. The KB method actually starts with a small data set size and increasing the size finds the data set that gives the minimum error. The method was also tested using different NNs with less or more hidden neurons. In all cases one data set was found that gave the best result for each specific NN. Further enhancement will include as parameters under selection the NN size and the maximum number of training epochs at each training versus the size of the training data set and the mean absolute error. The diagrams in Figure 6 show successive forecasting results corresponding to several data sets selected by the KBS and the way the forecasting results have matched the real demand at each month of the forecasting period.

5.2.2. Fault Diagnosis

The neuro-expert FD technique of Figure 4 has been applied to several plants. Here a discussion is provided on its application to the diagnosis of specific subprocesses of a Rayon manufacturing plant where a particular SCADA system (GENESIS™) was

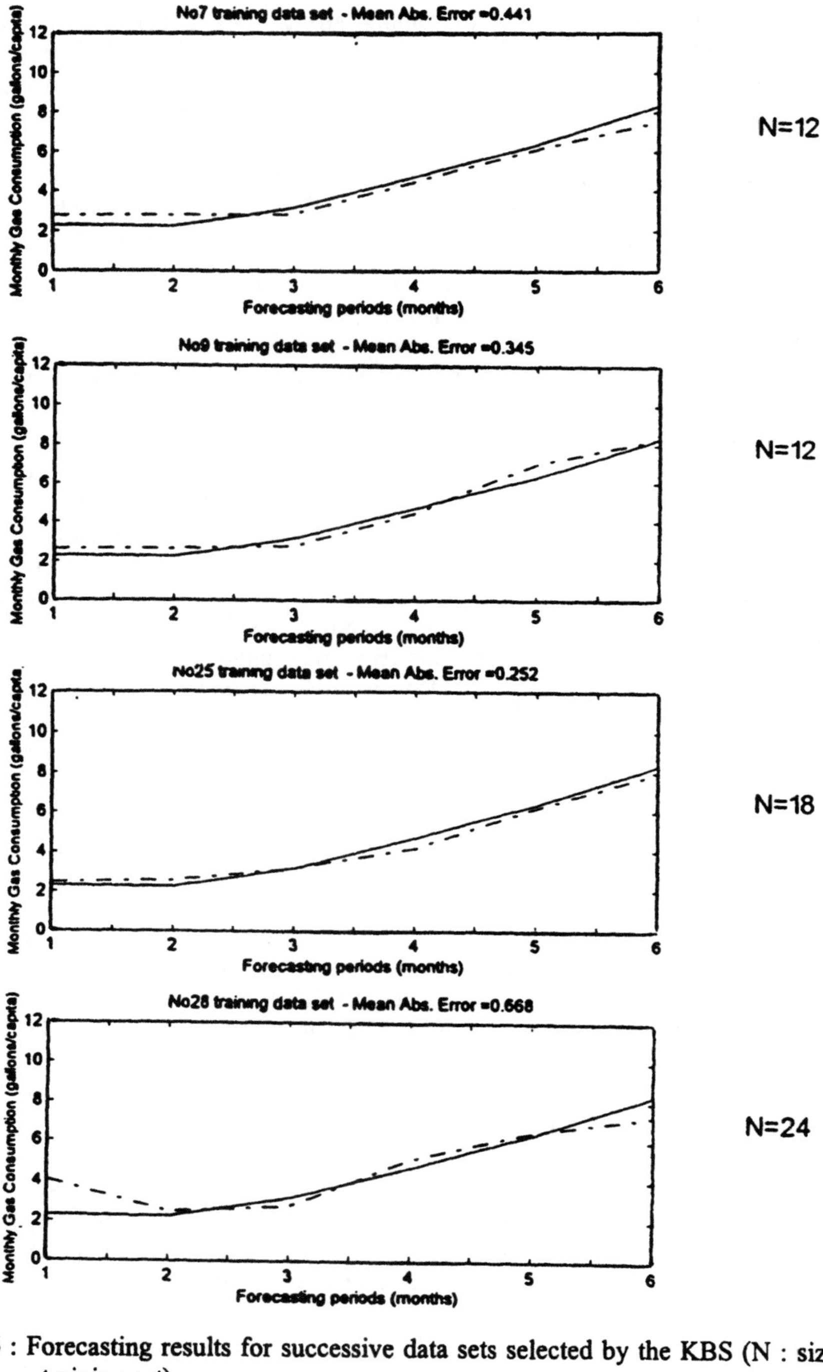

Figure 6 : Forecasting results for successive data sets selected by the KBS (N : size of training set).

installed with the collaboration of the authors' group. Two classes of experiments were performed; the first with the EDBP network and the second with the RBF network.

To find out the best EDBP network size for a specific case study, experiments were carried out with 20 to 150 hidden nodes, as initial parameters. It was found that the algorithm performs better when the initial hidden nodes are more than 100. Then the EDBP network reduced them to 60-80 nodes, depending on the situation. The supervisor was adapted to the requirements of this type of network. Every 342 loops during training the adaptation error was measured and if it was found smaller than 20%, all the connections with magnitudes less than 10^{-7} were pruned. After this the training was continued. The experiments showed that the EDBP approach performs well in most cases : over 95% for known faults, between 85% and 90% when noise was present. Although the training time seems yet to be very considerable, the EDBP algorithm is much faster than the standard BP algorithm and provides lower error rates.

Now, concerning the RBF network-based FD we first conducted experimental work to find the best number h of RBF units, since the RBF performance depends on this number. The network was implemented in the usual way and the k-means clustering algorithm was used to find a set of h RBF unit centers and a partition of the data set into h subsets, where h was determined via the p-nearest neighbor technique. The training was again supervised by the expert system, in which some rules regarding the task were incorporated. The training was again much faster than standard BP. The smaller misclassification value (diagnostic error) was found to be 15% and was obtained with an RBF network having 120 hidden nodes. When noise was added, it showed a performance of 83% successful diagnosis, and only 77% of the faults were detected when dealing with multiple fault cases. Summarizing, the EDBP technique provides an average of 95% success in fault identification, while RBF goes near 85%. However RBF still remains faster, especially because of the less training effort.

5.2.3. Fault Detection and Accommodation

The hybrid NN / model-based FD technique of Figure 5 was applied to an interconnected power system example with three stations (Yu and Siggers, 1971). The initial (start-up) conditions were assumed to have zero values and the optimal decentralized controllers were used to bring each subsystem to its associated setpoint. At t=28, parameter disturbances (representing faults) were introduced at subsystems 1 and 2, and the outputs of the three subsystems begun to deviate. After some time steps, the neural estimators derived a model of sufficient accuracy which was able to bring the system back to set point (fault accommodation). Representative performance plots are shown in Figure 7. The occurrence of long transient phenomena are due to both model mismatch and the time delay of neural estimation. However, although the learning period takes a little time, satisfactory tuning is achieved after acceptable time periods. Further work will be conducted to investigate the effectiveness of the technique for repeated parameter variations (faults).

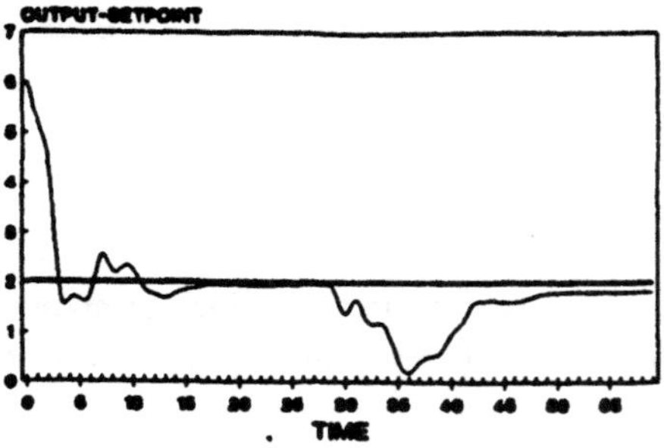

(a) Subsystem 1

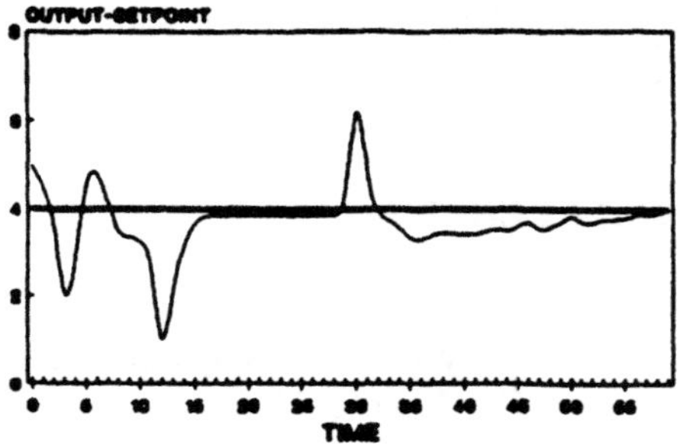

(b) Subsystem 2

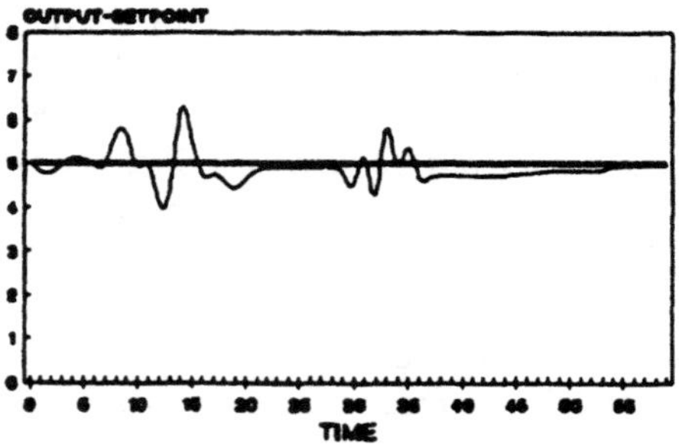

(c) Subsystem 3

Figure 7 : State trajectories of the three subsystems.

6. Conclusions

The work carried out by the author's group within the IFS European Thematic Network involves the investigation of neural networks, fuzzy reasoning and genetic/evolutionary algorithms for state forecasting and fault diagnosis. On the theoretical side the following have been completed (besides the issues presented in this report) :

- Adaptive-predictive control design for LSS using neuro-fuzzy identifiers for the interconnection variables (Tzafestas et. al., 1998).
- Robust fuzzy controller design for industrial plants (including power systems) (Tzafestas and Rigatos, 1998).
- Development of a general fuzzy inference method for system modelling, forecasting and decision making (Stamou and Tzafestas, 1998).
- Development of a meta-algorithm combining genetic and neural techniques, suitable for evolutionary (forecasting/fault diagnosis) computation. (Raptis and Tzafestas, 1998).

Jointly with HRA (Hellenic Aspropyrgos Refineries) NTUA has collected initial operational data and worked on the preprocessing of them (erroneous data rejection, data reduction) in view of their use for forecasting and fault diagnosis purposes. These data are currently used for testing the methodologies already developed or to be developed by the authors' group.

REFERENCES

[1] Baldi, P. and Hornik, K. (1989). Neural Networks and principal component analysis : Learning from examples without local minima, *Neural Networks*, Vol. 2,pp 53-58.

[2] Chen,L-H. and Chang,S. (1995). An adaptive learning algorithm for principal component analysis, *IEEE Trans. on Neural Networks*, Vol. 6, No. 5,pp1255-1263.

[3] Gonzales, L. (1992). Report on Temperature Effect on Gas Volumes, *Dept. of Finance and Qualitative Analysis*, Otago Univ., New Zeland

[4] Patton, R.J., Frank, P.M. and Clark, R.N.,eds. (1989). *Fault-Diagnosis in Dynamic Systems : Theory and Applications*, Prentice Hall.

[5] Powell, M.J. (1987). Radial Basis Functions for Multivariable Interpolation; A review, In : J.C. Mason and M. G. Cox (eds), *Algorithms for Approximation*, Oxford, UK, pp 143-167.

[6] Raptis, S.N. and Tzafestas, S.G. (1998). A Blueprint for a Genetic Meta-Algorithm, *EUFIT'98 : Fuzzy Logic Neural Networks-Evolutionary Computation Conf.*, Aachen, Germany, Sept. 7-10.

[7] Stamou, G.B. and Tzafestas, S.G. (1998). A method for fuzzy system modelling and fuzzy decision making, AII'98 : *Intl. Conf. On Artificial Intelligence in Industry : From Theory to Practice*, High Tatras, Slovakia, April 22-24.

[8] Tzafestas, S.G.,ed. (1997 a). Methods and Applications of Intelligent Control, *Kluwer*, Dordrecht/Boston

[9] Tzafestas, S.G., ed. (1997 b). Knowledge Based Systems : Advanced Concepts, Techniques and Applications, *World Scientific*, Singapore/London.

[10] Tzafestas, S.G., Dalianis, P.J. and Anthopoulos, G. (1996). On the overtraining phenomena of backpropagation neural networks, *Math. & Computers in Simul.*, Vol. 40, No. 5-6, p 507-521.

[11] Tzafestas, S.G. and Tzafestas E.S. (1997). Knowledge Acquisition : A Comprehensive Overview of Methods and Tools, *Studies in Informatics and Control*, Vol. 6, No. 2, pp. 181-214.

[12] Tzafestas, S.G., Kyriannakis, E. and Apostolikas, G. (1998). Decentralized predictive control of large scale systems using neuro-fuzzy identifiers for their interconnections, *CAEA'98 : First IFAC Workshop on Intelligent Control Applications of Ergonomics in Agriculture*, Athens, Greece, June 14-17

[13] Tzafestas, S.G. and Mekras, N. (1998). Forecasting using knowledge based techniques and artificial neural networks, *EURISCON'98 : Third European Robotics, Intell. Systems and Control Conf.*, Athens, Greece, June 22-25

[14] Tzafestas, S.G. and Rigatos, G.G. (1998). A simple robust fuzzy-logic sliding-mode controller of the diagonal type, *BASYS'98 : Third IEEE/IFIP Intl. Conf. on Balanced Automation Systems*, Prague, Aug. 26-28

2

PARAMETER IDENTIFICATION OF DYNAMICAL SYSTEMS USING NEURAL NETWORKS

N. YADAIAH*, L. SIVAKUMAR** and B.L.DEEKSHATULU***

** SCDE, JNT University*
Mahaveer Marg
HYDERABAD-500 028
Andhra Pradesh, INDIA
***BHEL(R&D), Vikas Nagar*
HYDERABAD-500093
Andhra Pradesh, INDIA

**** CSSTEAP*
IIRS Campus
No. 4, Kalidas Road
Dehradun - 248 001
Utter Pradesh, INDIA

1.0 Introduction:

The application of neural networks for parameter identification of linear dynamical systems is considered. For the past few years extensive research is going on in the area of application of neural networks to Control Systems Problems, such as System Identification, Adaptive Control and Function Approximation etc. Satyendra Bhama etal. [2], used a single layer neural network for parameter identification. Kumpati S. Narendra etal. [3] demonstrated that neural networks can be used effectively for the identification and control of nonlinear dynamical systems. S. Reynold Chu, etal. [4], developed Hopfield Network for System Identification. It is used to minimize the least mean square of error rates of estimated state variables. John Kuschenski etal. [5] proposed adalines , two - layer and three – layer feed forward neural networks for identification and control of dynamical systems. Sung – Woo Kim etal [6] , suggested a multi-layered neural network for unknown parameter identification. K.J. Hunt , etal. [7], presented some basic ideas and techniques of artificial neural networks for applications of a variety of neural networks architectures in control. Kavin Warwick [9] proposed a multi – layer perception and Radial basis function neural networks for system identification and control.

The main aim of this chapter is to identify parameters of a dynamical system using single layer neural networks. An improved technique over [2] is proposed for identification. It can be used for on line identification.

2.0 Statement of the Problem:

Consider the following dynamical system

$$\dot{X}(t) = A X(t) + B U(t) \qquad (1)$$

S.G. Tzafestas (ed.), Advances in Intelligent Systems, 17–26.

Where u(t), X(t) $\in R^n$ are the input and state at time t.

Our aim is to determine the elements of A and B, i.e., a_{ij} and b_i where i = 1,2, .. n and j = 1,2, ...n, of the unknown system with given values of state vector X(t), derivative of state vector X(t), and input u(t). A single layer neural network with fast learning algorithm is used for its training.

3.0 Mathematical Formulation:

3.1 Formulation of Error Function:

Let us consider the state variable model of linear dynamical system, which is described by equation (1). The problem is to determine the elements of A and B, of the unknown system with given values of state vector X(t), derivative of X(t) and input u(t).

In order to optimize the network, its error performance index is formulated in such way that it is quadratic in terms of the parameters estimated. If A_e and B_e represent the estimated values of A and B, then the dynamics of the estimated system can be expressed as

$$\dot{X}_e(t) = A_e X(t) + B_e U(t) \tag{2}$$

Let us define an error e(t) between the state vector and estimated state vector as

$$e(t) = X(t) - X_e(t) \tag{3}$$

Assuming a single input u, we are interested only in the error rate. which is defined as

$$\dot{e}(t) = \dot{X}(t) - \dot{X}_e(t) \tag{4}$$

$$\dot{e}_i = \dot{x}_i - \sum_{j=1}^{n}(a_{eij}x_j + b_{ei}u) \tag{5}$$

$$\dot{X}_e(t) = W_i^T X(t) \tag{6}$$

$$W_i = [a_{ei1}(t) \quad a_{ei2}(t) \quad a_{ei3}(t) \; . \quad . \quad . \; a_{ein}(t) \quad b_{ei}]^T \tag{7}$$

Equation (5) can be expressed as

$$\dot{e}_i(t) = \dot{x}_i(t) - W_i^T X(t) \tag{8}$$

Where $X = [x_1(t) \quad x_2(t) \quad x_3(t) \quad . \quad . \quad . \quad x_n(t) \quad u(t)]^T$ (9)

Now we define the error function of our network for over a period of time as

$$E = \frac{1}{2T}\int_0^T (\dot{x}_i - \dot{x}_{ei})^2 \, dt$$

$$= \frac{1}{2T}\int_0^T (\dot{e}_i)^2 dt = \frac{1}{2T}\int_0^T (\dot{e}_i)^{\mathrm{T}}(\dot{e}_i) dt \qquad \text{or}$$

$$E = \mathrm{tr}\left(\frac{1}{2T}\int_0^T (\dot{e}_i)\,(\dot{e}_i)^T dt \right) \tag{10}$$

Where tr represents the trace of the matrix.

3.2 Estimation of parameters of a dynamical systems:

Using the above error function , substituting the value of $\dot{e}_i(t)$ from equation (8) into the equation (10), we will get the following expression

$$E = \mathrm{tr}\left(\frac{1}{2T}\int_0^T \left(\dot{x}_i - W_i^T X\right)\left(\dot{x}_i - W_i^T X\right)^T dt\right) \tag{11}$$

The purpose of Single Layer Neural Network(SLNN) is to optimize the values of coefficients a's and b's i.e., W_i's.

Simplifying the equation (11) , we get

$$E = \mathrm{tr}\left(\frac{1}{2T}\int_0^T \left(\dot{x}_i \dot{x}_i^T + W_i^T X X^T W_i - 2\dot{x}_i X^T W_i \right) dt \right) \tag{12}$$

The gradient of error function w.r.t W_i written as

$$\nabla_i = \frac{1}{T}\int_0^T X X^T W_i \, dt - \frac{1}{T}\int_0^T X \dot{x}_i^T dt \; ; \quad \forall_i \tag{13}$$

In discrete form the equation is expressed as

$$\nabla_i = \frac{1}{N}\sum_{l=1}^{N} X X^T W_i - \frac{1}{N}\sum_{l=1}^{N} X \dot{x}_i^T \; : \quad \forall_i \tag{14}$$

Where N represents the number of samples considered for the purpose of simulation.

$$W_i(l+1) = W_i(l) + \lambda \nabla_i \tag{15}$$

Where λ is constant and it is less than 1.0.

3.3 Proposed Method:

A single layer neural networks is used for parameter identification based on equations (11) to equations (15) by [2]. The significant differences between their's and our proposed method are as follows:

(1) In a dynamical system if there are N discrete samples for each of derivative state vector, state vector and inputs, the parameters of the system are updated based on mean square error computed using N pairs of estimated and observed values. Where as in the proposed method the parameters are updated based on the mean square error values computed using P pairs of estimated and observed values. The mathematical relation for the proposed method as follows:

$$\nabla_{ik} = \frac{1}{P}\sum_{l=1}^{P} XX^T W_i - \frac{1}{P}\sum_{l=1}^{P} X \dot{x}_i^T \; ; \qquad \forall_i \qquad P = \frac{N}{K} \tag{16}$$

Where K is the number of subintervals.

The condition of optimization of the SLNN is determined by setting of gradient of error function equal to zero, i.e.,

$$\nabla_i = \sum_{k=1}^{K} \nabla_{ik} = 0 \tag{17}$$

(2) In the earlier method, the learning rate is kept constant, where as in the proposed method the learning rate is updated based on the mean square error value computed using all observed and estimated values. The revised update rule is as follows:

$$W_i(l+1) = W_i(l) - \lambda(t)\,\nabla_{ik} \tag{18}$$

$\lambda(t)$ is the positive definite matrix.

(3) Instead of merely updating the parameters based on mean square error values, the proposed method provides an intelligent check over the rate of at which the parameters are converging and the parameters are updated with the quantum increments using the gradients of the parameter trends. Let us say, the checks are at 10th iteration, W_i^1, at 20th iteration W_i^2 and at 30th iteration W_i^3. Suppose the direction of change in parameters are same, then the following relation is used for updating parameters as

$$W_i(l) = W_i(l) + g\,(W_i^3 - W_i^2) \tag{19}$$

Where 'g' is a scalar quantity. The repetition will be continued with the following relations.

The mean square error (MSE) is computed as $\left\|\dot{x} - \dot{x}_e\right\|$, where $\|\,.\,\|$ represent the norm. All the weight components are computed this way by each network N_i and the networks are allowed to iterate till MSE is less than a threshold ε, i.e.,

$$\left\| \dot{X} - \dot{X}_e \right\| \leq \varepsilon. \qquad (20)$$

If the condition in (20) is not satisfied, update the weight as

$$W_i(l) = W_i(l+1) \qquad (21)$$

4.0 Architecture of Single Layer of Neural Network:

As shown in Fig.1, there are n, i.e., N_1, N_2,. . N_n, neural networks circuits, each neural network circuit N_i consisting of (n+1) states . A simple analog processor could represent each neuron. There will be n summers (analog adders), one buffer, and one comparator. For example, a second-order (n=2) system will have two parallel networks N_1 and N_2, each consisting of n+1=3 neurons; total of 6 neurons {n(n+1)=6} in the whole system with zero bias. Each neuron represents the adaptable weight corresponding to one of the parameters to be determined (Fig.2). There are two summers, one buffer, and one comparator.

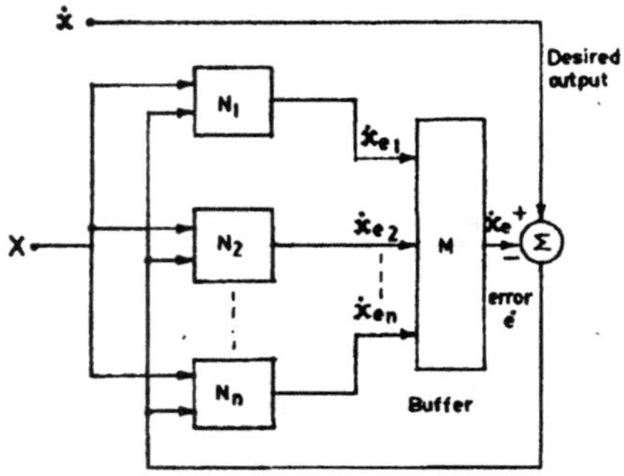

Fig .1. Architecture of a single layer neural network (SLNN)

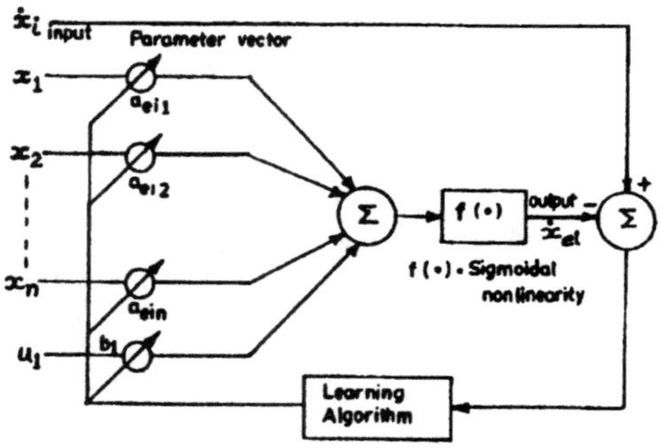

Fig.2. Internal structure of an neural network N_I in SLNN.

5.0 Flow Chart:

Flow chart is given in Fig.3 illustrating the calculation procedure for parameter identification with fast convergence.

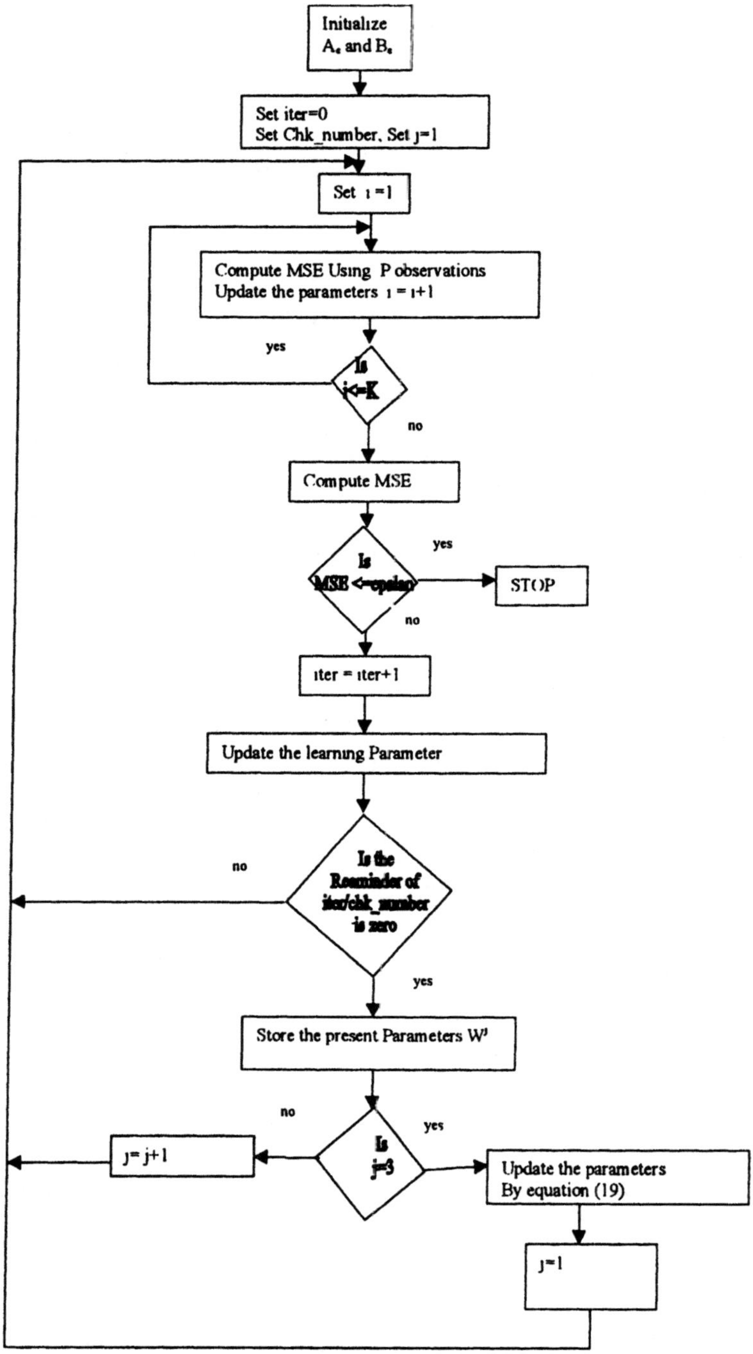
Initialize
A and B
Set iter=0
Set Chk_number, Set j=1
Set i=1
Compute MSE Using P observations
Update the parameters i = i+1
yes
Is
i<=K
no
Compute MSE
yes
Is
MSE <=epsilon
STOP
no
iter = iter+1
Update the learning Parameter
no
Is the
Reaminder of
iter/chk_number
is zero
yes
Store the present Parameters W
no
yes
j= j+1
Is
j=3
Update the parameters
By equation (19)
j=1

Fig. 3

6.0 Simulation Results:

In this section, we show that our algorithm is faster than already existing algorithm for determining the parameters of A and B of unknown second order dynamical systems whose state vector X(t), derivatives of state vector $\dot{X}(t)$ and the input u(t) are given. We have considered two different types of systems with different inputs for simulation. The algorithms were tested for different types of systems. The function of SLNN is to minimize the least square error between the $\frac{dx(t)}{dt}$ and $\frac{dx_e(t)}{dt}$ and thus ensure the optimization of unknown parameters. We observed that, whatever be the initial conditions and learning rate our algorithm works well

Example 1: Consider the following second order system:

$$\dot{X}(t) = A X(t) + B U(t)$$

Where $A = \begin{bmatrix} -0.94248 & 12.566 \\ -12.566 & -0.94248 \end{bmatrix}$, $B = \begin{bmatrix} 1.0 \\ 2.0 \end{bmatrix}$

In general in order to apply this technique, u, x, $\dot{x}$ should be known in advance However, in order to verify the computed results, we have selected different inputs, such as unit ramp and unit step with same values A and B for generation of the state vector X(t), derivatives of state vector $\dot{X}(t)$. Later, with known values of u, X(t) and $\dot{X}(t)$ we have applied our technique to determine the parameter of the system. The simulation results are shown the Fig .4.

Example 2: Consider the following second order system

$$\dot{X}(t) = A X(t) + B U(t)$$

Where $A = \begin{bmatrix} 0 & 1 \\ -2 & -3 \end{bmatrix}$ and $B = \begin{bmatrix} 2.0 \\ 0.0 \end{bmatrix}$

For ramp and step inputs, state vector, derivative of state vector were generated with zero initial conditions. With this information we applied our technique to identify the parameters of system and compared the results with original parameters. The simulation results are shown in Fig.5.

7.0 Conclusions:

A fast gradient decent technique for the parameter identification of a linear dynamical system has been considered. The mathematical formulation of the technique has been illustrated. The computer simulation results show that our method is works well in terms of faster convergence and irrespective of the choice of initial conditions.

	With Unit Step Response	With Unit Ramp Response
a_{11}	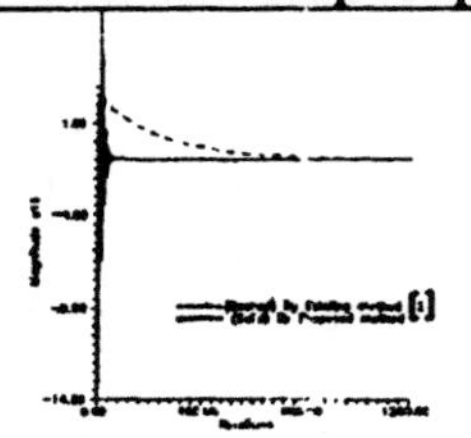	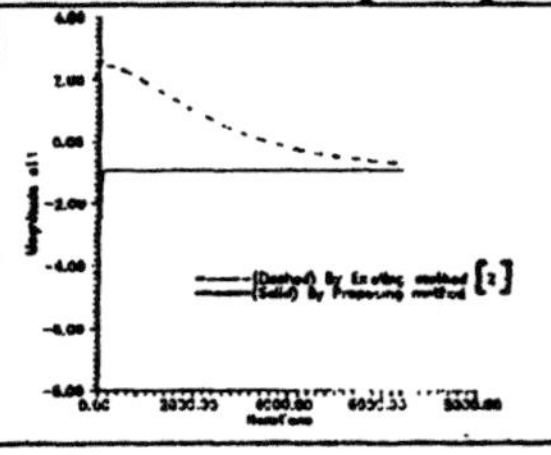
a_{12}	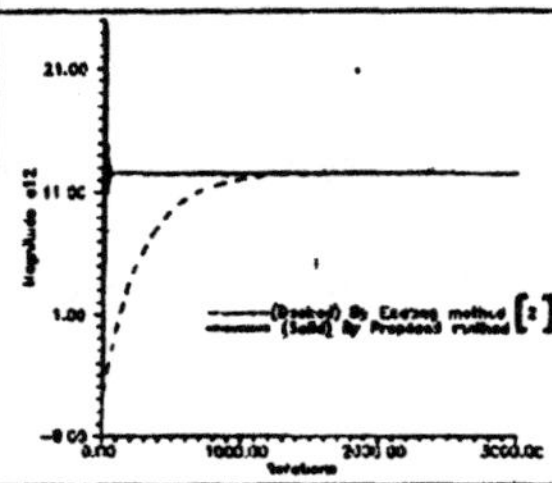	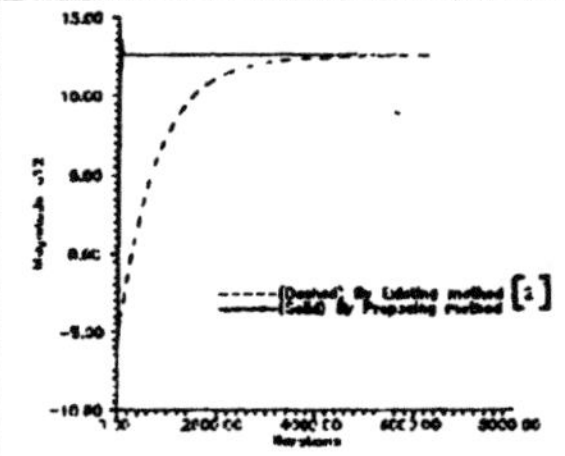
a_{21}	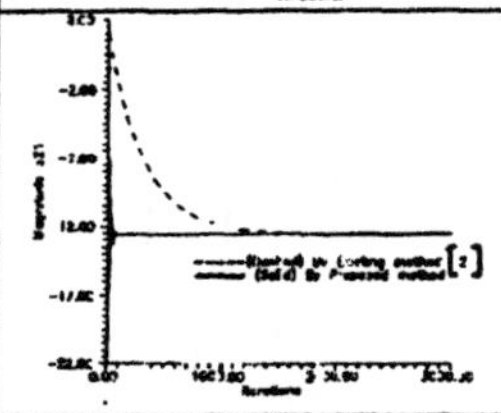	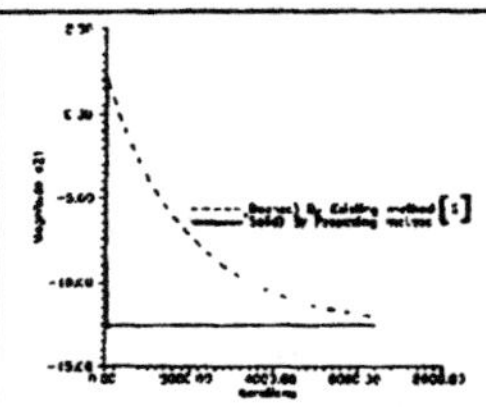
a_{22}	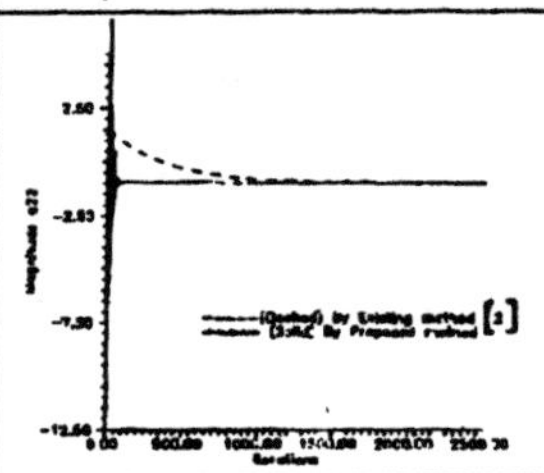	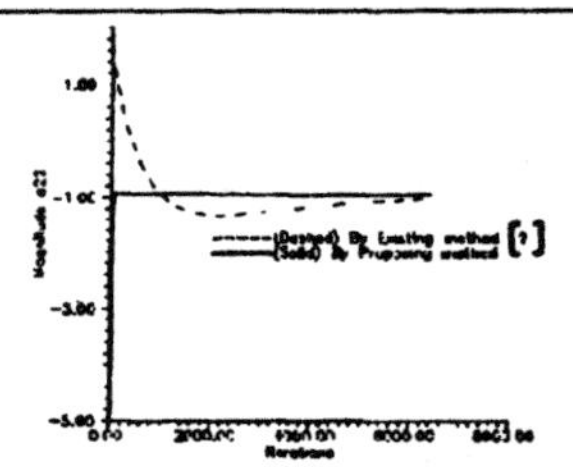
b_1	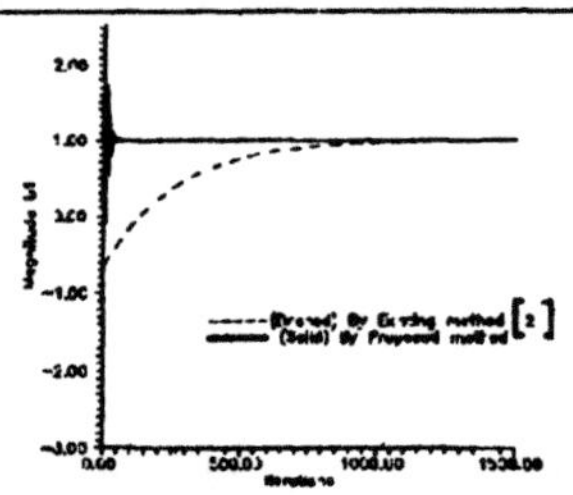	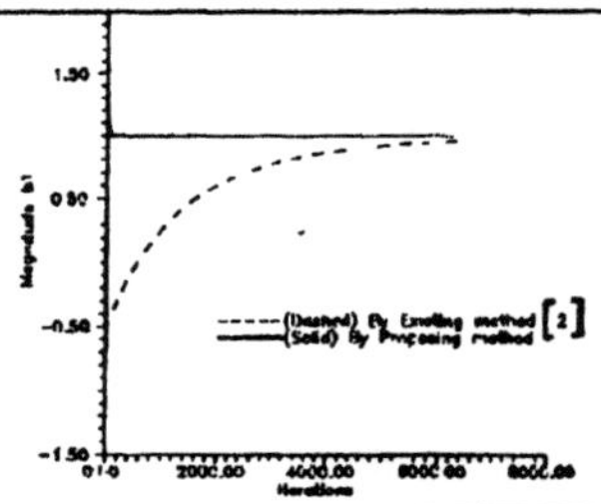
b_2	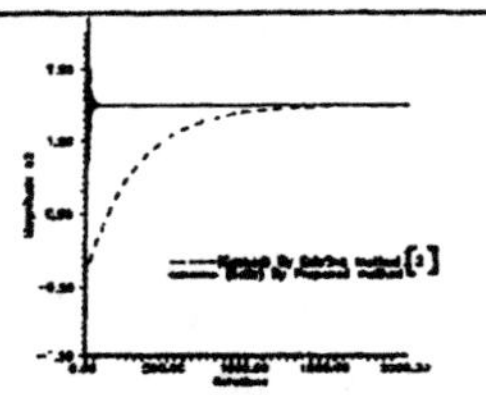	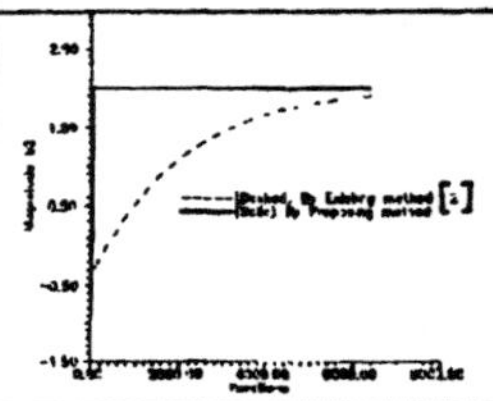

Fig.4Parameter curves of Example1

	With Unit Step Response	With Unit Ramp Response
a_{11}	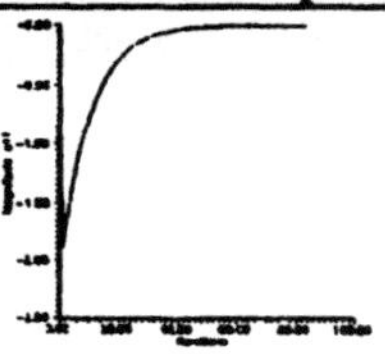	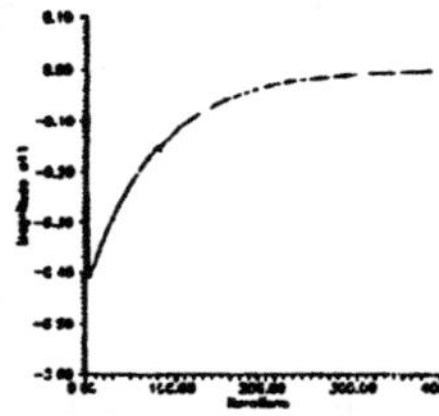
a_{12}	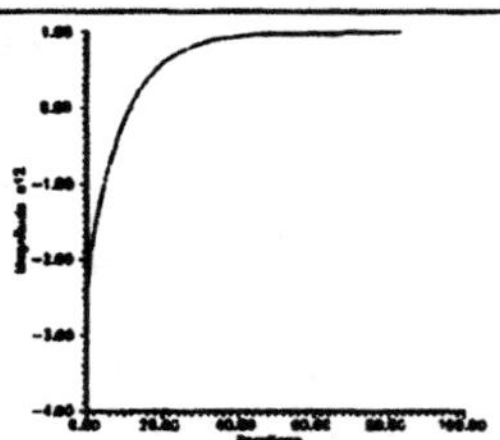	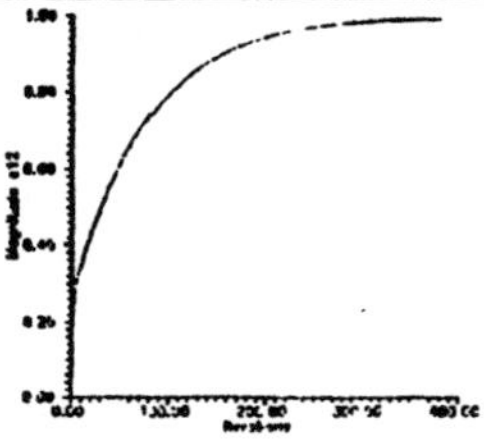
a_{21}	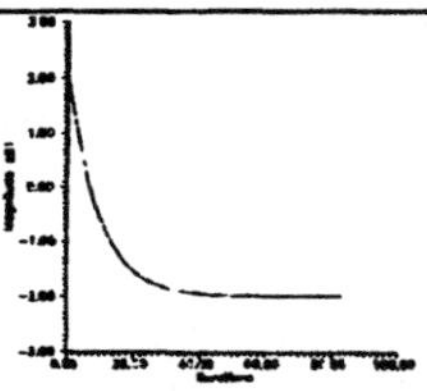	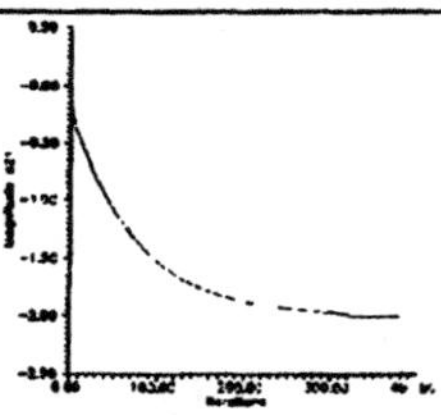
a_{22}	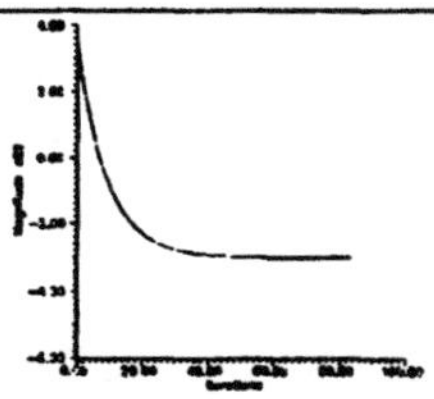	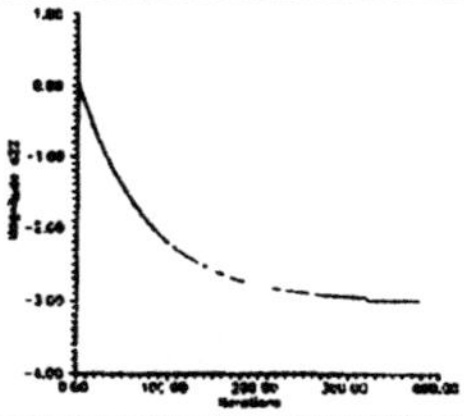
b_1	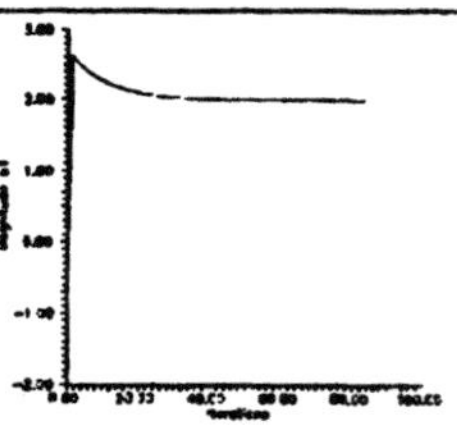	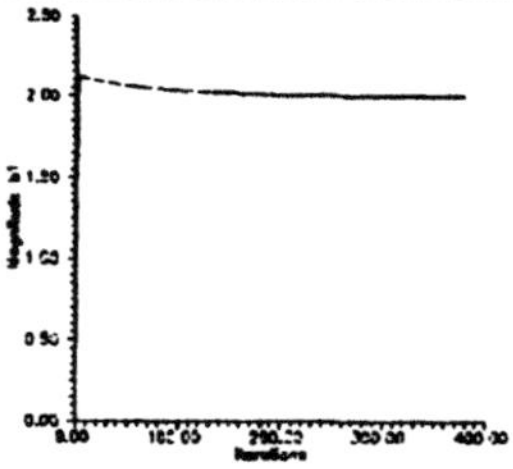
b_2	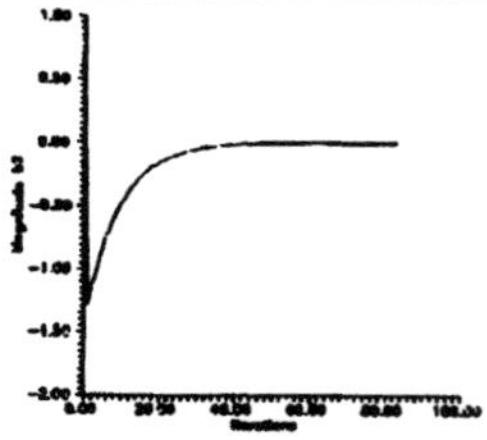	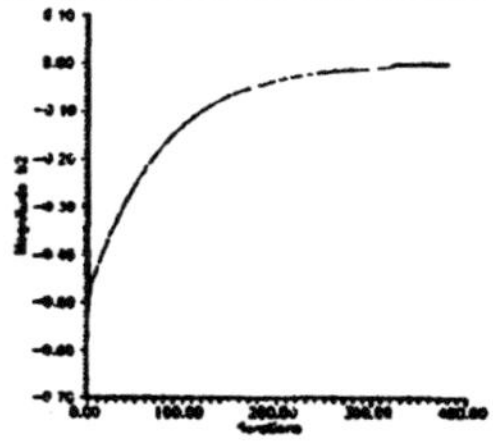

Fig.5 Parameter curves of example2

8.0 References:

1. W.T.Miller, R.S.Sutton, and P.J. Werbos, (eds), Neural Networks for Control", MIT Press,1990.
2. Satyendra Bhama and Harpreet Singh, "Single Layer Neural Networks for Linear Systems Identification Using Gradient Descent Technique", IEEE Trans. On Neural Networks, Vol.4, No.5, September 1993.
3. Kumpati S. Narendra and Khannan Parthasarathy, "Identification and Control of Dynamical Systems Using Neural Networks", IEEE Trans. On Neural Networks, Vol.1,No.1,pp.4-27,March 1990.
4. S. Reynold Chu, Rahmat Shauresh, and Manoel Tenorio, "Neural Networks for System Identification", IEEE Control Systems Magazine, Vol.10, No.3, April 1990.
5. Jhon Kuschenski , Stefen Hui, and Stanislaw H. Zak, "Application of Feedforward Neural Networks to Dynamical System Identification and Control "- IEEE Trans. On Control Systems Technology, Vol.1,No.1, March 1993.
6. Sung – Woo Kim and Ju-Jang Lee, "Unknown Parameter Identification of Parameterized System Using Multi-Layered Neural Networks",. Proc. Of IEEE Int. Conf. on Neural Networks, 1993.
7. K.J.Hunt, D. Sbarbaro , R. Zbikowski and P.J.Gawtrop, "Neural Networks for Control Systems –Survey", Automatica Vol.28,No.6, pp 1083-1112,1992.
8. N.Yadaiah and B.Sanjeev Rao, "Control of Dynamical Systems Using Neural Networks",Proc. Of National Conf. On Neural Networks and Fuzzy Systems,1995.
9. Kevin Warwick, " The Control Of Dynamical Systems by Neural Networks, " Proc. Of IEEE Int. Conf. On IA&C ,1995.
10. Jian-Kang Wu, "Neural Networks and Simulation Methods", Marcel Dekker Inc., 1994.

***

3

RECENT DEVELOPMENTS IN NEURAL NETWORK PID AUTOTUNING

A.E.B RUANO
UCEH, University of Algarve
Algarve, Portugal

1. Introduction

PID controllers are widely used in industrial applications. Because the plant can be time variant, methods of autotuning of this type of controllers, are of great economical importance, see (Astrom, 1996). Since 1942, with the work of Ziegler and Nichols (Ziegler and Nichols, 1942), several methods have been proposed in the literature.

Recently, a new technique using neural networks was proposed (Ruano *et al.,* 1992). This technique has been shown to produce good tunings as long as certain limitations are met.

2. Existing technique

The work described in this chapter arises from previous work in neural network PID autotuning, and, for this reason, it is briefly reviewed in this section. Two approaches were introduced (Ruano *et al.*, 1992), (Mamat *et al.*, 1995), differing mainly in the identification measures employed. Basically, the neural networks, previously trained off-line, are used to supply, in real-time, the PID parameters to a standard PID controller, according to the identification measures obtained.

Concentrating here in the first (chronologically) of the two approaches (Ruano et al., 1992), in this technique three MultiLayer Perceptrons (MLPs) are used, each one of them supplying one of the PID parameters to a PID controller. As the identification measures employed may be obtained in open and in closed loop, this technique is applicable to both, with similar tunings being achieved in both situations. A schema of this technique is presented in Fig. 1.

S.G. Tzafestas (ed.), Advances in Intelligent Systems, 27–38.

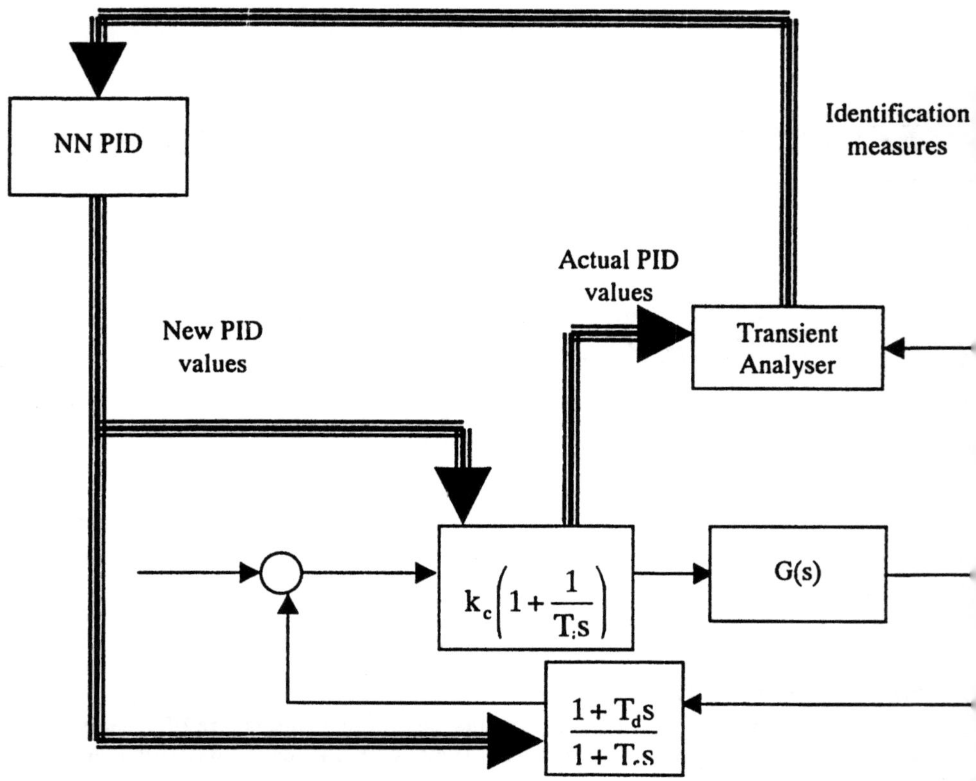

Fig. 1 – PID autotuning scheme

2.1. *Neural network inputs*

Let us consider a plant with transfer function:

$$G_p(s) = k_p e^{-d_p s} \frac{\prod_{i=1}^{n_z}\left(1+\tau_{z_i} s\right)}{\prod_{i=1}^{n_p}\left(1+\tau_{p_i} s\right)} \tag{1}$$

The identification measures $F(\alpha)$:

$$F(\alpha)=\frac{G(\alpha)}{k_p}, \alpha \in \Re, \tag{2}$$

for a range of values of α, will be used to identify the plant. In order to serve this purpose, α must be related with the time delay (d_p) and time constants (τ) of the plant. An algebraic sum of these factors, T_T:

$$T_T = d_p + \sum_{i=1}^{n_p} \tau_{P_i} - \sum_{i=1}^{n_z} \tau_{Z_i} , \tag{3}$$

is used as a scaling factor for α:

$$\alpha = \frac{\sigma}{T_T}, \sigma \in [0.1, 10] \tag{4}$$

For the sake of the following explanation it is convenient to express the identification measures in terms of σ:

$$F(\sigma) = e^{-\frac{d_p \sigma}{T_T}} \frac{\prod_{i=1}^{n_z} \left(1 + \frac{\tau_{Z_i} \sigma}{T_T}\right)}{\prod_{i=1}^{n_p} \left(1 + \frac{\tau_{P_i} \sigma}{T_T}\right)} \tag{5}$$

These last identification measures, $F(\sigma)$, are the inputs of the MLPs. The advantage of using them is related to the fact that they can be obtained in real time, both in open and in closed loop, through the use of integral measures of the output and/or the control signal (please see Ruano *et al.*, 1992).

2.2. *Neural networks outputs*

The outputs of the MLPs (k_c', T_i', T_d') are the parameters of the PID controller (k_c, T_i, T_d), conveniently scaled. Specifically:

$$k_c = \frac{1}{k_c' k_p}, T_i = T_i' T_T, T_d = T_d' T_T \tag{6}$$

2.3. *Neural network training*

Before being used in real-time control, the MLPs are trained off-line. In order to obtain good results from this technique, two conditions must be fulfilled:

- The examples employed for training must cover the whole operational range of the controller. An indirect way to solve this problem is to have an *a priori* knowledge of the types of transfer functions, as well as the range of its time constants and delay times that are exhibited during the lifetime of the process. Discretizing this multidimensional space, suitable examples required for training can be obtained.

- For every transfer function in the training set, it is necessary to determine the corresponding PID values. Although any tuning criterion can be employed, which accounts for the versatility of the proposed technique, a suitable criterion that produces well-damped responses should be selected. Thanks to its selectivity, the Integral of Time multiplied by the Absolute Error (ITAE) is used in our method.

After having obtained the training data, the MLPs are trained using a fast training algorithm (Ruano *et. al.*, 1991).

2.4. Results

Three MLPs were trained using systems with transfer functions with time delay and one and two poles, and systems with three and four poles, belonging to the following set† :

$$G_p(s) = k_p \frac{e^{-d_p s}}{(s+p)} \qquad d_p \in \{1:.1:3\}, p = -1 \tag{7 i}$$

$$G_p(s) = k_p \frac{e^{-d_p s}}{(s+p_1)(s+p_2)} \qquad d_p \in \{1:.3:2.8\}, p_1 = -1, p_2 \in \{-1:-1:-6\} \tag{7 ii}$$

$$G_p(s) = \frac{k_p}{(s+p_1)(s+p_2)(s+p_3)} \tag{7 iii}$$

$$p_1 \in \{-.1,-.2,-.4,-.8\}, p_2 \in \{-1,-2,-4,-8\}$$
$$p_3 \in \{-1.5,-3,-6,-12\}$$

$$G_p(s) = \frac{k_p}{(s+p_1)(s+p_2)(s+p_3)(s+p_4)} \tag{7 iv}$$

$$p_1 = -1, p_2 \in \{-.1,-.2,-.4,-.8\},$$
$$p_3 \in \{-1,-2,-4,-8\}, p_4 \in \{-1.5,-3,-6,-12\}$$

In total, the training set was composed of 191 examples. Figures 2a) and b) show some examples of the tunings obtained, for different plants, in situations where no noise was considered.

The solid line denotes the output obtained with the neural network autotuner, while the dashed line represents the results obtained using Ziegler Nichols tuning rule (frequency response). The plant is identified during the first step using the open loop output. As

---

†In eqs. (7), the notation {*x*:*y*:*z*} means a set of values with *x* as lower limit, *z* as upper limit, and with *y* increment.

soon as the next step is detected, the PID parameters are computed and the loop is closed. Due to the derivative kick introduced by the automatic loop closing, the second step is not considered for closed loop identification, and the tuning obtained from open loop identification is observed better in the third step. The output shown in the fourth step is obtained with closed loop identification, using the control signal recorded in the third step. As it can be observed, the tunings obtained using open and closed loop identification are similar.

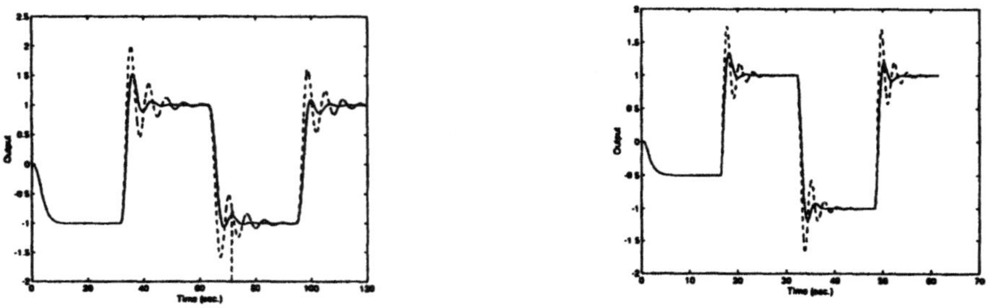

a) $G(s) = \frac{1}{(s+1)^4}$ b) $G(s) = \frac{e^{-0.5s}}{(s+1)(s+2)}$

Fig. 2 – Neural Network PID control of two different plants

3. Current work

The existing technique works well provided two conditions are fulfilled: the training data adequately spans the input space of the neural networks, and the actual plant to be controlled lies within the training space. The neural PID autotuners can be somehow envisaged as *blind experts,* who were previously taught how to control a certain range of plants, and subsequently perform their task in a very efficient manner. Unfortunately their performance is not improved as a result of continuously on-line operation.

To extend the capabilities of the neural PID autotuners, current research is targeted to improve (on-line training) the mapping capabilities of the neural networks by making use of the on-line experimental results, and also to address the use of multiple criteria to on-line tune the PID controllers.

In order to address the first objective (adapting the neural networks mappings on-line) the desired output data must be know for the actual controlled plant. Unfortunately this can not be know directly, and we propose here to employ additional neural networks to model the objective functions employed in the calculation of the optimal PID parameters.

In what concerns the second aim (multiple criteria) we propose here the application of genetic algorithms (GA) in the search of the optimal solution.

The following figure illustrates our overall autotuning schema.

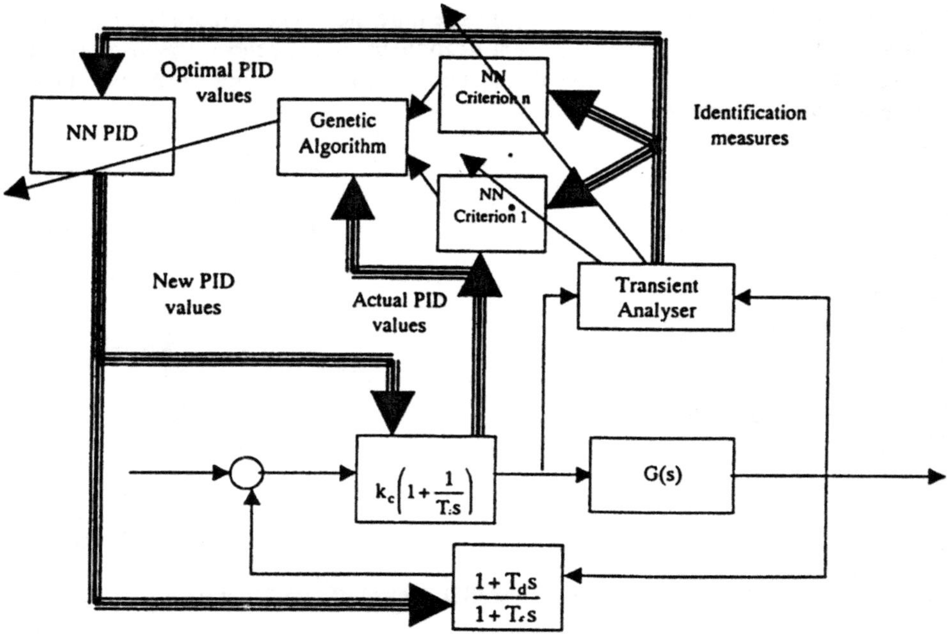

Fig. 3 – Overall autotuning schema

This chapter describes current work that is being developed towards that final objective.

3.1. Time invariant case

The plant considered in this case was chosen from a set of representative transfer functions representing a great number of industrial processes, and it is given by:

$$G(s)=\frac{e^{-s}}{s+1} \tag{8}$$

In this phase of work, we consider only two objectives for the tuning of the PID, namely, reference following and output disturbance rejection. This way, we define the following criteria that we pretend minimise, (9) and (10). In what concerns reference following, the ITAE criterion is employed:

$$\text{ITAE}=\int t|e(t)|dt \tag{9}$$

where: $e(t)=y(t)-r(t)$.

Relatively to output disturbance rejection, considering a null reference and a unit step added to the G(s) output, the criteria to minimise is (10):

$$ITAY = \int t\left|y(t)\right|dt \tag{10}$$

Two MLPs are used to approximate the mappings between the PID parameters and the criteria (ITAE and ITAY) described above.

The NN inputs are PID normalised values defined by (6). In the current application, with the transfer function (8), we have $T_T = 2$ and $k_p=1$.
The NN outputs are normalised criteria values given by (11):

$$ITAE' = \frac{T_T^{\,2}}{ITAE} \quad ITAY' = \frac{T_T^{\,2}}{ITAY} \tag{11}$$

After having determined that the best topology for this application was $\begin{bmatrix}3 & 9 & 5 & 1\end{bmatrix}$, (3 input neurons, 9 neurons in the first nonlinear hidden layer, 5 neurons in the second hidden layer, and one output neuron), the MLPs were trained using a fast training algorithm (Ruano *et al*, 1991), and used to output the objective functions to the GA algorithm. The GA is employed to minimise a function that takes into account the two objectives defined by (9) and (10). Representing the function to minimise by f(), we have:

$$f(k_c,T_i,T_d) = w_1 ITAE(k_c,T_i,T_d) + w_2 ITAY(k_c,T_i,T_d) \tag{12}$$

In Fig. 3 and Fig. 4, we compare the closed-loop responses in three different situations:

a) the optimal PID values are obtained employing a common gradient method for the ITAE minimization. In this case the ITAE values are computed by dynamic simulation, i.e., no neural network is employed;

b) the optimal PID values are obtained by a GA minimizing the ITAE criterion. The ITAE values are obtained using the trained neural network;

c) the PID is tuned by Ziegler & Nichols tuning rules.

In Fig. 4 y(t) represents the closed loop step output response with a null disturbance. Fig. 5 represents the closed-loop output response when the reference is set to zero, and an unit step disturbance is added to the plant output.

In both figures the upper part represents the response y(t), and the lower part represents the evolution of the ITAE and ITAY, respectively. In both figures the dashed-dot line represents situation a), the solid line b), and the crossed (heavy) line situation c).

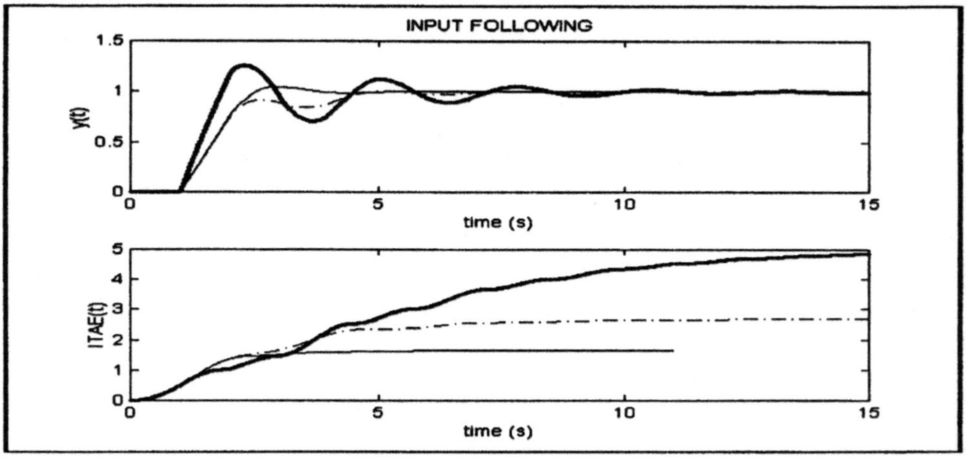

Fig. 4 – Reference tracking: comparison between situations a), b) and c)

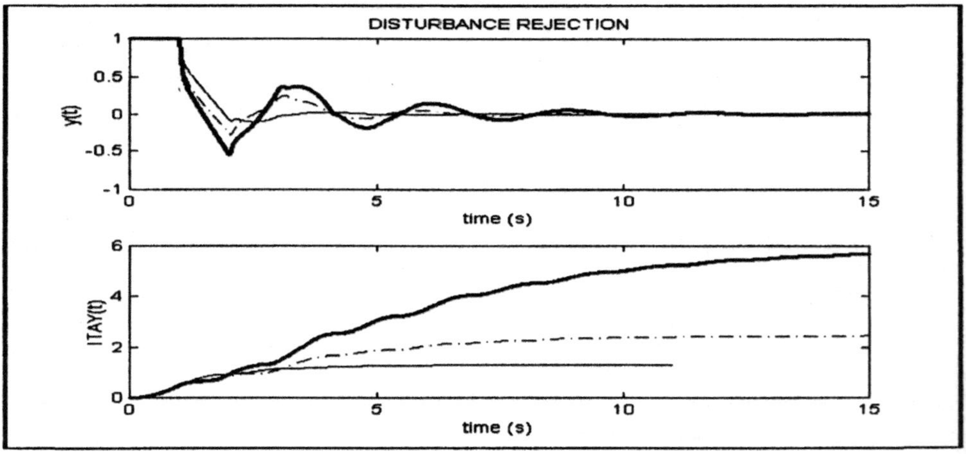

Fig. 5 – Disturbance rejection: comparison between situations a), b) and c)

Analysing these figures, we can conclude that, by employing the neural network as a model for the ITAE, together with the GA, better results than employing a common gradient minimization technique together with dynamic simulation, and much better than the Ziegler-Nichols results, are achieved. More important, situation b) takes an insignificant fraction of the time taken by approach a), and, since there is no need to perturb the control loop to compute the ITAE, it can be employed in real-time.

In Fig. 6 and Fig. 7, the optimisation was just performed using the GA, and the neural networks as models of the objective functions. In each figure we can compare the results of three different optimizations, corresponding to different combination of weights w_1 and w_2 in eq. (12). In these figures, the dashed-dot line represents $w_1 = 1$, $w_2 = 1$, the solid line $w_1 = 1$, $w_2 = 0$, and the crossed (heavy) line $w_1 = 0$, $w_2 = 1$

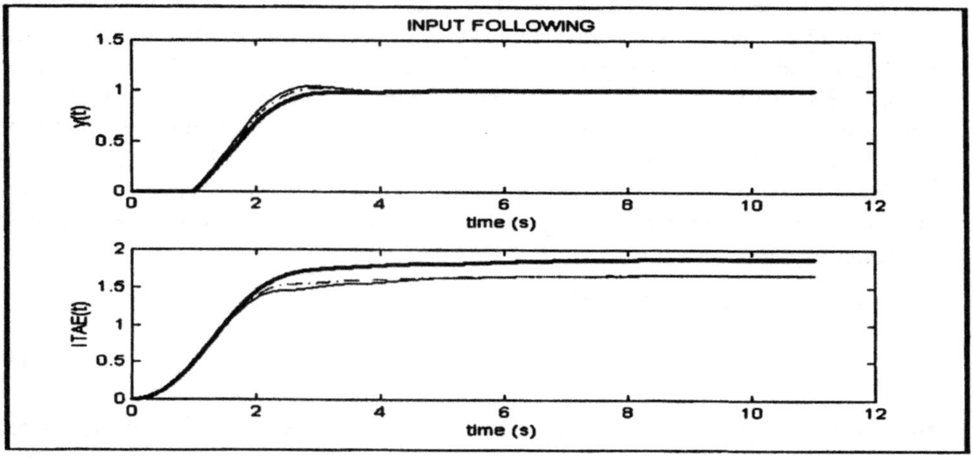

Fig. 6 – Reference tracking. GA optimisation: comparison between three objective functions.

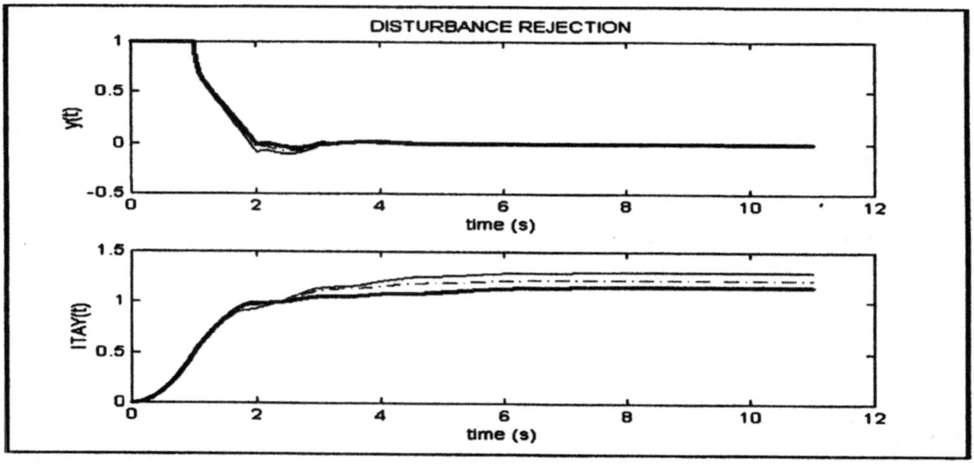

Fig. 7 – Disturbance rejection. GA optimisation: comparison between three objective functions.

We conclude that the neural networks proved to be effective as ITAE models, and, for this particular plant, the results obtained for tracking and disturbance rejection do not differ very much.

3.2. *Time-varying plants*

We consider, at this stage, only one objective, and that the plant is time-varying. If MLPs or RBFs are employed, as these are memoriless networks, on-line training within one region of the input space can modify the mapping achieved in different regions of the input space, which is an undesirable feature. For this reason, the neural networks used here, and in our final objective, are B-spline neural-networks (Brown and Harris, 1994). B-spline networks are a kind of Associative Memory Networks which differs from the others because they store information locally ("learning" about one part of the input space minimally affects the rest of it). To overcome the problem known as *curse of dimensionality* and in order to obtain, during off-line training, a good network structure for later on-line training, we employ the ASMOD (Weyer *et al.*, 1995) construction algorithm.

The plant considered here has a transfer function

$$G(s) = \frac{e^{-Ls}}{s+1}, \text{ with } L \in \{0.1 : 0.1 : 3\}, \tag{13}$$

from which we obtain the identification measures $F(\sigma)$, given by (5), considering only $\sigma = 1$. Some of the identification measures from this set (6 out of the 30 examples) are used to off-line train three neural networks. As in this case we are considering PID control, tuned by the ITAE criterion, their outputs are optimal scaled PID values given by (6). After that, one iteration of the on-line training algorithm was performed, considering all the 30 examples in the training set. Figs 8a) to c) illustrate the desired mapping (solid line), the off-line results (dashed line) and the on-line results (dashed-dotted line).

We can observe that slight improvements were obtained as a result of on-line learning. It should be however noticed that only one iteration of the on-line training algorithm was performed.

Another neural network will be trained to produce the mapping of the ITAE performance criterion measure, as a function of the actual PID parameters and the identification measures, i.e.:

$$ITAE = f(k_c, T_i, T_d, F(1)) \tag{14}$$

After off-line training, this network will be used by a minimisation algorithm to obtain the scaled optimal PID parameters, which will be employed as target data by the other networks for on-line training. This network is also going to be trained on-line using ITAE measures obtained, during the identification procedure, in conjunction with the identification measures.

In what concerns this network, we are currently looking at it. As the other networks have just got one input, the neural network can be implemented with univariate sub-networks. This network (the ITAE model) has 4 inputs, and we are at the moment adapting the training algorithm to accommodate multi-variable sub-networks. The training algorithm that is employed is the ASMOD algorithm, which needs an address

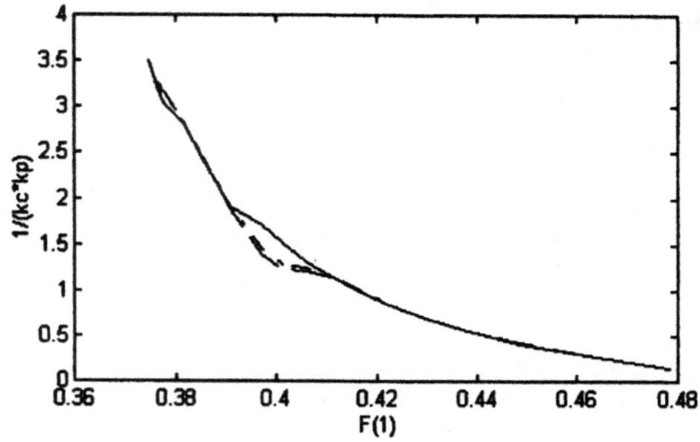

a)

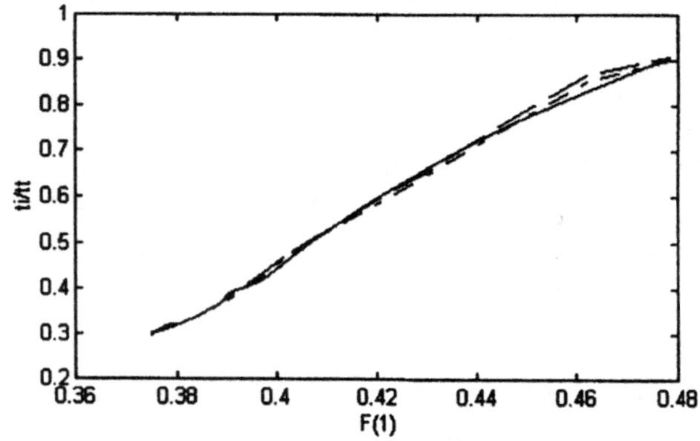

b)

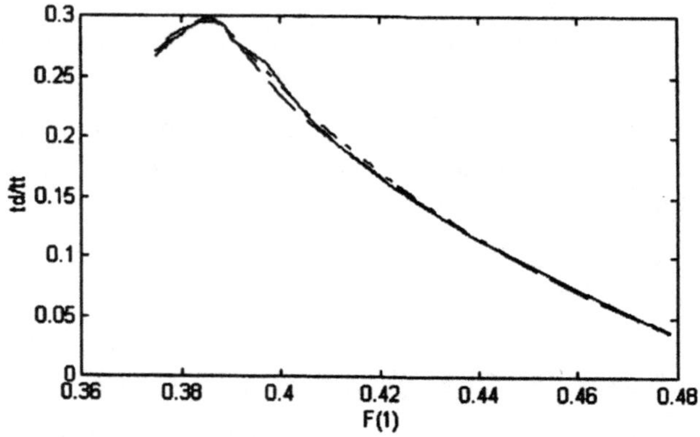

c)

Fig. 8 – Neural Networks Mappings

a) Proportional gain, b) Integral Time constant, c) Derivative Time Constant calculation routine, and that is the computing block that must be changed to this multivariable situation.

4. Conclusions

In this chapter we have introduced a new scheme for PID autotuning, involving neural networks and genetic algorithms. As this is work in progress, we have described here two situations, a time-invariant case with two tuning objectives, and preliminary results concerning the time-variant approach.

With respect to the former, it was verified that neural networks can accurately model the criteria under study and that they generalize well. By employing these neural networks as models, together with the GA, the PID tunings obtained were better than the ones achieved using a gradient technique with dynamic simulation. Of great importance, the formers were obtained faster and without the need of perturbing the loop, which are key factors considering a real-time application. Neural networks and Gas proved also to be suitable for multiobjective minimization.

With respect to the latter B-splines neural networks were employed, first trained off-line, and subsequently adapted on-line, while performing on-line control. Preliminary results show that this will be a promising approach.

Future work will address the implementation of the ITAE modelling network and the incorporation of this block in the overall PID autotuning approach.

References

Astrom, K, (1996), From Tuning to Adaptation, 13th IFAC World Congress (Plenary Lecture), San Francisco, USA.

Brown, Martin and Chris Harris, (1994), Neurofuzzy Adaptive Modelling and Control, Prentice Hall, London.

Mamat, R., P.J. Fleming, and A.E. Ruano, (1995), Neural Networks assisted PID autotuning, 2nd Int. Conf. on Industrial Automation, Nancy, France, pp. 849-854

Ruano, A.E., D.I. Jones and P.J. Fleming, (1991), A New Formulation of the Learning Problem for a Neural Network Controller, Proc. IEEE Conference on Decision and Control, Brighton, U.K., pp. 865-866

Ruano, A.; Fleming, P.; Jones, D.; (1992) A Connectionist Approach to PID Autotunung, IEE Proceedings, Part D., Vol.139, No3, , pp.279-285.

Weyer, E. and T. Kavli, (1995), The ASMOD algorithm. Some new theoretical and experimental results, SINTEF report STF31 A95024., Oslo.

Ziegler J, Nichols N. (1942), Optimum settings for automatic controllers. Trans ASME; **65**: 759-768.

4

ADAPTIVE SELF-TUNING NEURAL-NETWORK-BASED CONTROLLER

P. POTOČNIK and I. GRABEC
University of Ljubljana,
Faculty of Mechanical Engineering
Aškerčeva 6, SI-1000 Ljubljana
Slovenia

1 Introduction

The control of complex dynamical systems is not always sufficiently addressed by conventional control theory. The problem of environmental uncertainty often requires human-like decision making with heuristic reasoning and learning from past experiences. Learning is required when the complexity of the problem or the uncertainty thereof prevents *a priori* specification of a satisfactory solution. Adequate solutions can be obtained through accumulating information about the problem and using this information to dynamically generate an acceptable solution. Such approach is referred to as "intelligent control" and denotes systems having an ability to emulate human capabilities, such as planning, learning and adaptation [1]. The field combines conventional control theory with techniques such as neural networks, fuzzy logic, genetic algorithms and a variety of optimization approaches.

The problem of intelligent control is related to reinforcement learning (RL), where an agent learns a proper behavior through trial-and-error interactions with a dynamic environment [2]. One of the central issues of RL is the exploitation versus exploration dilemma [3]. The agent must sufficiently explore the environment in order to identify an optimal controller and on the other hand, the accumulated knowledge must be exploited during learning to reduce the costs. The tradeoff between exploration and exploitation demands efficient exploration capabilities, maximizing the effect of learning while minimizing the costs of exploration.

In our research we investigate a class of NN-based self-tuning control of unknown time-dependent nonlinear plants. NN-based direct self-tuning control of nonlinear plants is discussed in [4], where the controller is directly identified as a mean square inverse estimator of a pseudo-plant. In [5] self-tuning NN-based approach

S.G. Tzafestas (ed.), Advances in Intelligent Systems, 39–46.

is presented for automatically determining the most appropriate parameter values of the controller. The approach is based on employing several identical NN, each of which uses the description of the process to be controlled for estimating a single parameter of the controller. The RBFN was proposed in [6] for the implementation of reinforcement learning in continuous spaces. In our previous research [7, 8], we investigated the application of empirical modeling of natural phenomena to the optimal self-control of an autonomous system in a chaotic environment.

In this chapter, we propose a novel approach to NN-based self-tuning control, where the self-tuning principle and reinforcement learning are combined with special emphasis on solving the exploration–exploitation dilemma. The control scheme is composed of a controller, a utility estimator, an exploration module, a learning module and a rewarding module. The controller and the utility estimator are implemented together in a single radial basis function network (RBFN). The learning method involves structural adaptation (growing RBFN) and parameter adaptation. The exploration–exploitation dilemma is solved with smooth transitions between the two modes. This enables fast exploration response to novel plant dynamics and stable operation in the absence of changes in plant dynamics.

The organization of the chapter is as follows. A growing RBFN which implements the controller and the utility estimator is presented in section 2. The proposed control scheme is discussed in section 3 and results of a simulation study are presented in section 4.

2 Growing RBFN

A radial basis function network (RBFN) is a three-layered neural network [9] which has a universal approximation property for continuous real-valued functions [10, 11]. The output of RBFN is defined by the equation

$$\widehat{y}(x) = \sum_{k=1}^{K} w_k \, \phi_k(x). \tag{1}$$

The symbols denote: input vector, x, network output vector, $\widehat{y}$, basis function, ϕ, weight, w, and number of basis functions, K. A common choice of basis function is a Gaussian which is given by the equation

$$\phi_k(x) = \exp\left(-\frac{||x - q_k||^2}{2\sigma^2}\right). \tag{2}$$

Here q_k represents the center of the basis function and σ denotes its width. The norm $||\cdot||$ in equation (2) can be expressed by Euclidean distance. Construction of a RBFN involves determination of the parameters w_k, q_k and σ. Since we use a growing RBFN structure, the number of centers, K, is not fixed *a priori* but is adjusted during the course of the learning process. The only predetermined parameter is σ which can be regarded as a modeling resolution.

The learning process of the growing RBFN is initiated by the first input–output sample, (x_1, y_1), where the first RBFN center, q_1, obtains the value of x_1. When new samples, $(x_t, y_t;\ t = 2, 3, \ldots)$, are presented to the network, the subsequent centers are added when the excitation of the existing basis functions, $(\phi_k;\ k = 1, \ldots, K)$, is below the threshold, θ:

$$\max_k \phi_k(x_t) < \theta \quad \Longrightarrow \quad q_{k+1} = x_t. \tag{3}$$

In our research we used the threshold value, $\theta = 0.6$, which guarantees for proper overlapping of basis functions. The output weights are adjusted according to the gradient learning rule

$$w_k(t+1) = w_k(t) + \alpha\,(y(x_t) - \widehat{y}(x_t))\,\phi_k(x_t) \tag{4}$$

where α denotes a learning rate in interval $[0, 1]$. The learning process of growing RBFN is local. Consequently, only output weights of the most excited neurons are changed considerably without detrimental effects on other operating areas. Growing RBFN is suitable for on-line modeling of time-dependent systems where the model structure must continuously adapt to the variable operating conditions.

3 Control scheme

In this section we present a model-free NN-based self-tuning control scheme which is suitable for control of nonlinear systems for which very little *a priori* knowledge is available and controlled systems may also change with time. Figure 1 shows the control scheme which consists of a controller, a utility estimator, an exploration module and a learning module. A rewarding module is a part of the environment which must provide for a suitable reinforcement signal, U. The particular parts of the control scheme perform the following tasks:

Controller and the utility estimator are implemented together in a single RBFN. The inputs of the RBFN are the current state of the plant, y, and the future reference signal, r, which defines the desired plant trajectory. If necessary, delayed values of y, r and u can also be added to the input vector. Based on these inputs, the controller generates control action, u.

Utility estimator is designed to estimate the expected suitability of the proposed control action, u. The estimated suitability, $\widehat{U}$, is fed to the exploration module and to the learning module.

Exploration module is responsible for a proper exploration strategy. If the estimated suitability, $\widehat{U}$, is low, a random exploration component, δu, is added to the proposed control action in the exploration module. The modified action, $u + \delta u$, is then executed, the system proceeds to a new state and the environment (rewarding module) generates a new reinforcement signal, U.

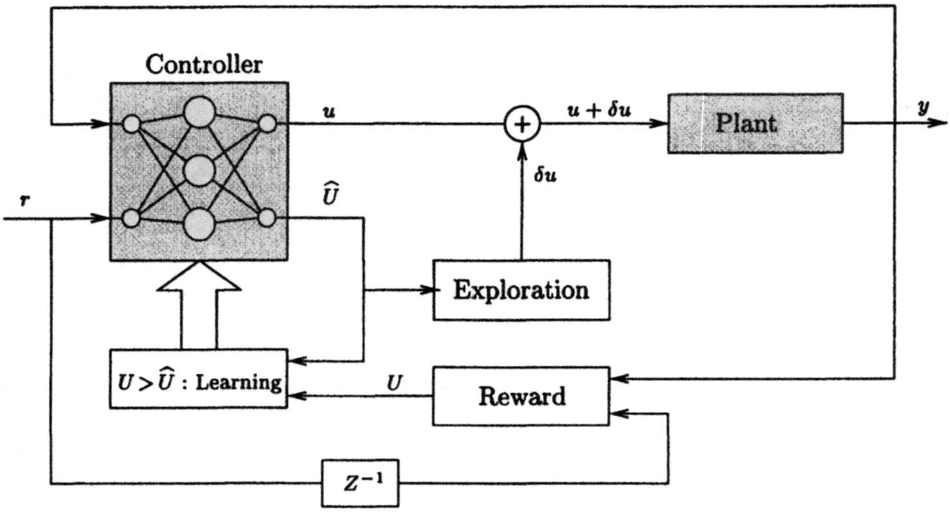

Figure 1: Adaptive self-tuning neural-network-based control scheme, composed of a controller, a utility estimator, an exploration module and a learning module. A rewarding module is a part of the environment which must provide for a suitable reinforcement signal. The labels denote: state of the plant, y, reference, r, control, u, perturbation of control, δu, estimated suitability, $\widehat{U}$, and reinforcement signal U.

Learning module compares the received reinforcement, U, with the estimation, $\widehat{U}$, and in the case of positive difference, $\Delta = U - \widehat{U}$, learning of the controller is activated. The utility estimator is continually updated to approximate the expected utility. Learning of the RBFN involves structural adaptation (growing neural network) and parameter adaptation. The network is supplemented with new neurons if necessary, and the output weights of the controller and the utility estimator are adapted according to the following equations:

$$w_{k,\hat{U}}(t+1) = w_{k,\hat{U}}(t) + \alpha\,(U(t) - \hat{U}(t))\,\phi_k(t) \tag{5}$$

$$w_{k,u}(t+1) = \begin{cases} w_{k,u}(t) + \alpha\ \delta u\ \phi_k(t) & \text{if } U(t) > \hat{U}(t) \\ w_{k,u}(t) & \text{otherwise} \end{cases} \tag{6}$$

The learning method for the RBFN is local and therefore only specific operation domain is adapted without detrimental effects on other operating areas.

The controller is capable of asymptotically approaching the desired reference trajectory. The exploration–exploitation dilemma is solved with smooth transitions between the two modes. This enables fast exploration response to novel plant dynamics and stable operation in the absence of changes in plant dynamics.

4 Simulation study

An application of the proposed control scheme is presented in a simulation study. The objective is to control the time-dependent nonlinear system according to the reference trajectory. The system is described by the equation:

$$y(t+1) = \begin{cases} \dfrac{1.5y(t)}{1+y^2(t)} + 0.3\cos(y(t)) + 1.2u(t) & \text{for } t \le 4000 \\ \\ \dfrac{1.5y(t)}{1+y^2(t)} + 0.3\sin(y(t)) + 0.6u(t) & \text{for } t > 4000 \end{cases} \qquad (7)$$

and the reference signal, r, is given as a filtered periodic step-function which is switched between ± 1 after every 100 time steps. The reference is altered to lower frequency at time $t = 8000$. The controller and the utility estimator are implemented as a 2-input 2-output growing RBFN

$$[u(t), \hat{U}(t+1)] = f_{\text{RBFN}}[r(t), y(t)]. \qquad (8)$$

We applied the learning rate, $\alpha = 0.1$, which was found to be convenient during the experiment. The reinforcement signal, U, was generated in the interval $[0, 1]$ with value 1 in a case of perfect tracking. A random exploration component of the control action, $\delta u(t)$, was generated in the interval $[-(1 - U(t))\delta u_{\max}, (1 - U(t))\delta u_{\max}]$, where $\delta u_{\max}$ denotes the maximal value of random perturbation δu.

Figure 2 shows results of control with the proposed control scheme. System output, y, reference signal, r, and control, u, are shown in the following time intervals: $[0 : 1000]$, $[3500 : 4500]$ and $[7500 : 8500]$. Intensive exploration phase is initiated at time $t = 0$. The controller is investigating the effect of various control actions on the behavior of the system. After several hundred time steps, tracking accuracy improves considerably and the exploration component δu is automatically reduced. After several hundred subsequent time steps, a stable control law with accurate tracking is established.

The system is changed at time $t = 4000$ according to Eq. (7) and an operation of the controller is again switched to intensified exploration. Soon, output error is reduced and the changed control law stabilizes. The last change was introduced at time $t = 8000$ when the reference, r, was modified. The dynamics of the controlled system were not changed and therefore no additional exploration was necessary for accurate control.

Figure 3 shows the number of neurons, K, for the RBFN implementing the controller and the utility estimator. A rapid increase in the number of neurons is observed at the beginning of operation at time $t = 0$ and at time $t = 4000$ when the system has changed.

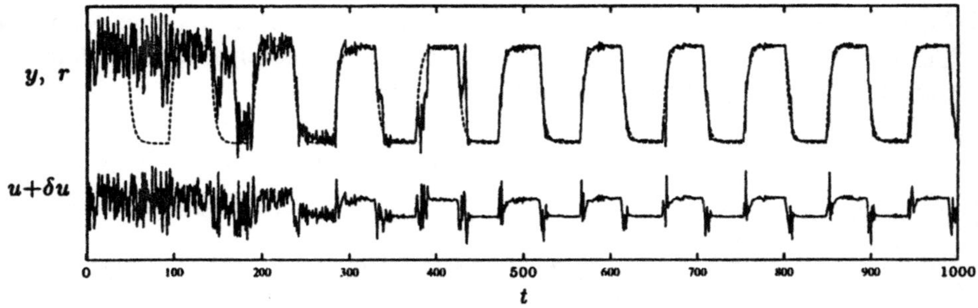

(a) At time $t = 0$ learning of the controller begins with exploration of the unknown system.

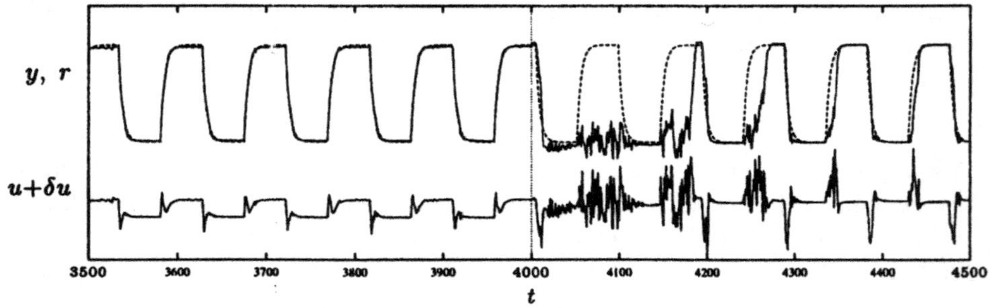

(b) Properties of the controlled system are changed at time $t = 4000$ and therefore the new period of exploration is initiated.

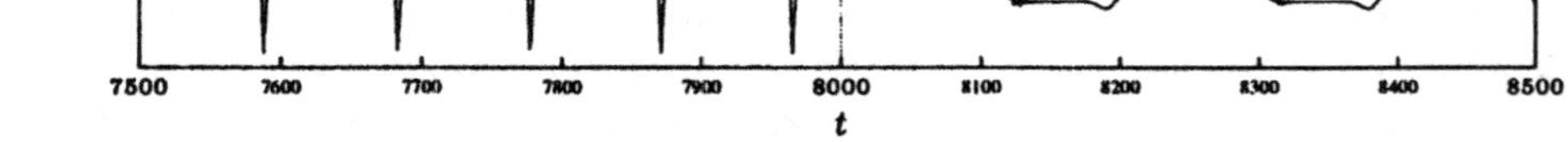

(c) Reference signal, r, is altered at time $t = 8000$; dynamics of the controlled system is not changed. Consequently, no additional exploration is necessary for accurate control.

Figure 2: Application of adaptive self-tuning neural-network-based controller for control of time-dependent nonlinear system. Reference trajectory $r(t)$ (dashed line), plant output $y_p(t)$ and control signal $u(t)$ are plotted for three time intervals: **(a)** $[0 : 1000]$, **(b)** $[3500 : 4500]$ and **(c)** $[7500 : 8500]$.

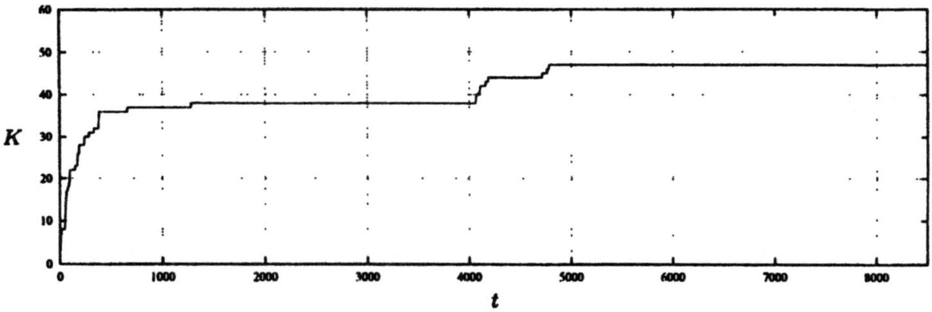

Figure 3: Application of adaptive self-tuning neural-network-based controller for control of time-dependent nonlinear system. Number of neurons, K, is plotted during the evolution of growing RBFN which implements the controller and the utility estimator.

5 Conclusion

Neural-network-based direct self-tuning control of nonlinear plants has been discussed in this chapter. The method proposed is model-free and uses adaptive exploration/exploitation strategy to investigate the properties of the system. The controller and the utility estimator are implemented together in a single growing RBFN. The method was shown to successfully control the simulated nonlinear system, but there are still many questions to be addressed in the future work:

- The stability of the method was not investigated yet. Particularly the exploration strategy may cause instability and therefore needs careful consideration.

- A simple exploration strategy with generation of random perturbations, δu, was used in this study. Only maximal amplitude of the perturbation, $\delta u_{\max}$, was adapted on-line. A more elaborated exploration strategy could be utilized to improve the convergence properties. One of the possibilities is to upgrade the method proposed with reactive search strategies, such as the reactive tabu search [12].

- A growing RBFN was used to implement the controller and the utility estimator. For real-world applications it would be necessary to add pruning capacity in order to limit the size of the neural network. A strategy should be devised to determine and eliminate obsolete neurons which were generated during the intensive exploration phase, before the stabilization of the control.

Acknowledgement—The authors would like to acknowledge the National Ministry of Science and Technology, Slovenia, for their financial support.

References

[1] D. A. Linkens and H. O. Nyongesa, "Learning systems in intelligent control: an appraisal of fuzzy, neural and genetic algorithm control applications," *IEE Proceedings. Control Theory and Applications*, vol. 143, no. 4, pp. 367–386, 1996.

[2] L. P. Kaelbling, M. L. Littman, and A. W. Moore, "Reinforcement learning: A survey," *Journal of Artificial Intelligence Research*, vol. 4, pp. 237–285, 1996.

[3] S. B. Thun, "The role of exploration," in *Handbook of Intelligent Control. Neural, Fuzzy, and Adaptive Approaches* (D. A. White and D. A. Sofge, eds.), ch. 14, pp. 527–559, New York: Van Nostrand Reinhold, 1992.

[4] M. S. Ahmed and M. F. Anjum, "Neural-net-based direct self-tuning control of nonlinear plants," *International Journal of Control*, vol. 66, no. 1, pp. 85–104, 1997.

[5] B. Freisleben and S. Strlen, "Design of self-tuning controllers using neural networks," in *Proceedings of the 5th European Congress on Intelligent Techniques and Soft Computing EUFIT'97*, Volume 1, (Aachen, Germany), pp. 424–428, 1997.

[6] R. Valverde and M. A. Salichs, "Neurocontrol of continuous processes through reinforcement learning," in *Proceedings of the 5th European Congress on Intelligent Techniques and Soft Computing EUFIT'97*, Volume 1, (Aachen, Germany), pp. 389–393, 1997.

[7] I. Grabec, J. Gradišek, and P. Potočnik, "An empirical approach to optimal self-control," *Engineering Applications of Artificial Intelligence*, vol. 10, no. 1, pp. 73–85, 1997.

[8] I. Grabec and W. Sachse, *Synergetics of Measurements, Prediction and Control.* Heidelberg, Germany: Springer Verlag, Series in Synergetics, 1997.

[9] J. Moody and C. J. Darken, "Fast learning in networks of locally-tuned processing units," *Neural Computation*, vol. 1, pp. 281–294, 1989.

[10] E. J. Hartman and J. D. Keeler, "Layered neural networks with gaussian hidden units as universal approximations," *Neural Computation*, vol. 2, pp. 210–215, 1990.

[11] F. Girosi and T. Poggio, "Networks and the best approximation property," *Biological Cybernetics*, vol. 63, pp. 169–176, 1990.

[12] R. Battiti, "Reactive search: Toward self-tuning heuristics," in *Modern Heuristic Search Methods* (V. J. Rayward-Smith, I. H. Osman, C. R. Reeves, and G. D. Smith, eds.), ch. 4, pp. 61–83, John Wiley & Sons, 1996.

5

NEURO-CONTROL FOR MULTIVARIABLE SYSTEMS

S. OMATU and M. YOSHIOKA
Department of Computer and Systems Sciences
College of Engineering, Osaka Prefecture University
Sakai, Osaka 599-8531, Japan

1. INTRODUCTION

Computational intelligence approaches like fuzzy theory, neural network, genetic algorithm, artificial life, etc. have been well-developed and applied to many real control problems in an efficient way[1]-[4]. In real control processes, the PID controller has been used as the major control methods and the operators select suitable PID gains from time to time based on their experience and knowledge about the plant dynamics. However, it is required to find a suitable PID gains automatically. The main reason is that its structure is simple and PID controllers are robust with noise and parameter variations. To use the PID controller, we must tune the PID gains which have been determined by trial and error based on experience and knowledge of experts. In this paper, we propose a tuning method of the PID gains by using neural networks to control a single-input multi-output system and apply it to stabilization of the inverted pendulum.

2. NEURO-CONTROL APPROACH

Recently, many neuro-control schemes have been proposed [1]-[4] where neural networks can be trained to realize good controllers by learning an inverse model of the plant or a nonlinear emulator for the plant. Among many neural network learning methods, the error back-propagation(BP) method[1] is the most widely used in a wide variety of applications. In control problems, there has been development on neuro-control for robots control problems[4]. Although various intelligent control methods may be applied to his problem, PID control is the major approach since it is robust to noise and stable for parameter change.

Using the back-propagation method, various kinds of neuro-controllers could be

S.G. Tzafestas (ed.), Advances in Intelligent Systems, 47–56.

trained in such a way that the desired plant output is attained as much as possible. The basic control scheme is depicted as Fig.1 where feedforward controller(FFC) and feedback controller(FBC) are adjusted. The neuro-controller is used in FFC or FBC instead of a conventional controller. Three typical neuro-controllers are a series type, a parallel type, and a self-tuning type. In this paper, we consider the self-tuning type when FFC is neglected and the PID structure is adopted for FBC.

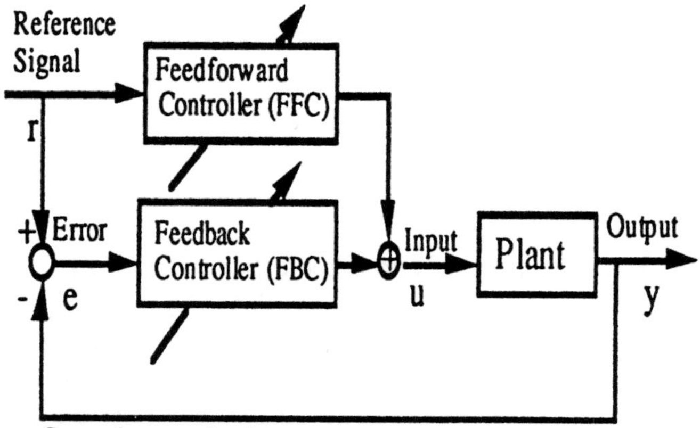

Fig. 1. General control scheme.

3. SELF-TUNING NEURO-PID CONTROLLER

The control structure of the self-tuning neuro-PID controller can be shown in Fig.2 where the outputs of the neural network are proportional(P), integral(I), and derivative(D) gains and the inputs are selected in a suitable way according to the specific problem.

In order to derive the self-tuning neuro-PID control algorithm for a single-input single-output(SISO) system, we will define mathematical notation of neuron models. Total input to neuron j is denoted by net_j and is given by

$$net_j = \sum_{i=1}^{n} w_{ji} O_i + \theta_j$$

where θ_j denotes a threshold. Then the output O_j of neuron i becomes

$$O_j = f(net_j), \quad f(x) = \frac{1}{1+e^{-x}}$$

By connecting many neurons in a layered array, we can derive the BP algorithm based on the gradient method which is described as follows:

<u>*Step* 1</u>. Find the generalized error δ_k in the output layer

$$\delta_k = (t_k - O_k) f'(net_k) = (t_k - O_k) O_k (1 - O_k)$$

and the generalized error in the hidden layer

$$\delta_j = f'(net_j) \sum_{k=1}^{K} \delta_k w_{kj} = O_j (1 - O_j) \sum_{k=1}^{K} \delta_k w_{kj}$$

where t_k is the target pattern, O_k is the actual output of neural networks, and w_{ji} denotes the connection weight from neuron i to neuron j.

Step 2. Change the connection weights by Δw_{ji} in the following way: In the output layer

$$\Delta w_{kj}(n+1) = \eta \delta_k O_j + \alpha \Delta w_{kj}(n) + \beta \Delta w_{kjj}(n-1)$$

and in the hidden layer

$$\Delta w_{ji}(n+1) = \eta \delta_k O_j + \alpha \Delta w_{ji}(n) + \beta \Delta w_{ji}(n-1)$$

where a modified BP method has been used to avoid the local minimum problem.

Step 3. If the error becomes smaller than a predetermined value, then stop and otherwise, n is replaced by n+1 and go to *step* 1.

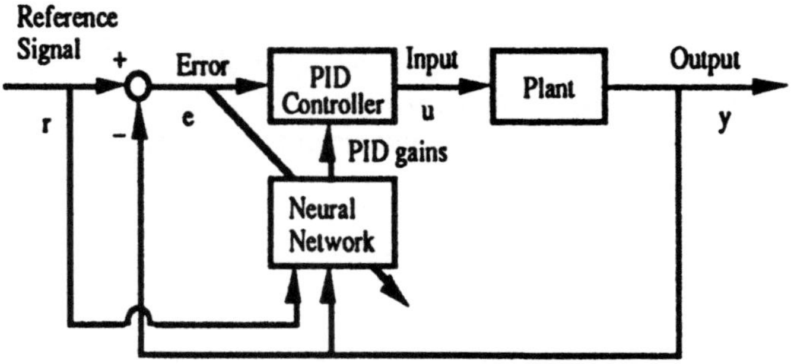

Fig. 2. Self-tuning neuro-PID control scheme.

Using the above BP algorithm, we can derive the self-tuning neuro-PID control algorithm. First, note that the PID controller in discrete-time is described as

$$u(n) = u(n-1) + K_P(e(n)-e(n-1)) + K_I\, e(n) + K_D e(n)-2e(n-1)+e(n-2)) \quad (1)$$

where K_P, K_I, and K_D are proportional, integral, and derivative gains, respectively, u(n) denotes the plant input at nT, and e(n) is the error at nT between a desired value r(n) and the actual output y(n), that is,

$$e(n) = r(n) - y(n).$$

Here, T denotes a sampling interval.

In order to adjust K_P, K_I, and K_D adaptively, we use a three layered neural network based on the BP method. Each layer consists of N_1, N_2, and N_3 neurons where N_1 and N_2 can be selected by trial and error according to complexity of the plant and N_3 is three which are equal to the number of PID gains. The BP method requires a cost function E which should be minimized. Here, the following E is adopted such that the PID controller can decrease the squared error E at (n+1)T,

$$E = \frac{1}{2}e(t+1)^2.$$

The connection weights w_{kj} at the output layer are updated by the following relation.

$$\Delta w_{kj}(n+1) = \eta\delta_k O_j + \alpha\Delta w_{kj}(n) + \beta\Delta w_{kjj}(n-1)$$

where

$$\delta_k = -\frac{\partial E}{\partial net_k}.$$

Using the chain rule for derivative, we get

$$\delta_k = e(n+1)\frac{\partial y(n+1)}{\partial u(n)}\frac{\partial u(n)}{\partial O_k}\frac{\partial O_k}{\partial net_k}$$

Furthermore, we have from (1)

$$\frac{\partial u(n)}{\partial O_k} = \begin{cases} e(n)-e(n-1), & for\, k=1 \\ e(n), & for\, k=2 \\ e(n)-2e(n-1)+e(n-2), & for\, k=3 \end{cases}$$

where $O_1 = K_P$, $O_2 = K_I$, and $O_3 = K_D$.

If the plant is unknown, two approximation algorithms to calculate the system Jacobian $\frac{\partial y(n+1)}{\partial u(n)}$ have been proposed.

The first one is to use the following relation:

$$\frac{\partial y(n+1)}{\partial u(n)} = \text{sign}\left(\frac{\partial y(n+1)}{\partial u(n)}\right)\left|\frac{\partial y(n+1)}{\partial u(n)}\right|.$$

Thus, if we know only the sign of the Jacobian, we can adjust the learning rate η such that $\left|\frac{\partial y(n+1)}{\partial u(n)}\right|$ is included into η.

The second way is to build the emulator which can simulate the input/output map of the plant. Regarding the emulator as the true plant and using the BP method, we can adjust the connection weights by combining the neural network to tune the PID gains with a neural network to simulate the plant. The total configuration for the second approach is given by adding the emulator to the plant in a parallel way. Using the control scheme, we have obtained fine tuning of the PID gains in many cases. The total configuration of the neuro-PID controller with an emulator is shown in Fig.3 where inputs of the emulator are past input and output data of the plant. But it takes much time to get convergence results even though the results obtained are similar to those by human operators who have engaged with control operation of the process for a long time and obtained much experience how to tune the PID gains when unknown circumstances came out. Therefore, the only problem for the above approach is to shorten the computational time to find the fine tuning values. The convergence time depends on the initial values of the PID gains and connection weights, we can use the genetic algorithm to find them roughly.

4. SIMO CASE

We consider a single-input two-output PID controller by extending the previous one for the SISO case. Let us consider the situation of Fig.4 where single-input two-output system is described. This is the case of stabilization of an inverted pendulum. In order to construct the self-tuning PID controller in this system, we introduce the tuning structure of Fig.5. In Fig. 5, one neural network is used to tune the PID gains for two PID

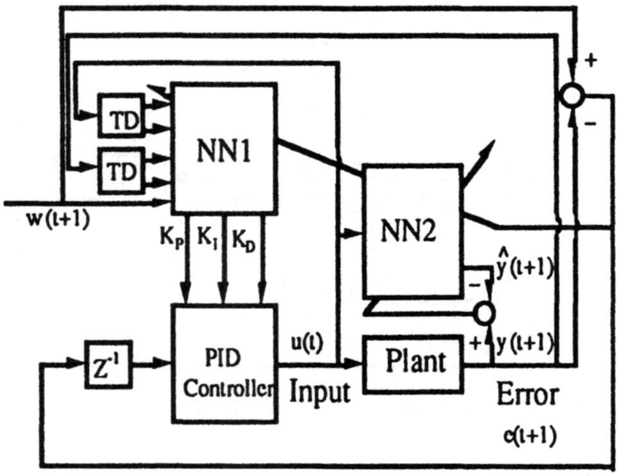

Fig. 3. Neuro-PID controller configuration.

controllers. The tuning structure will be described through the example of stabilization of inverted pendulum. The basic principle is to tune the total PID gains as the output of the neural networks. In that case, we need the system Jacobian as in the single PID control case. Thus, we must construct the system emurator as in the preceding section.

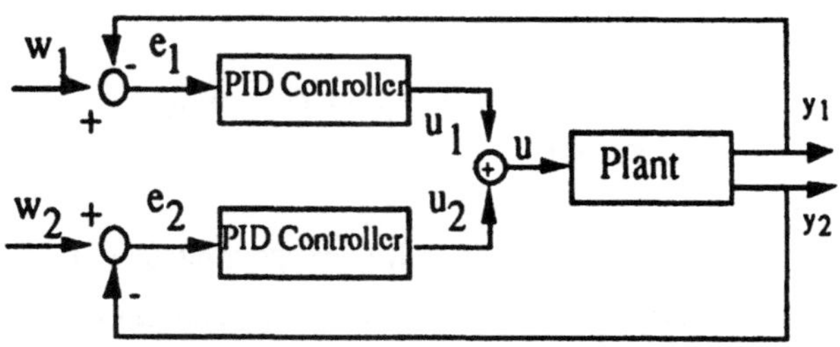

Fig.4 SIMO PID control system.

5. APPLICATION TO AN INVERTED PENDULUM

We consider the application of the single-input two-output self-tuning PID controller to stabilization of the inverted pendulum. Control system of the inverted pendulum has been described in Fig. 6. The aim is to move the cart of the pendulum to the center of the bar(x=0) and stand it (θ =0). To construct the PID controller to the control

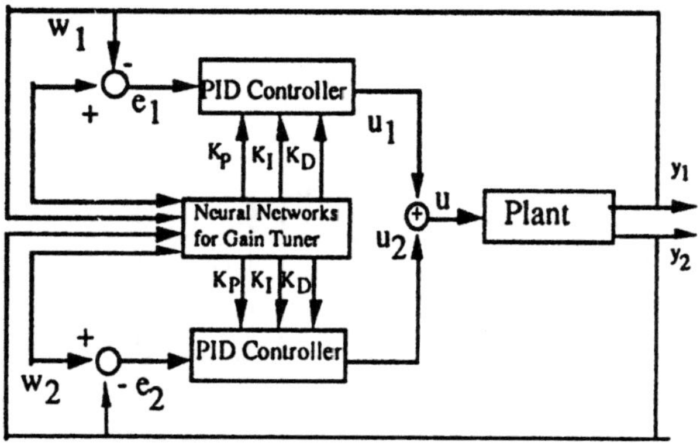

Fig.5. Self-tuning PID for SIMO case.

problem, the PID gains are adjusted by the neural network as shown in Fig.7 where the minimum cost function for tuning is the squared error between the actual output and the desired values. In this case, the desired values are x=0 and θ =0. As in the SISO case stated in Section 3, we need the system Jacobian of the inverted pendulum which is derived by using the emulator given by Fig. 8. Here, the input layer consists of 6 units and the number of the output layer is 2 as shown in Fig.8. Using this emulator, we can derive the system Jacobian by the following equations:

$$\frac{\partial \hat{x}(n+1)}{\partial u(n)} = \sum_{j=1}^{3} \dot{w_{kj}} O_j(1-O_j) w'_{j1}$$

$$\frac{\partial \hat{\theta}(n+1)}{\partial u(n)} = \sum_{j=3}^{6} \dot{w_{kj}} O_j(1-O_j) w'_{j1}$$

where suffix 1 of w_{j1} denotes the neuron unit corresponding to u(n) at the input layer and j=1,2,3 mean the three neurons in the hidden layer which are connected to $\hat{x}(n+1)$.

Similarly, j=4,5,6 means the remaining neurons in the hidden layer which are connected to the $\hat{\theta}(n+1)$.

Using the above relations, we have obtained the control results of the inverted pendulum.

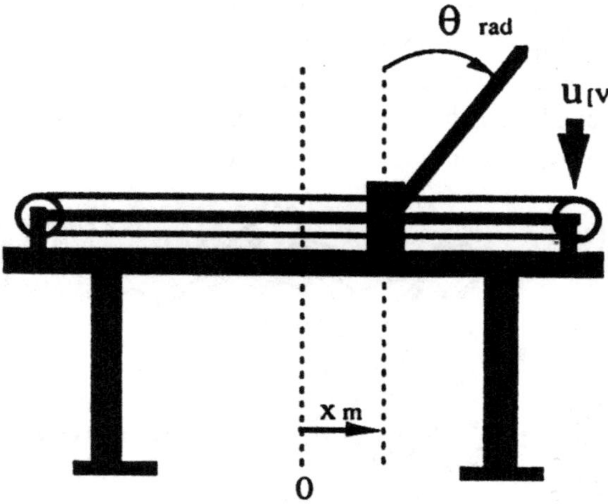

Fig. 6. Control system of inverted pendulum.

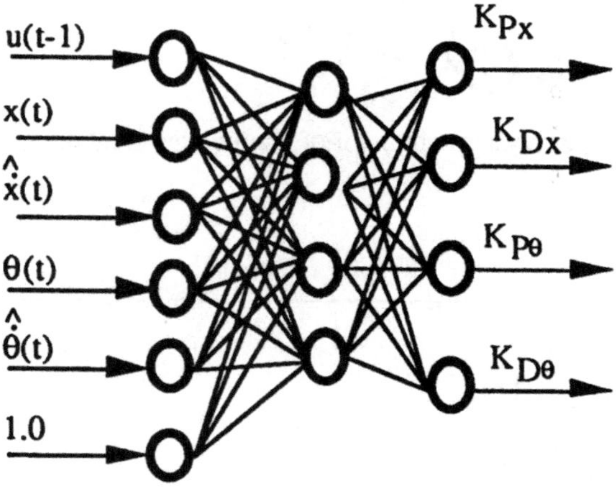

Fig. 7. Gain tuning by a neural network.

The control results were shown in Figs. 9 and 10. Figure 9 shows the control results of the inverted pendulum where x=0 and θ =0 are attained after 2 seconds. Figure 10 shows the corresponding inputs where u is the total input and u_x and u_θ denote the corresponding components for x and θ.

6. CONCLUSIONS

We have considered the neuro-PID control systems and their applications to stabilization problems of the inverted pendulum. In this case we must control the position of the pendulum and stabilize the pendulum by adjusting the cart, which means single-input two-output system. From our experimental results we can see that the single-input multi-output system can be treated by the present neuro-PID control algorithms. The other application examples have been discussed in [6]-[12]. From these application examples, the future perspective of the neuro-control would be preferable in various application fields. Furthermore, by combining the present approaches to vision systems, the neuro-approach will become more powerful in intelligent robotics area.

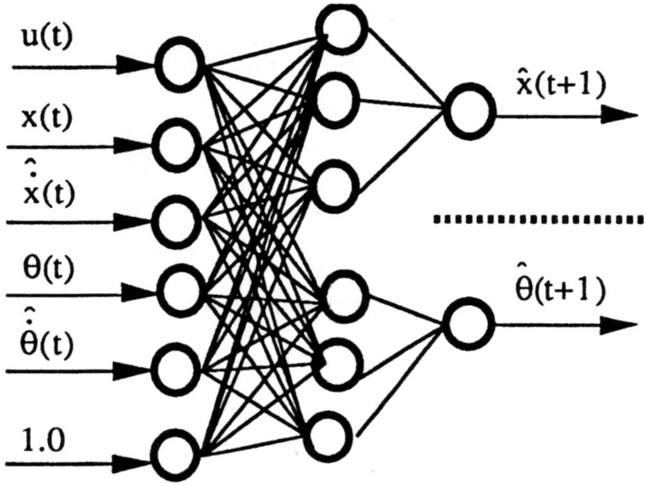

Fig.8. System emulator of the inverted pendulum.

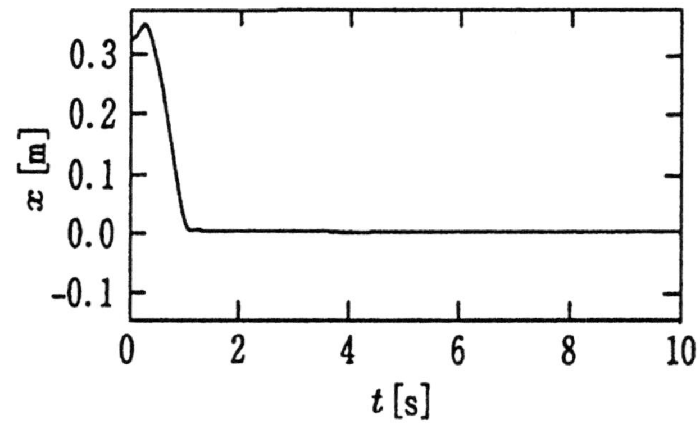

(a)

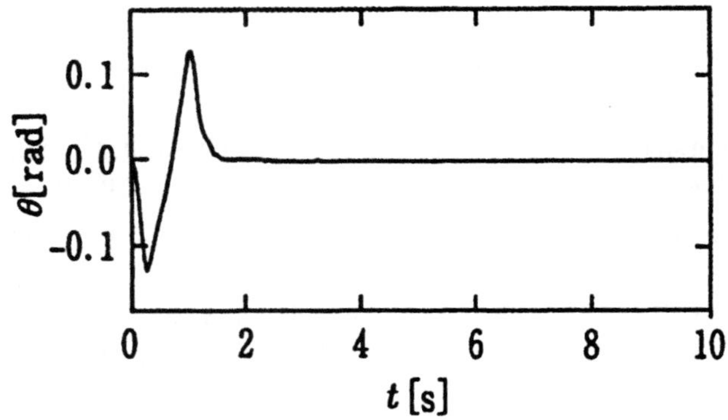

(b)

Fig.9. Control results of inverted pendulum.
(a) The cart position x.
(b) Angle of the pendulum.

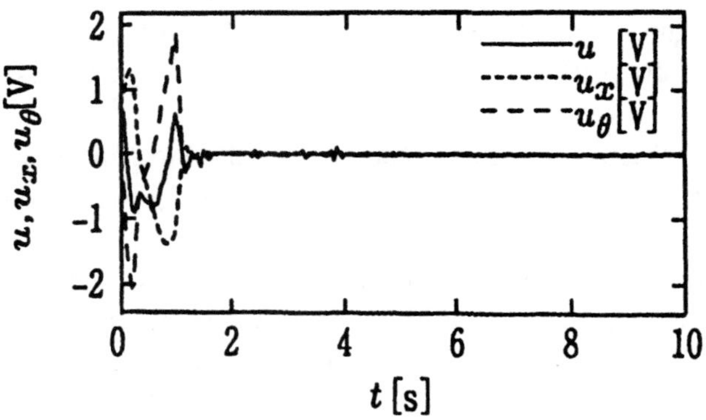

Fig. 10. Control inputs for the inverted pendulum.

References

[1] W. T. Miller III, R. S. Sutton, and P. J. Werbos, "Neural Networks for Control", MIT Press, (1990).
[2] D. A. White and D. A. Sofge Edited, "Handbook of Intelligent Control", Neural, Fuzzy, and Adaptive Approaches, Van Nostrand Reinhold, (1992).
[3] S. Omatu, M. Khalid, and R. Yusof, " Neuro-Control and Their Applications", Springer- Verlag, (1995).
[4] M. Kawato, Y. Uno, M. Isobe, and R. Suzuki, "Hierarchical neural network model for voluntary movement with application to robotics", IEEE Control Systems Magazine, Vol.8, pp.8-16, (1988).
[5] T. Aoyama and S. Omatu, "Design of a Self-Tuning PID Control System by Neural Networks", Trans. on IEE of Japan, Vol.116-C, No.11, pp.1197-1201, (1996) (in Japanese).
[6] T.Aoyama,S.Omatu,and Y.Kishida, "Neuro-PD Control for Modeling and Stabilization of Inverted Pendulum", Trans. on the Institute of Systems, Control and Information Enginners, Vol.1, No.1, pp.10 -11, (1998) (in Japanese).
[7] S. Omatu,"Intelligence Based on Neuro-Control", Proc. of the Third Internartional Symposium on Artificial Life and Robotics, Vol.1, pp.137-140, Beppu, Japan, (1998).
[8] S. Omatu," Neuro-Approach to Motion Control", Proc. of the Third Internartional Symposium on Artificial Life and Robotics, Vol.2, pp.528-531, Beppu, Japan, (1998).
[9] S. Omatu, Y. Kishida and M. Yoshioka, "Neuro-Control for Single-Input Multi-Output Systems", Proc. of 1998 International Conference on Knowledge-Based Intelligent Electronic Systems, Vol.1, pp.202 -205, Adelade, Australia, (1998).
[10] T. Iwasa, N. Morizumi, and S. Omatu, "Temperature Control in a Batch Process by Neural Networks", Proc. of 1998 IEEE International Joint Conference on Neural Networks, Vol.2, pp.992-995, Anchorage, USA, (1998).
[11] S. Omatu and M. Yoshioka," Stability of Inverted Pedulum by Neuro-PID Control with Genetic Algorithm", Proc. of 1998 IEEE International Joint Conference on Neural Networks, Vol.3, pp.2142-215, Anchorage, USA, (1998).
[12] S. Omatu," Neuro-Control Applications in Real-World Problems", Proc. of the Tenth Yale Workshop on Adaptive and Learning Systems, pp.92-97, New Haven, USA, (1998).

PART II

FUZZY LOGIC-BASED INTELLIGENT ESTIMATION AND CONTROL

6

ORTHOGONAL LEAST-SQUARES BASED FUZZY MODEL FOR SHORT TERM LOAD FORECASTING

P. A. MASTOROCOSTAS, J.B. THEOCHARIS,
S.J.KIARTZIS and A.G.BAKIRTZIS
Department of Electrical and Computer Engineering
Aristotle University of Thessaloniki, Greece

1. Introduction

Accurate short-term load forecasting (STLF) is a challenge to power engineers operating modern Energy Management Systems (EMS). Reliable load forecasts with lead times from a few hours to seven days, provide a very important input to EMS fundamental operating functions such as unit commitment, hydro-thermal coordination and interchange evaluation.

Power market competitive environment, especially under the newly established energy policy based on deregulation, is a most promising area for cost saving applications relying on operating reconstruction and accurate forecasts.

A variety of models for STLF have been reported in the literature during the last few decades. Traditional STLF approaches include regression models [1] or time series models [2]. Recent developments on artificial intelligence techniques provided new tools for the solution of the STLF problem. Expert systems have been successfully applied to STLF [3], although the existence of an expert to train the system is required. Artificial neural networks are currently established as a useful and much promising approach to a variety of power systems engineering problems, with impressive performance in STLF [4]. Artificial neural networks are able to cope with the non-stationary nature of the load and provide accurate forecasts, learning by themselves the functional relationship between system inputs and outputs, through a training process [5-8].

Fuzzy systems is another recently introduced approach for STLF [9-12]. Fuzzy systems seem to have certain advantages over neural networks [9], such as: (a) faster training, (b) model structure and parameters which have a physical meaning as contrasted to the "black-box" representation of the neural networks and (c) ability to incorporate linguistic IF-THEN expert rules.

A new approach for STLF is presented in this chapter employing an Orthogonal Least Squares based Fuzzy Inference System. In section 2, the fuzzy system is briefly described while in section 3 the Orthogonal Least Squares method is presented. In the next two sections the training algorithm is presented followed by the proposed

S.G. Tzafestas (ed.), Advances in Intelligent Systems, 59–70.

application of the model to STLF. Finally, in the last section simulation results are reported.

2. Fuzzy Inference System

The fuzzy inference system considered in this chapter consists of Takagi-Sugeno's IF-THEN rules [13]. A fuzzy rule for a Multiple-Input-Single-Output system can be written in the form:

$$\begin{aligned} R^{(j)}: \ & \textbf{IF} \ u_1 \text{ is } A_1^j \ \textbf{AND} \ldots \textbf{AND} \ u_m \text{ is } A_m^j \\ & \textbf{THEN} \ g_j = w_0^j + w_1^j u_1 + \ldots + w_m^j u_m \end{aligned} \tag{1}$$

where m is the number of inputs $u_1 \ldots u_m$, g_j denotes the j-th rule output, w_i^j are the polynomial coefficients that form the consequent parameter set and A_i^j are labels of fuzzy sets pertaining to the j-th rule. Each linguistic label A_i^j is associated with a membership function. $\mu_{A_i^j}(u_i)$ which specifies the degree to which a given u_i satisfies the quantifier A_i^j. The membership functions considered here are of Gaussian type (bell-shaped) and are presented in (2)

$$\mu_{A_i^j}(u_i) = \exp\left[-\frac{1}{2} \cdot \frac{(u_i - m_{ij})^2}{\sigma_{ij}^2}\right] \tag{2}$$

where m_{ij} and σ_{ij} are the mean value and the standard deviation of the Gaussian membership function, respectively, forming the premise parameter set. Given an input vector $\underline{u}$, the output of the fuzzy system is derived as follows:

$$y = \frac{\sum_{j=1}^{n} [g_j \prod_{i=1}^{m} \mu_{A_i^j}(u_i)]}{\sum_{j=1}^{n} [\prod_{i=1}^{m} \mu_{A_i^j}(u_i)]} \tag{3}$$

where n is the total number of rules.

The above fuzzy inference method was suggested by Takagi and Sugeno in [13]. It is also referred to in the literature as type-3 reasoning [14]. The output of each rule is a linear combination of the input variables plus a constant term and the output of the fuzzy system is the weighted average of each rule's output.

3. The Fuzzy Modeling Approach

3.1 STRUCTURE IDENTIFICATION

Having determined the fuzzy inference system, the next task is to perform the structure identification of the fuzzy model, which involves the selection of the input variables and the partitioning of the input space to fuzzy subspaces. The premise part of a fuzzy rule corresponds to a fuzzy hyper-cell and the respective fuzzy sets can be regarded as the rule co-ordinates. In that sense, the degree of fulfilment of each rule, presented in (4), can be treated as the membership function of the rule's hyper-cell.

$$\mu_j = \prod_{i=1}^{m} \mu_{A_i^j}(u_i) \tag{4}$$

Therefore, given a training set of input/output training data pairs, the partitioning of the input space leads to the determination of the number of hyper-cells and their location within the input space.

In the literature, the premise space partitioning has been based either on experience or on a number of methods such as fuzzy clustering [15] or combinatorial techniques [13]. In this chapter, an alternative approach is proposed, based on the Orthogonal Least Squares algorithm (OLS). Once the OLS partitions the input space, the premise parameters are determined. Since the rules' outputs are linear with respect to the consequent parameters, the latter can be determined by applying the Recursive Least Squares Estimation method (RLSE) [13-14].

3.2 THE ORTHOGONAL LEAST SQUARES METHOD

Let P be the total number of input/output pairs that constitute the training data set. The objective of the OLS is to determine the number of hypercells necessary to efficiently cover the input space and their location (their centers) within the input space. The method is an iterative procedure: Initially, each data point is considered as the center of a hypercell, forming a set of P hypercells. At each iteration, the algorithm extracts the hypercell that covers most of the input data set. This cell is the most significant among the whole set of cells. When a hypercell is selected, it is extracted from the set of hypercells and the algorithm is applied to the rest of the set. Thus, a series of hypercells is created, each of which is more significant than the proceeding ones. In order to highlight this notion, a brief mathematical description is given below: Substituting (4) to (3), it is derived

$$y = \frac{\sum_{j=1}^{n} \mu_j g_j}{\sum_{j=1}^{n} \mu_j} \tag{5}$$

Defining

$$\frac{\mu_j(\underline{u})}{\sum_{i=1}^{n} \mu_j(\underline{u})} = p_j(\underline{u}) \tag{6}$$

(5) becomes

$$y = \sum_{j=1}^{n} p_j(\underline{u}) \cdot g_j \tag{7}$$

Considering the functions $p_j(\underline{u})$ as regressors, the expansion of the regressors in (7) can be viewed as a special case of the linear regression model

$$d^{[p]} = \sum_{j=1}^{n} p_j^{[p]} \cdot g_j + e^{[p]} \tag{8}$$

where $d^{[p]}$ is the desired system output, g_j are real parameters, for a certain input vector, and the regressors are fixed functions of system inputs $\underline{u}^{[p]}$, that is $p_j^{[p]} = p_j(\underline{u}^{[p]})$. The error signal $e^{[p]}$ is assumed to be uncorrelated with the regressors.

As mentioned above, given P input/output pairs, $p=1,...,P$, in the beginning we consider $n = P$ regressors. The centers of each membership function $\mu_j(\underline{u})$ are equal to an input data vector, $m_{ij} = u_i^{[j]}$, where $u_i^{[j]}$ denotes the i-th component of the j-th input data vector. The standard deviations of the membership functions are selected in a way that there will be sufficient overlapping between the resulting fuzzy hyper-cells. The OLS method aims at selecting a subset of the most significant regressors from the total set of P regressors. Once the regressors are extracted, the partitioning of the input space is complete. A diagrammatic representation of the resulting partitioning is given in section 5.3, where a load forecasting test-case is presented.

The OLS method is fully presented in [16] and [18] and will not be discussed any further. The procedure is terminated at the M_sth step when the criterion $1-\sum_{j=1}^{M_s}[err]_j<\rho$ is fulfilled, where $0<\rho<1$ is a chosen tolerance. The term $[err]_j$ is called *error reduction ratio* and is defined in [16]. Thus, the OLS determines the number of rules and the premise part structure, providing a partition of the input space to M_s hyper-cells as well as their centers, for given values of the standard deviations. Since the rules' outputs are linear with respect to the consequent parameters, the latter can be determined by applying the RLSE method [13-14].

3.3 COMMENTS ON THE OLS

It should be noted that the OLS has been employed before in [16-18] for determining the parameters of a system. However, there are major differences between the present application of the OLS and the preceding ones:

a) In this chapter, the OLS is used to perform the premise partitioning and, consequently, provide the initial premise parts of the fuzzy rules. In [16-18] they are employed to determine the final model.
b) In the preceding references, the g_j's are constant coefficients, while in the inference system employed here they constitute the consequent parts and are linear polynomials of the inputs.

c)In [17] and [18], the number of rules is predetermined. In our case, the rule extraction is terminated when a criterion, introduced in [16], is fulfilled.

4. The Training Algorithm

The modeling technique presented in the previous section leads to the formulation of an efficient initial fuzzy model. Using the OLS algorithm, the centers of the most significant hyper-cells, for given σ's, are selected, while the RLSE provides a consequent parameter vector that corresponds to the global optimum point in the consequent parameter space [14]. However the standard deviations of the membership functions are considered to be fixed, thus reducing the system's degrees of freedom. Additionally, as stated in [17], the choice of σ's has a significant influence on the approximation results. In order to tune the standard deviations along with the rest of the model parameters so as to further improve the performance of the model, a hybrid training method is applied. At this point, the fuzzy model can be considered as a Fuzzy Neural Network, entirely described by its parameter set. The method, which is fully presented in [19], is an iterative two-stage learning algorithm and proceeds as follows: At each iteration, the training data are supplied to the network, producing the current membership values and the rule outputs. Then, the consequent parameters are identified using the RLSE (the consequent parameter learning stage). Throughout this process the premise parameters are considered to be fixed. In the sequel, the premise parameters are updated by employing a Steepest Ascent Method (the premise parameter learning stage). This procedure is repeated until an acceptable level of performance is achieved.

4.1 PREMISE PARAMETER LEARNING STAGE

The premise part is nonlinear with respect to the premise parameters. Therefore, the adjustment of these parameters is carried out by means of a Steepest Ascent Method. This method is an iterative procedure which aims at achieving two objectives, simultaneously: First, it is desired that an error measure function, E, should be minimized, in order to perform the input-output mapping, successfully. The Mean Square Error (MSE) is selected here and defined as follows:

$$E = \frac{1}{P}\sum_{p=1}^{P}(d^{[p]} - y^{[p]})^2 \tag{9}$$

where $d^{[p]}$ is the desired output.

Second, a nonlinear multivariable function Φ., which relates the current parameters with those of the preceding iteration, should be maximized. This function is called the *pay-off function* and is introduced as follows:

$$\Phi = (\underline{\theta} - \underline{\Theta})^T (\underline{\theta} - \underline{\Theta}) \tag{10}$$

$\underline{\theta} = [\underline{m}^T, \underline{\sigma}^T]^T$ is the premise parameter vector at the current iteration, called the *control vector*. and $\underline{\Theta}$ is the average of the control vector at the previous iteration. As reported in [20], maximization of this pay-off function at each epoch implies that the current and previous weight updates are highly aligned, thus avoiding zig-zag trajectories in the parameter space and improving the convergence speed.

Optimization of the above two functions is iteratively performed with respect to the decision variables, $\underline{\theta}$ and the vector $\underline{x} = [\underline{\mu}^T, y]^T$, called the *state vector*, and should satisfy the constraints imposed by the *state equations* (4). (5). In that sense, the learning process can be regarded as a constrained optimization problem. This problem has been solved in [21], using methods based on optimal control theory and the calculus of variations.

Following the procedure dictated in [21], at each iteration the control variables are changed by a small amount $d\underline{\theta}$, lying on the surface of a hypersphere of radius *dp*, which is considered as the step size:

$$(dp)^2 = d\underline{\theta}^T \cdot d\underline{\theta} \tag{11}$$

Thus, at each iteration. the search in the parameter space is restricted to the surface of a small hypersphere of radius dp, centered at the point defined by the current control vector. The quantity dp represents the step size of the algorithm. These perturbations of the control vector induce changes in the state variables $\underline{x}$ and in the functions E and Φ. If the step size is small enough, these changes can be approximated by the first differentials dE and $d\Phi$. For steepest ascent, we wish to find the optimal parameter changes $d\underline{\theta}$ that maximize $d\Phi$, for given value of dE. The update of the premise parameter vector is given by the following formula, where a detailed derivation can be found in [19]:

$$\begin{aligned} \underline{\theta}(t+1) = \underline{\theta}(t) + d\underline{\theta} = \underline{\theta}(t) + \sqrt{\frac{(\nabla E)^2 (dp)^2 - (dE)^2}{(\nabla\Phi)^2 \cdot (\nabla E)^2 - (\nabla\Phi)\cdot(\nabla E)}} \cdot \\ \cdot \{\nabla\Phi - \nabla E \cdot \frac{(\nabla\Phi)\cdot(\nabla E)}{(\nabla E)^2}\} + \nabla E \cdot \frac{dE}{(\nabla E)^2} \end{aligned} \tag{12}$$

∇E and $\nabla\Phi$ are the gradients of E and Φ with respect to the control vector.

5. Fuzzy Model For Short Term Load Forecasting

5.1 SELECTION OF MODEL INPUTS

A six-input single-output model has been developed to perform hourly based short term load forecasting. The input vector comprises the following variables:

$\hat{T}_{min,d}$ minimum daily temperature forecast of day d

$\hat{T}_{max,d}$ maximum daily temperature forecast of day d

$L_{d-1,h-1}$	average load during hour h-1 of day d-1
$L_{d-1,h}$	average load during hour h of day d-1
$L_{d-1,h+1}$	average load during hour h+1 of day d-1
$L_{d-7,h}$	average load during hour h of day d-7

where d=1,...,365 is the day index and h=1,...,24 is the hour index. The output of the model is the load at the h-th hour of day-d, $\hat{L}_{d,h}$.

The first two inputs are selected because they were found to be highly correlated with the desired load. The next three inputs are selected to supply information concerning the dynamics of the load curve at each hour, which is quite crucial, particularly at the extreme points (peak and valley loads). Finally, the last input associates the load at the same hour of the previous week with the desired one, which was found to be essential, especially for the forecasting of weekend loads.

5.2 DEFINITIONS - NOTATIONS

One fuzzy model is assigned to each day of the week for every season. Hence a total of $4*7=28$ fuzzy models are required for the forecasting of the daily load curves of all seasons.

Two data sets are employed. The training data set is used during the structure identification process as well as the learning process. It contains historical data from six consecutive past years. The evaluation set, used for testing the forecasting capabilities of the fuzzy model, contains the load data of the year 1995.

The scope of the training process is to adjust the network parameters so that an error measure is minimized. The average percentage error (APE) is selected to be the error measure in this chapter, and is defined as follows:

$$APE=\frac{1}{N_w}\sum_{d=1}^{N_w}[\frac{1}{24}\sum_{h=1}^{24}\frac{\left|L_{d,h}-\hat{L}_{d,h}\right|}{L_{max,d}}]\cdot 100\%$$

where N_w is the number of days in the data set, $L_{d,h}$ and $\hat{L}_{d,h}$ are the actual and predicted loads, respectively, during the h-th hour of the d-th day, and $L_{max,d}$ is the maximum actual load of the day.

5.3 PRESENTATION OF A TEST-CASE

In order to highlight the fuzzy modeling process, the prediction of the daily load for Summer Fridays is presented, as an illustrative example. The training and evaluation sets comprise 2088 and 432 input/output pairs, respectively. First, the OLS is applied to determine the number of rules and to perform the premise space partition. The tolerance ρ is set to 0.2 and is attained at the fourth step, leading to a four-rule rule-base. The premise parameter set comprises of $4*12=48$ parameters. In the sequel, the RLSE is employed to determine the $4*7=28$ consequent parameters. The resulting rule base is cited in Table I.

The scatter diagrams of $L_{d-1,h}$ versus $\hat{T}_{max,d}$ and $L_{d-1,h-1}$ versus $L_{d-1,h}$ are shown in Fig.1 and 2, respectively. Moreover, the 0.4-cuts (α-cuts) of the hyper-cells' membership functions, which correspond to the rules' degrees of fulfilment are also plotted. As it can be seen, the OLS has performed a very efficient partition of the input space, since most of the data points are covered by a rule with a degree of fulfilment greater than 0.4.

Next, the training method is applied; dp is set to 0.008. Simulation lasts for 90 iterations and an APE of 1.28% is attained. Then the evaluation set is fed to the fuzzy system, giving a prediction APE of 1.49%.

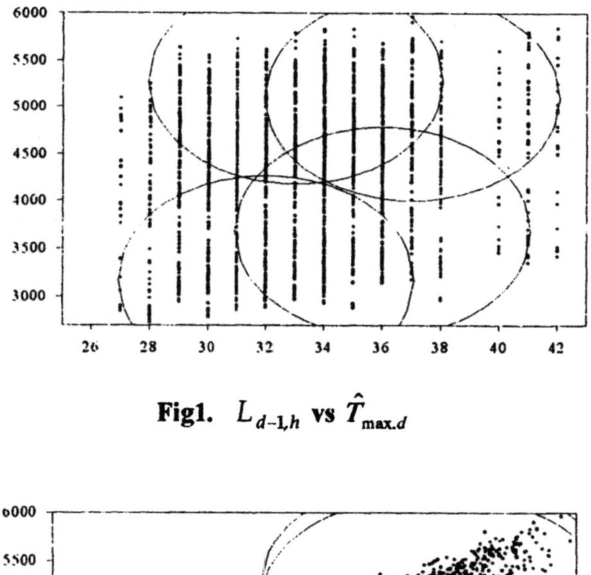

Fig1. $L_{d-1,h}$ **vs** $\hat{T}_{max,d}$

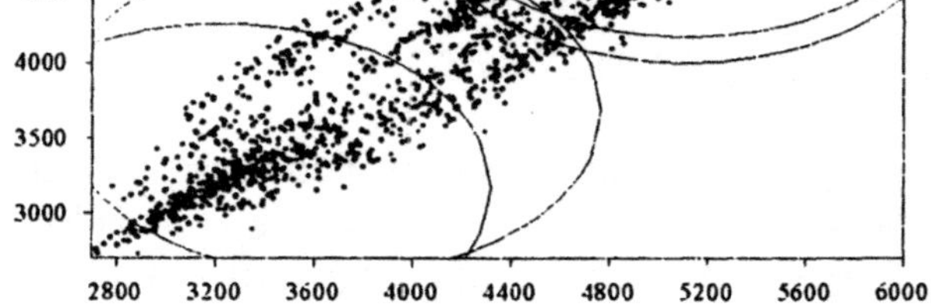

Fig2. $L_{d-1,h-1}$ **vs** $L_{d-1,h}$

Table I. Fuzzy Rule Base for Summer Fridays

	Rule 1			Rule 2		
	m	σ	w	m	σ	w
u_0			366.7			805.47
u_1	23.28	1.71	-29.67	24.22	2.21	-2.31
u_2	30.67	2.23	9.76	31.29	2.08	30.26
u_3	3471.1	807.3	0.34	5134.8	1156	0.14
u_4	3402.8	784.2	0.25	5476.2	696.5	0.46
u_5	3301.2	754.8	0.33	5398.1	919.3	0.18
u_6	3709.1	805.2	0.05	5439.1	963.7	-0.07
	Rule 3			Rule 4		
	m	σ	w	m	σ	w
u_0			-2268			569.4
u_1	24.82	2.01	2.62	25.23	1.38	-7.83
u_2	32.99	2.61	51.1	32.92	2.03	12.28
u_3	5194.5	871.1	0.16	3901.7	843.2	0.81
u_4	5124.8	1022	0.48	3782.2	778.7	-0.54
u_5	5128.7	717.7	0.24	3598.4	419.3	0.62
u_6	5036.3	746.7	0.19	3793.8	791.2	0.01

6. Simulation Results

In order to evaluate the learning and forecasting qualities of the developed fuzzy model. hourly based prediction for the whole year has been performed. Structure identification has been carried out via the OLS method, leading to rule bases of 3 or 4 rules. Therefore. each fuzzy model contains either 57 or 76 parameters. The seasonal numbers of parameters along with the total numbers are shown in Table II. It is evident that the OLS method has provided fuzzy models with a relatively small number of parameters and. consequently. with reduced complexity. Table III summarises the predicted APE for all days of all seasons and the seasonal APE's. The yearly APE is **1.82**.

The statistical analysis regarding the difference in the actual and the predicted load is reported in Table IV. The yearly mean absolute forecast error is **91.1 MW**. It should be noted that the standard deviation of the absolute forecast error is 77.6 MW for the whole year and takes similar values for all seasons, leading to the conclusion that the predicted load does not differ more than 170 MW for most of the 8760 hours, comprising a year.

Finally, the forecast error duration curve is presented in Fig.3. It gives the percentage of the hours of the year when the absolute forecast error is greater than the MW value shown on the MW-axis. It can be easily observed that nearly 64% of the hourly predictions has a forecast error of less than 100 MW. Moreover, only 15 of the predicted loads (0.017%) have an error in excess of 500 MW.

Based on the results presented above and to the fact that the standard deviation of the forecast error is small, we can conclude that the suggested method successfully trains the fuzzy system to perform efficient and robust forecasting.

Table II. Fuzzy model parameters

# of parameters	Winter	Spring	Summer	Autumn	Total
Premise	*288*	*252*	*336*	*312*	*1188*
Consequent	*168*	*147*	*196*	*182*	*693*

Table III. Seasonal and yearly APE

Day of the week	Winter	Spring	Summer	Autumn	Year
Monday	*2.05*	*2.07*	*1.68*	*1.85*	*1.90*
Tuesday	*1.93*	*1.86*	*1.36*	*2.09*	*1.75*
Wednesday	*1.84*	*1.73*	*1.48*	*1.52*	*1.64*
Thursday	*1.91*	*1.77*	*1.83*	*1.81*	*1.83*
Friday	*1.84*	*1.64*	*1.49*	*1.57*	*1.63*
Saturday	*1.97*	*2.05*	*1.70*	*1.85*	*1.88*
Sunday	*2.54*	*2.19*	*1.83*	*2.00*	*2.13*
Week	*2.01*	*1.90*	*1.63*	*1.82*	***1.82***

Table IV. Absolute load difference in MW

	Winter	Spring	Summer	Autumn	Year
Aver.	*101.8*	*89.6*	*84.1*	*91.8*	*91.1*
Std. Dev.	*85.4*	*74.7*	*75.5*	*72.0*	*77.6*

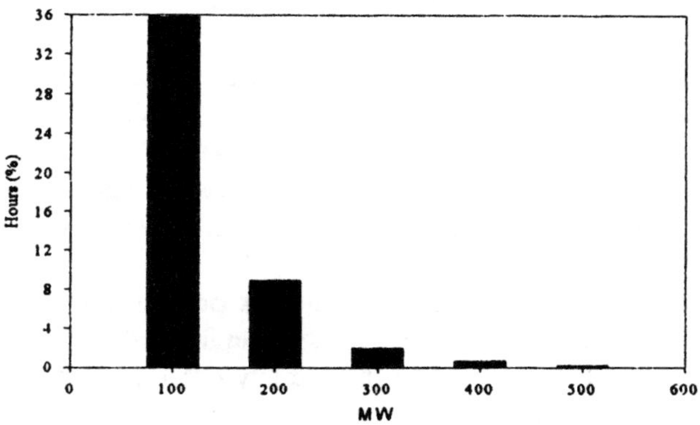

Fig.3 Forecast error duration curve

7. Conclusions

A fuzzy model for short-term load forecasting has been developed. The OLS method has been employed to perform the premise partitioning, thus to provide the initial premise parts of the fuzzy rules. The RLSE method has been used to determine the consequent parameters. Different fuzzy model have been assigned to each day of the week for every season. Test results show that the developed STLF model produces accurate load predictions.

Acknowledgement

This research was funded by the Secretary General of Research and Technology of the Greek Ministry of Industry, Energy and Technology and the Public Power Corporation (PPC) of Greece. The authors would like to thank Mr Katsigiannakis of PPC for supporting this research.

References

[1] A.D. Papalexopoulos and T.C. Hesterberg, "A Regression-Based Approach to Short Term System Load Forecasting", *IEEE Trans. on Power Systems*, Vol. 5, No. 4, pp. 1535-1547, 1990.

[2] J.H. Park, Y.M. Park and K.Y. Lee, "Composite modeling for adaptive short-term load forecasting", *IEEE Trans. on Power Systems*, Vol. 6, No. 2, pp. 450-457, 1991.

[3] S. Ranman and R. Bhatnagar, "An expert system based algorithm for short term load forecast", *IEEE Trans, PWRS*, Vol. 3, pp. 392-399, 1988.

[4] IEEE/Power Engineering Society, "A tutorial course on Artificial Neural Networks with Applications to Power Systems", edited by M. El-Sharkawi and D. Niebur, 96TP112-0, 1996.

[5] D.C. Park, M.A. El-Sharkawi, R.J. Marks, L.E. Atlas and M.J. Damborg, "Electric Load Forecasting Using an Artificial Naural Network", *IEEE Trans. on Power Systems*, Vol. 6, No. 2, pp. 442-449, 1991.

[6] D. Niebur et al., "Artificial neural networks for Power Systems", CIGRE TF38.06.06 Report, ELECTRA, No. 159, pp. 77-101, April 1995.

[7] T.M. Peng, N.F. Hubele and G.G. Karady, "Advancement in the Application of Neural Networks for Short Term Load Forecasting", *IEEE Trans. on Power Systems*, Vol. 7, No. 1, pp. 250-258, 1992.

[8] A.G. Bakirtzis, V. Petridis, S.J. Kiartzis, M.C. Alexiadis and A.H. Maissis, "A Neural Network Short Term Load Forecasting Model For The Greek Power System", *IEEE Trans. on Power Systems*, Vol. 11, No. 2, pp. 858-863, 1996.

[9] A.G. Bakirtzis, J.B. Theocharis, S.J. Kiartzis and K.J. Satsios, "Short Term Load Forecasting Using Fuzzy Neural Networks", *IEEE Trans. on Power Systems*, Vol. 10, No. 3, pp. 1518-1524, 1995.

[10] D. Srinivasan, C.S. Chang and A.C. Liew, "Demand Forecasting Using Fuzzy Neural Computation, with special emphasis on weekend and public holiday forecasting", *IEEE Trans. on Power Systems,* Vol. 10, No. 4, pp. 1897-1903, 1995.

[11] H. Mori and H. Kobayashi, "Optimal Fuzzy Inference for Short-Term Load Forecasting", *IEEE Trans. on Power Systems,* Vol. 11, No. 1, pp. 390-396, 1996.

[12] A. Piras, A. Germond, B. Buchenel, K. Imhof and Y. Jaccard, "Heterogeneous Artificial Neural Network for Short Term Electrical Load Forecasting", *IEEE Trans. on Power Systems,* Vol. 11, No. 1, pp. 397-402, 1996.

[13] T. Takagi and M. Sugeno, "Fuzzy Identification of Systems and Its Applications", *IEEE Trans. on Syst., Man, and Cybern.*, Vol.15, No. 1, pp. 116-132, 1985.

[14] J.S.R. Jang, "ANFIS: Adaptive-Network-Based Fuzzy Inference System", *IEEE Trans. on Syst., Man, and Cybern.*, Vol. 23, No. 3, pp. 665-685, 1993.

[15] M. Sugeno and T. Yasukawa, "A Fuzzy-Logic-Based Approach to Qualitative Modeling", *IEEE Trans. on Fuzzy Systems,* Vol. 1, No. 1, pp. 7-31, 1993.

[16] S. Chen, C.F.N. Cowan and P.M.Grant, "Orthogonal Least Squares Learning Algorithm for Radial Basis Functions Networks", *IEEE Trans. on Neural Networks*, Vol. 2, No. 2, pp. 302-309, 1991.

[17] J. Hohensohn and J.M. Mendel, "Two-Pass Orthogonal Least-Squares Algorithm to Train and Reduce Fuzzy Logic Systems", *Proc. 1994 Int. Fuzzy Systems Conf,* pp. 696-700.

[18] L. X. Wang and J.M. Mendel, "Fuzzy Basis Functions, Universal Approximation, and Orthogonal Least-Squares Learning", *IEEE Trans. on Neural Networks*, Vol. 3, No. 5, pp. 807-814, 1992.

[19] P. Mastorocostas and J. Theocharis, "A Fast Learning Hybrid Algorithm for Training Fuzzy Neural Networks", *Proc. 1996 IASTED Int. Conf. on Systems, Modeling, Identification and Control*, pp. 129-132.

[20] D.A. Karras and S.J. Perantonis, "An Efficient Constrained Training Algorithm for Feedforward Networks", *IEEE Trans. on Neural Networks*, Vol. 5, No. 6, pp. 1420-1434, 1995.

[21] A. E. Bryson and W. F. Denham, "A Steepest-Ascent Method for Solving Optimum Programming Problems", *Journal of Applied Mechanics*, 29, pp.247-257, June 1962.

7

AN UNSUPERVISED FUZZY CLASSIFICATION ALGORITHM FOR NON ELLIPTIC CLASSES

P. BILLAUDEL, A. DEVILLEZ
and G. VILLERMAIN LECOLIER
Laboratoire d' Automatique et de Micro-électronique
Faculté des Sciences, Moulin de la Housse, B.P. 1039
51687 REIMS Cedex 2, France

1. Introduction

The non elliptic classes cause problem in unsupervised classification. It's often necessary to use a multiprototype approach. The set of points is divided into a large number of subclasses. The main difficulty consists in determining the subclasses to merge in order to restore the real classes. Another problem is the automatic determination of the number of classes.

In this chapter we have put forward an unsupervised method allowing, in a first stage, to create the subclasses. To that effect we use the Fuzzy C-Means algorithm. In a second stage we have defined the concept of neighbourhood between two classes by using the principle of ambiguity rejects. This has allowed us to achieve the merging of subclasses, thus entailing the restoring of real classes.

2. Problems of the Fuzzy C-Means Method

2.1. Problem of non elliptic classes

Unsupervised classification methods like Fuzzy C-Means [1,2] work satifactorily if the shape of classes is spherical or elliptical. Figure 1 shows the result of clustering achieved through this method. Two classes of parabolic shape and one class of elliptic shape are included in the set of points. Each point is represented by the number of the class to which it was allocated. We can note that the three classes resulting from this clustering don't correspond to the real classes. Therefore, owing to the shape of the classes this set can't be divided by the fuzzy C-Means algorithm.

S.G. Tzafestas (ed.), Advances in Intelligent Systems, 71–80.

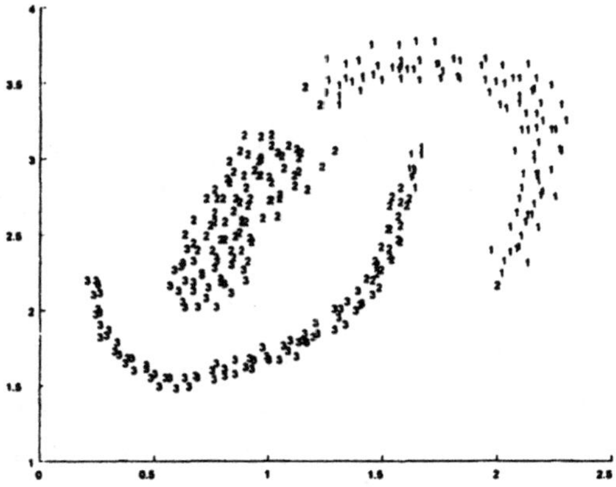

Figure 1. *Clustering obtained by the Fuzzy C-Means Algorithm, for a set of points containing classes of non elliptic shape. Points belonging to the same class are marked by a similar number.*

2.2. Determination of the number of classes

The classical Fuzzy C-Means algorithm requires the a priori knowledge of the number of classes. The easiest simple way to solve this problem consists in achieving the clustering on an increasing number of classes. The stop is made on the convergence of a criterion. These methods among which ISODATA is best known [3], requires important computing times.

Other methods derived from the Fuzzy C-Means divide the set into a large number of subclasses. These ones are then merged following a similarity criterion [4]. This relatively performing method uses several parameters which are not easy to determine.

3. Multiprototype approach

3.1. Creation of the prototypes

The first stage of our algorithm consists in using the Fuzzy C-Means algorithm to divide the set into c subclasses in order to obtain c prototypes. The result of this algorithm is a fuzzy membership matrix U. The element u_{ij} represents the membership degree of sample j to subclass i. Each subclass must belong to one real class only. The number of subclasses must be greatly higher than the number of real classes. In our example c must be higher than 3. It's possible to use some heuristic criteria to choose c [5]. We used the *fuzzy hyper volume, compactness and separability* criteria defined in chart 1.

Criteria	Expressions
compactness	$\pi = \frac{1}{N} \cdot \sum_{i=1}^{c} \sum_{k=1}^{N} \left[u_{ik} \cdot \left\| X_k - V_i \right\| \right]^2$
séparability	$s = \left[\min_{i=1,c;j=1,c;j \neq i} \left\| V_i - V_j \right\| \right]^2$
fuzzy hypervolume	$F_{HV}(U,c) = \sum_{i=1}^{c} [\det(\sum_i)]^{(1/2)}$

Chart 1. Criteria of classes validity

In this chart :

N is the total number of samples,
X_k is the k^{ist} sample,
V_i is the centroïd of the sublass i,
$\sum_i$ is the fuzzy covariance matrix defined by :

$$\sum_i = \frac{S_i}{\sum_{k=1}^{N} (u_{ik})^2}$$ where S_i is the fuzzy dispersion matrix :

$$S_i = \sum_{k=1}^{N} (u_{ik})^2 * (X_k - V_i)(X_k - V_i)^T \qquad i=1,c$$

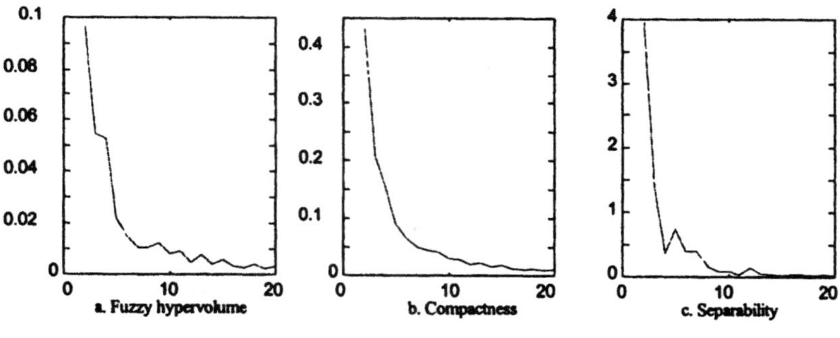

Figure 2. *Evolution of criteria according to the number of classes*

The results we obtained are presented in figure 2. We can note that the evolution of the three criteria includes an instability phase and a stability phase. We can say that when this last one is attained the classes are sufficiently compact and dividing them again won't favour the separability. We can estimate that in stable phase, each created subclass belongs to one real class only. For our example the stable phase is attained for a number of subclasses higher than 12. The division we obtained into 12 subclasses, is presented on figure 3a and the corresponding centroids appear on figure 3b. We can note that each subclass belongs to one real class only.

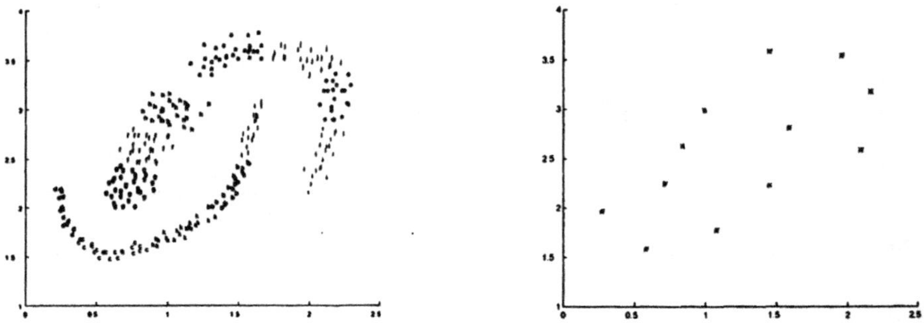

a b

Figure 3. *Division of the set into 12 subclasses. All points marked by a similar letter belong to the same subclass.*

3.2. Neighbourhood of classes and ambiguous points

The second stage of our algorithm uses the concept of neighbourhood of classes. We can consider that two classes are neighbouring if they have an ambiguity zone in common, that is to say a zone containing ambiguous points. The definition of an ambiguous point is given in [6, 5]. In our study we have used the following concept : a point is ambiguous if the difference between its first two belonging maximums is lower than a defined threshold. According to this principle and in defining the ambiguity threshold to 0.1, we have presented the obtained results in figure 4. We can note that the ambiguous points marked by the symbol « a » are situated in zones of class neighbourhood. In fact in the Fuzzy C-Means algorithm, for a given point i belonging to class j, the proximity of a class k induces the decreasing of the coefficient u_{ij} on behalf of the coefficient u_{kj}, the sum of all the coefficients being always equal to 1. The ambiguous points don't appear in zones situated between non neighbouring classes. So we are in conditions where ambiguity and distance rejects can not be confused.

Ambiguous points will serve as a basis to define the neighbourhood concept. Two classes j and k are considered as neighbouring if one of the following conditions is respected :

$$u_{ji}=\max(u) \text{ AND } u_{ki}= \max(u-\{u_{ji}\}),$$
$$\text{OR}$$
$$u_{ki}=\max(u) \text{ AND } u_{ji} = \max(u-\{u_{ki}\}),$$

and if the following relation is true for at least one point i :

$$|u_{ji}-u_{kj}|<s$$

where s is the ambiguity threshold,
and $u=(u_{1i}, \ldots, u_{zi}, \ldots, u_{ci})$ is the belonging vector of point i.

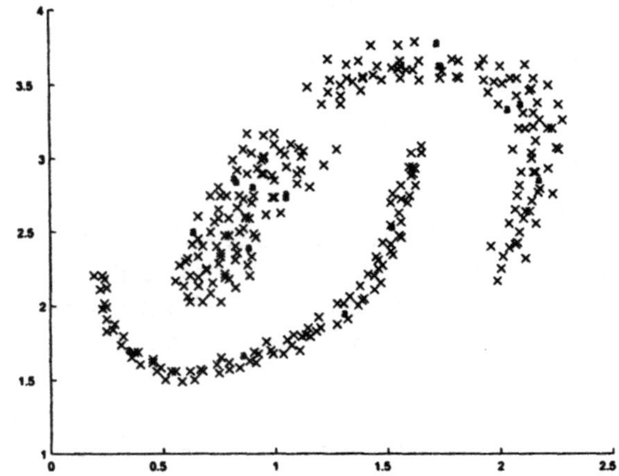

Figure 4. *Position of the ambiguous points between two or several classes*

3.3. Design of a proximity graph

We can build a proximity graph from ambiguous points. Summits represent subclasses and arcs represent links between them. A link is created between two summits if the corresponding subclasses are neighbouring. The graph relative to our example is presented on figure 5. Real classes are obtained by merging subclasses which exist in the connected components. A connected component is a whole of summits linked between one another [7]. For our example the algorithm reveals the following connected components :

C1=[a e i j]
C2=[f g h]
C3=[b c d k l]

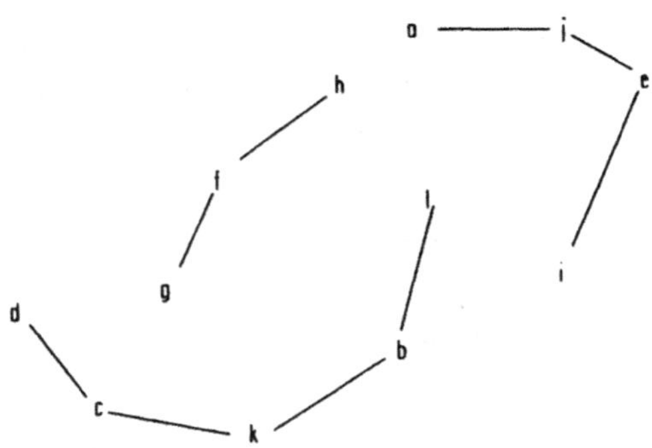

Figure 5. *Proximity graph*

The connected components correspond to the three starting classes. The method we develop allows us to find real classes and to determine the exact number of classes existing in a set of points.

3.4. Fusion of prototypes

For each point we have a membership degree for each subclass at our disposal. It is necessary to realise a fusion in order to obtain a membership degree for each real class. The fusion is made by the operator used in [8]. When a number k of subclasses SC_j exists within a class i, the membership function to this class is defined by :

$$u_i(x) = u_{SC_1}(x) \cup .... \cup u_{SC_{k_i}}(x) = \min\left[1, \sum_{j=1}^{k_i} u_{SC_j}(x)\right]$$

For our example, we obtained the following results :

$u_1=\min(1, u_a+u_e+u_i+u_j)$
$u_2=\min(1, u_f+u_g+u_h)$
$u_3=\min(1, u_b+u_c+u_d+u_k+u_l)$

The global result is presented on figure 6. The developed method enables us to divide sets of points containing classes of non elliptic shape into unsupervised mode. Yet, the starting set mustn't contain any ambiguous points. This condition is generally respected if the cluster is used as a learning whole for a supervised classification method.

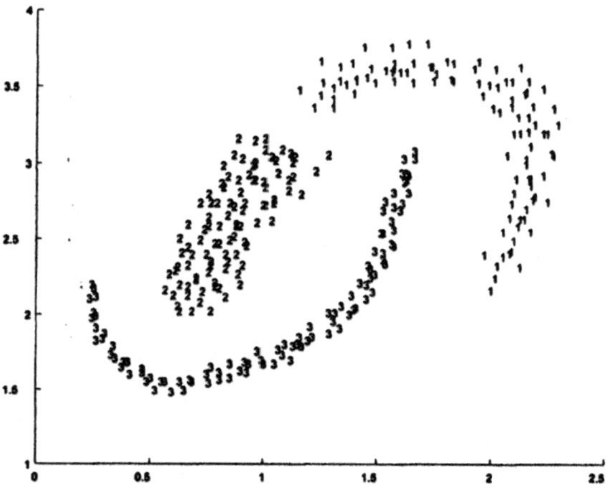

Figure 6. *Global clustering after fusion of subclasses. Points marked by a similar number belong to the same class.*

3.5. Algorithm

The algorithm is presented on figure 7. We can find all the phases described previously.

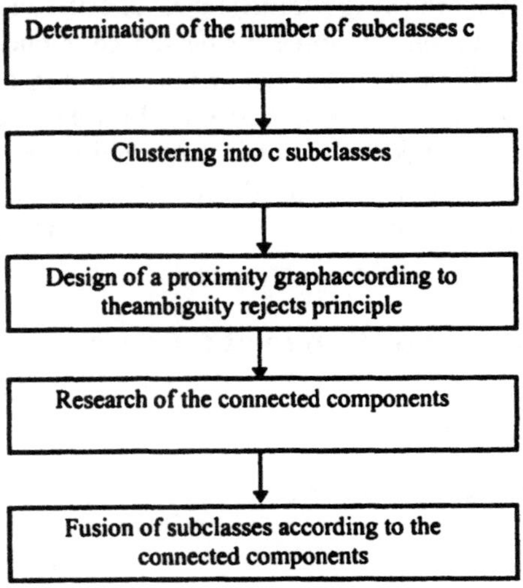

Figure 7. *Algorithm*

4. Application of the algorithm to IRIS data

To test and validate our method we used the IRIS data. The set contains three classes of 50 points corresponding to three types of flowers : IRIS setona versicolor and virginia. The four attributes which were retained are the length and width of petals and the length and width of sepals.

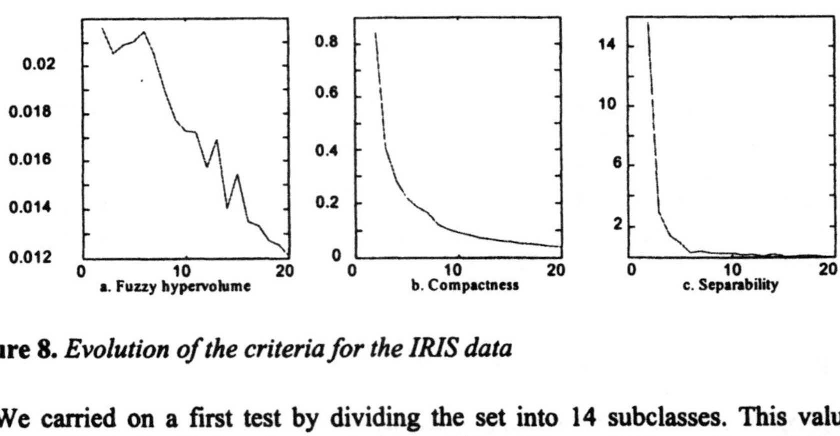

Figure 8. *Evolution of the criteria for the IRIS data*

We carried on a first test by dividing the set into 14 subclasses. This value was chosen with the heuristic criteria of class validity. Figure 8 shows its evolution. The result was not satisfactory because our algorithm couldn't distinguish the classes 2 and

3. Several classification errors appeared on the membership matrix.

The problem comes from the existence of several ambiguous points in the starting set of points. After the elimination of 20 ambiguous points , we applied the Fuzzy C-Means algorithm with the following parameters m=2, c=14 and ε=0.001. The use of our algorithm allowed the building of three connected components corresponding to the three classes of flowers. Four classification errors remain in the final clustering, which gives a classification of 96.9%. This example enables us to check the validity of our method when the starting set doesn't contain any ambiguous points. For an equivalent recognition rate, the algorithm determined the exact number of classes whereas this number must be imperatively known for the Fuzzy C-Means algorithm.

5. Application of the algorithm to other data

5.1. Presentation of the data

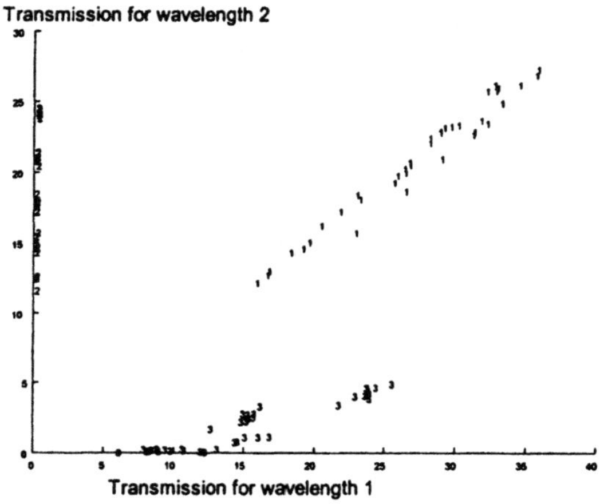

Figure 9. *Point set obtained for the transmission values of the plastic materials*

These data come from a study achieved on the sorting of plastic bottles composed of vinyl polychloride, therphtalate polyethylene, high density polyethylene. The analysis of the spectrums of each material with a statistic analysis allowed us to retain two wavelengths to represent the classes. We have 90 transmission values for each wavelength. These values are divided into three classes of 30 points and correspond to the three polymers as shown on figure 9.

5.2. Application of the algorithm and results

5.2.1. *Problems linked with the shape of classes*

We can note that the point sets have very long shapes, which is problematic when we use the Fuzzy C-Means with the euclidian distance. The recognition rate attains 97%. When using the Mahalanobis distance, we obtain a recognition rate of 100 %.

The presence of empty zones in classes constitutes a second problem to choose the ambiguity threshold which serves to build the graph.

5.2.2. *Résults*

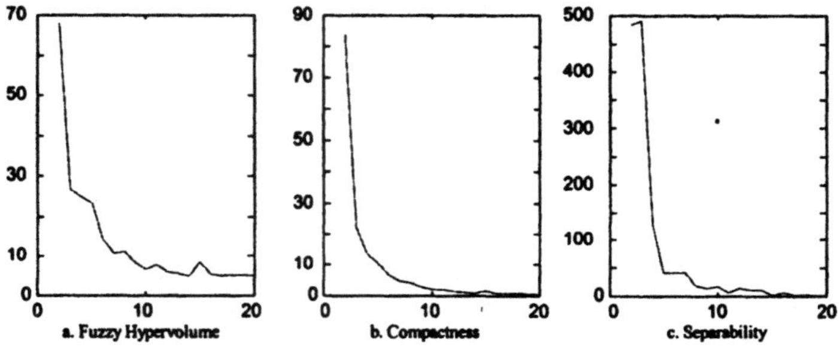

Figure 10. *Evolution of the criteria of class validity for plastic material data*

Figure 10 presents the evolution of the criteria according to the number of classes. They allow us to divide the set into 14 subclasses. The threshold to build the graph was fixed at 0.85. The obtained recognition rate is 100%, which shows that the algorithm finds the subclasses relative to each real class. From this result we can note that the algorithm functions correctly in spite of presence of empty zones and the long shape of classes. However in this case, it is not easy to determine the value of the threshold defining the neighbourhood notion.

6. Conclusion

The unsupervised method which we have developed, has enabled us to find subclasses relative to each real classes. With the fusion of these subclasses may be achieved a clustering respecting the shape of classes. This method provides a solution for the unsupervised classification of data offening complex shapes of classes. Unlike the Fuzzy C-Means method, it doesn't require the a priori knowledge of the number of classes.

Nevertheless our method contains some restrictions due to the separability of classes. The presence of empty zones in a class can generate problems as regards the choice of neighbourhood threshold.

It's possible to use the clustering achieved by this type of algorithm as a learning basis for supervised methods such as K Nearest Neighbours or Fuzzy Pattern Matching [9].

References

[1] J.C. BEZDEK, "Pattern recognition with fuzzy objective function algorithms", Plenum Press, New-York , 1981.

[2] H. CHOE et J.B. JORDAN, "On the optimal choice of parameters in the fuzzy C-means algorithm", Proc. of fuzzy IEEE 92, pp 349 354, 1992.

[3] J. C. DUNN, "A fuzzy relative of the Isodata process and it use in detecting compact well separated clusters", Journal of Cybernetics, Vol. 3, pp 32-57, 1974.

[[4] H. FRIGUI, R. KRISHNAPURAM, "A robust algorithm for automatic extraction of an unknown number of clusters from noisy data", Pattern Recognition Letters, Vol. 17, pp 1223-1232, 1996.

[5] C. FRELICOT, "Un système adaptatif de diagnostic prédictif par reconnaissance de formes floue", Thèse en contrôle des systèmes, Université Technologique de Compiègne, 1992.

[6] B. DUBUISSON, "Diagnostic et reconnaissance des formes", Traité des Nouvelles Technologies, série Diagnostic et Maintenance, HERMES, 1990.

[7] M. GONDRAN, M. MINOUX, "Graphes et algorithmes", Collection des Etudes et Recherches d'Electricité de France, Eyrolles, 1985, Paris.

[8] M. H. MASSON, B. DUBUISSON, C. FRELICOT, "Conception d'un module de reconnaissance de formes floues pour le diagnostic", RAIRO-APII-JESA, Vol. 30, n° 2-3, 1996.

[9] A. DEVILLEZ, P. BILLAUDEL, G. VILLERMAIN LECOLIER, "Use of the Fuzzy Pattern Matching in a diagnosis module based on the pattern recognition", CESA'98, IEEE Systems, Man and Cybernetics, Vol. 4, pp 902-907, April 1-4, Nabeul-Hammamet, Tunisia, 1998.

8

NECESSARY AND SUFFICIENT CONDITIONS FOR STABILITY OF DYNAMIC FUZZY SYSTEMS

F. MATÍA, B. M. AL-HADITHI and A. JIMÉNEZ
DISAM - Universidad Politécnica de Madrid
José Gutiérrez Abascal 2
E-28006, Madrid, Spain

1 Introduction

Although currently fuzzy control has become a widely used technique, stability issues are still under study when using Takagi-Sugeno Model [1]. Many research works have dealt with stability problems regarding this model [2], [3], [4], [5], [6],[7].

Takagi-Sugeno model has its main advantage in modeling unknown non-linear systems, but with known local dynamics. The consequent part of each rule is a linear function of the input

$$x^{(n)} = a_o + a_1 x + a_2 \dot{x} + \ldots + a_n x^{(n-1)}$$

In general, stability analysis of fuzzy systems must be approached by using standard techniques for nonlinear systems. Tanaka-Sugeno present several stability theorems for Takagi-Sugeno discrete model, by using Lyapunov stability theory [4]. But their stability criterion is only available for a particular case in which $a_0 = 0$.

As we will see, this is only applicable to the particular case in which the linearization of all the linear subsystems (consequent part of each rule) is made with respect to the origin, which has no sense in practical applications. Other results dealing with the stability of discrete and continuous fuzzy systems have also been presented [2], [6], but they have the same limitation . In this chapter, we introduce new theorems for the general case in which $a_o \neq 0$ by using Lyapunov method.

The organization of the chapter is divided into two parts. First, we review some recent stability theorems of Takagi-Sugeno model in sections 2 and 3. Then, new theorems for the stability of the general case of Takagi-Sugeno model, with examples, will be presented in sections 4 and 5.

S.G. Tzafestas (ed.), Advances in Intelligent Systems. 81–92.

2 The fuzzy dynamic model

In this section we describe the continuous fuzzy dynamic model, proposed by Takagi and Sugeno. Given an nth order free system in which x is the state, the model represents the local dynamics of each IF-THEN rule $R^{i_1 \dots i_n}$, by a linear subsystem, as follows:

$$R^{i_1 \dots i_n}: \; IF \; \left(x \; is \; M_1^{i_1}\right) \; AND \dots \; AND \; \left(x^{(n-1)} \; is \; M_n^{i_n}\right)$$
$$THEN \; x^{(n)} = a_0^{i_1 \dots i_n} + \sum_{l=1}^{n} a_l^{i_1 \dots i_n} x^{(l-1)}$$

where $M_1^{i_1} (i_1 = 1, 2, \dots, r_1)$ are the fuzzy sets for x, $M_2^{i_2} (i_2 = 1, 2, \dots, r_2)$ are the fuzzy sets for $\dot{x}$ and $M_n^{i_n} (i_n = 1, 2, \dots, r_n)$ are the fuzzy sets for $x^{(n-1)}$. Their membership functions are $\mu_{M_1^{i_1}}(x)$, $\mu_{M_2^{i_2}}(\dot{x})$ and $\mu_{M_n^{i_n}}(x^{(n-1)})$, respectively, as in Figure 1 (the shape may vary).

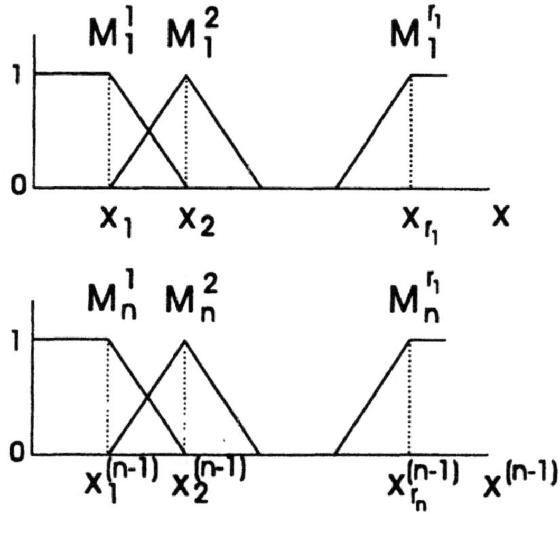

Fig. 1: Membership Functions Description

The total number of rules (and so of linear subsystems) is $r_1 r_2 \cdots r_n$. The previous notation can be simplified by using an array representation:

$$R^{i_1 \dots i_n}: \; IF \; \left(\mathbf{x} \; is \; \mathbf{M}^{i_1 \dots i_n}\right) \; THEN \; \dot{\mathbf{x}} = \mathbf{a_0}^{i_1 \dots i_n} + \mathbf{A}^{i_1 \dots i_n} \mathbf{x} \tag{1}$$

where $\mathbf{x}$ is the state array $\mathbf{x}^T = \left[x, \dot{x}, \dots, x^{(n-1)}\right]$, and

$$\left(\mathbf{x} \; is \; \mathbf{M}^{i_1 \dots i_n}\right) = \left(x \; is \; M_1^{i_1}\right) \; AND \dots \; AND \; \left(x^{(n-1)} \; is \; M_n^{i_n}\right)$$

If we restrict the analysis to the case in which we use the controllable canonical form

$$x^{(n)} = f(x, \dot{x}, \ldots, x^{(n-1)}) = a_0 + \sum_{l=1}^{n} a_l x^{(l-1)} \tag{2}$$

then the constant elements of equation (1) are

$$\mathbf{A}^{i_1 \ldots i_n} = \begin{bmatrix} 0 & 1 & 0 & \ldots & 0 \\ \vdots & \ddots & \ddots & \ddots & \vdots \\ \vdots & & \ddots & \ddots & 0 \\ 0 & \ldots & \ldots & 0 & 1 \\ -a_1^{i_1 \ldots i_n} & \ldots & \ldots & \ldots & -a_n^{i_1 \ldots i_n} \end{bmatrix} \qquad \mathbf{a_0}^{i_1 \ldots i_n} = \begin{bmatrix} 0 \\ \vdots \\ 0 \\ a_0^{i_1 \ldots i_n} \end{bmatrix}$$

The final free fuzzy system that results from blending all the linear subsystems becomes

$$\dot{\mathbf{x}} = \frac{\sum_{i_1=1}^{r_1} \cdots \sum_{i_n=1}^{r_n} w^{i_1 \ldots i_n}(\mathbf{x}) \left[\mathbf{a_0}^{i_1 \ldots i_n} + \mathbf{A}^{i_1 \ldots i_n} \mathbf{x}\right]}{\sum_{i_1=1}^{r_1} \cdots \sum_{i_n=1}^{r_n} w^{i_1 \ldots i_n}(\mathbf{x})} = \mathbf{a_0}(\mathbf{x}) + \mathbf{A}(\mathbf{x})\mathbf{x}$$

where it can be seen that it is a dynamic model in which its dynamics are expressed by the $\mathbf{A}(\mathbf{x})$ matrix. $\mathbf{A}(\mathbf{x})$ and $\mathbf{a_0}(\mathbf{x})$ may be calculated as follows:

$$\mathbf{A}(\mathbf{x}) = \frac{\sum_{i_1=1}^{r_1} \cdots \sum_{i_n=1}^{r_n} w^{i_1 \ldots i_n}(\mathbf{x}) \mathbf{A}^{i_1 \ldots i_n}}{\sum_{i_1=1}^{r_1} \cdots \sum_{i_n=1}^{r_n} w^{i_1 \ldots i_n}(\mathbf{x})}$$

$$\mathbf{a_0}(\mathbf{x}) = \frac{\sum_{i_1=1}^{r_1} \cdots \sum_{i_n=1}^{r_n} w^{i_1 \ldots i_n}(\mathbf{x}) \mathbf{a_0}^{i_1 \ldots i_n}}{\sum_{i_1=1}^{r_1} \cdots \sum_{i_n=1}^{r_n} w^{i_1 \ldots i_n}(\mathbf{x})}$$

where the weight of each rule is calculated as

$$w^{i_1 \ldots i_n}(\mathbf{x}) = \mu_{M_1^{i_1}}(x) \ldots \mu_{M_n^{i_n}}(x^{(n-1)}) = \prod_{l=1}^{n} \mu_{M_l^{i_l}}(x^{(l-1)})$$

In this equation, $w^{i_1 \ldots i_n}(\mathbf{x}) \geq 0$, $\forall x, \forall i_1, \ldots, i_n$, so we always have

$$\sum_{i_1=1}^{r_1} \cdots \sum_{i_n=1}^{r_n} w^{i_1 \ldots i_n}(\mathbf{x}) > 0$$

Furthermore, it can be demonstrated (see [8]) that if the membership functions for the variables in the premise part of each rule verify

$$\sum_{i_l=1}^{r_l} \mu_{M_l^{i_l}}(x^{(l-1)}) = 1, \ \forall x^{(l-1)}, \ \forall l = \{1, \ldots, n\}$$

then it can be stated that

$$\sum_{i_1=1}^{r_1} \cdots \sum_{i_n=1}^{r_n} w^{i_1 \ldots i_n}(\mathbf{x}) = 1$$

Special attention should be paid to the fact that $a_0(0) = \mathbf{0}$ must be satisfied to guarantee that the equillibrium holds at $\mathbf{x} = \mathbf{0}$.

3 Stability analysis

Tanaka-Sugeno have analyzed in [4] the stability of Takagi-Sugeno **discrete** model, considering only a special case in which the fuzzy system was represented by $\mathbf{x_{k+1}} = \mathbf{A}(\mathbf{x})\mathbf{x_k}$. This means that $\mathbf{a}_0^{i_1 \ldots i_n} = \mathbf{0}, \forall i_1, \ldots, i_n$. For this particular case, they demonstrate that the fuzzy system is asymptotically stable in the large, if there exists a common positive definite matrix $\mathbf{P}$ such that $\forall i_1, \ldots, i_n$,

$$(\mathbf{A}^{i_1 \ldots i_n})^T \mathbf{P} \mathbf{A}^{i_1 \ldots i_n} - \mathbf{P} < 0$$

where

$$\mathbf{A}(\mathbf{x_k}) = \frac{\sum_{i_1=1}^{r_1} \cdots \sum_{i_n=1}^{r_n} w^{i_1 \ldots i_n}(\mathbf{x_k}) \mathbf{A}^{i_1 \ldots i_n}}{\sum_{i_1=1}^{r_1} \cdots \sum_{i_n=1}^{r_n} w^{i_1 \ldots i_n}(\mathbf{x_k})}$$

Note that this result is reduced to the Lyapunov stability theorem for discrete linear systems when $r_1 = r_2 = \ldots = r_n = 1$.

The stability problem for Takagi-Sugeno **continuous** model, taking into account again the special case in which $\dot{\mathbf{x}} = \mathbf{A}(\mathbf{x})\mathbf{x}$ ($\mathbf{a}_0^{i_1 \ldots i_n} = \mathbf{0}, \forall i_1, \ldots, i_n$) was later developed by other authors. In [2] and [6] it was proved that the fuzzy system is asymptotically stable in the large, if there exists a common positive definite matrix $\mathbf{P}$ such that $\forall i_1, \ldots, i_n$,

$$(\mathbf{A}^{i_1 \ldots i_n})^T \mathbf{P} + \mathbf{P} \mathbf{A}^{i_1 \ldots i_n} < 0 \tag{3}$$

Again, note that this result is reduced to the Lyapunov stability theorem for continuous linear systems when $r_1 = r_2 = \ldots = r_n = 1$.

The stability condition for the **discrete** case shows that the fuzzy system may not be stable even if all the subsystems are stable [4]. This will not be the case in **continuous** fuzzy systems as we will see in section 4.

As we have seen, Takagi-Sugeno's model is based on the identification of the linear behaviour of the system around several linearization points. Of course, between each two linearization points, the model dynamic is only an approximation of the original non-linear system. Each linear subsystem has the form

$$\dot{\mathbf{x}} = \mathbf{a_0}^{i_1 \ldots i_n} + \mathbf{A}^{i_1 \ldots i_n} \mathbf{x}$$

In [4], Tanaka-Sugeno studied Takagi-Sugeno model when $\mathbf{a_0}^{i_1 \ldots i_n} = \mathbf{0}$, this means taht they formulate the subsystems as

$$\dot{\mathbf{x}} = \mathbf{A}^{i_1 \dots i_n} \mathbf{x}$$

which has no sense, as we can see in Figure 2. By eliminating the independent term, they are linearizing all the subsystems with respect to the origin, which does not represent a real system.

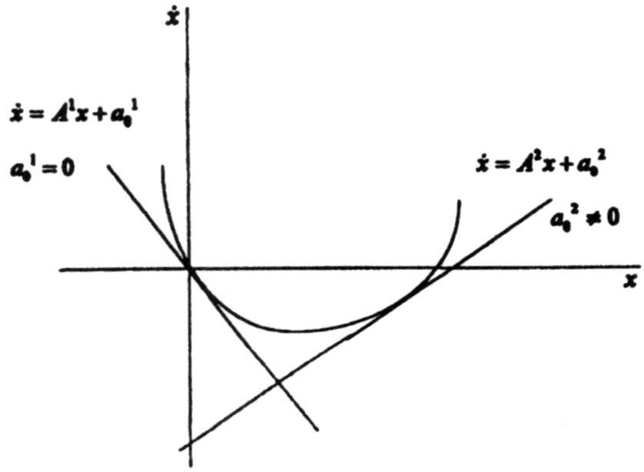

Fig. 2: Linearization in two points

4 Stability in the general case

Now, we present new stability theorems for the general case of Takagi-Sugeno model in which $\mathbf{a}_0^{i_1 \dots i_n} \neq 0$.

Let us assume that $\left(\mathbf{A}^{i_1 \dots i_n}\right)^{-1}$ exists (it can be demonstrated that this is true is we choose the canonical form shown in equation (2)). Then, if we define

$$\mathbf{B}^{i_1 \dots i_n} = -\left(\mathbf{A}^{i_1 \dots i_n}\right)^{-1}$$

$$\mathbf{b_0}^{i_1 \dots i_n} = -\left(\mathbf{A}^{i_1 \dots i_n}\right)^{-1} \mathbf{a_0}^{i_1 \dots i_n}$$

the rules of equation (1) may be rewritten in the following form:

$$R^{i_1 \dots i_n} : IF \ \left(\mathbf{x} \ is \ \mathbf{M}^{i_1 \dots i_n}\right) \ THEN \ \mathbf{B}^{i_1 \dots i_n} \dot{\mathbf{x}} + \mathbf{x} = \mathbf{b_0}^{i_1 \dots i_n}$$

with

$$\mathbf{B}^{i_1 \dots i_n} = \begin{bmatrix} b_1^{i_1 \dots i_n} & \dots & \dots & \dots & b_n^{i_1 \dots i_n} \\ -1 & 0 & \dots & \dots & 0 \\ 0 & \ddots & \ddots & & \vdots \\ \vdots & \ddots & \ddots & \ddots & \vdots \\ 0 & \dots & 0 & -1 & 0 \end{bmatrix} \qquad \mathbf{b_0}^{i_1 \dots i_n} = \begin{bmatrix} b_0^{i_1 \dots i_n} \\ 0 \\ \vdots \\ 0 \end{bmatrix}$$

and the resulting fuzzy system will be

$$\mathbf{B}(\mathbf{x})\dot{\mathbf{x}} + \mathbf{x} = \mathbf{b_0}(\mathbf{x}) \tag{4}$$

which is again a fuzzy dynamic system in which

$$\mathbf{B}(\mathbf{x}) = \frac{\sum_{i_1=1}^{r_1} \cdots \sum_{i_n=1}^{r_n} w^{i_1 \dots i_n}(\mathbf{x})\mathbf{B}^{i_1 \dots i_n}}{\sum_{i_1=1}^{r_1} \cdots \sum_{i_n=1}^{r_n} w^{i_1 \dots i_n}(\mathbf{x})}$$

$$\mathbf{b_0}(\mathbf{x}) = \frac{\sum_{i_1=1}^{r_1} \cdots \sum_{i_n=1}^{r_n} w^{i_1 \dots i_n}(\mathbf{x})\mathbf{b_0}^{i_1 \dots i_n}}{\sum_{i_1=1}^{r_1} \cdots \sum_{i_n=1}^{r_n} w^{i_1 \dots i_n}(\mathbf{x})}$$

and

$$\mathbf{b_0}(\mathbf{0}) = \mathbf{0}$$

It must be noted that

$$\mathbf{B}^{-1}(\mathbf{x}) \neq \frac{\sum_{i_1=1}^{r_1} \cdots \sum_{i_n=1}^{r_n} w^{i_1 \dots i_n}(\mathbf{x}) \left(\mathbf{B}^{i_1 \dots i_n}\right)^{-1}}{\sum_{i_1=1}^{r_1} \cdots \sum_{i_n=1}^{r_n} w^{i_1 \dots i_n}(\mathbf{x})}$$

Theorem 1 *The equillibrium of a free fuzzy system described by equation (4) is aymptotically stable in the large, if there exists a common positive definite matrix* $\mathbf{P}$ *such that,* $\forall \mathbf{x} \neq \mathbf{0}$,

$$\left[\mathbf{B}^{-1}(\mathbf{x})\right]^T \left[\frac{d\mathbf{b_0}(\mathbf{x})}{d\mathbf{x}} - I\right]^T \mathbf{P} + \mathbf{P}\left[\frac{d\mathbf{b_0}(\mathbf{x})}{d\mathbf{x}} - I\right] \mathbf{B}^{-1}(\mathbf{x}) < 0$$

Proof. If we define a continuous function $V(\mathbf{x})$ as

$$V(\mathbf{x}) = [\mathbf{x} - \mathbf{b_0}(\mathbf{x})]^T \mathbf{P}\,[\mathbf{x} - \mathbf{b_0}(\mathbf{x})] \tag{5}$$

where $\mathbf{b_0}(\mathbf{0}) = \mathbf{0}$ and $\mathbf{P} > \mathbf{0}$, then $V(\mathbf{x})$ satisfies the following two conditions:

$V(\mathbf{0}) = 0$

$V(\mathbf{x}) > 0, \ \forall \mathbf{x} \neq \mathbf{0}$

and must satisfy

$\dot{V}(\mathbf{0}) = 0$

$\dot{V}(\mathbf{x}) < 0, \ \forall \mathbf{x} \neq \mathbf{0}$

Taking the derivative of equation (5),

$$\begin{aligned}
\dot{V}(\mathbf{x}) &= \left[\dot{\mathbf{x}} - \tfrac{d\mathbf{b_0}}{d\mathbf{x}}\dot{\mathbf{x}}\right]^T \mathbf{P}\left[\mathbf{x} - \mathbf{b_0}(\mathbf{x})\right] + \left[\mathbf{x} - \mathbf{b_0}(\mathbf{x})\right]^T \mathbf{P}\left[\dot{\mathbf{x}} - \tfrac{d\mathbf{b_0}}{d\mathbf{x}}\dot{\mathbf{x}}\right] \\
&= \dot{\mathbf{x}}^T \left[I - \tfrac{d\mathbf{b_0}}{d\mathbf{x}}\right]^T \mathbf{P}\left[\mathbf{x} - \mathbf{b_0}(\mathbf{x})\right] + \left[\mathbf{x} - \mathbf{b_0}(\mathbf{x})\right]^T \mathbf{P}\left[I - \tfrac{d\mathbf{b_0}}{d\mathbf{x}}\right]\dot{\mathbf{x}} \\
&= \left[\mathbf{x} - \mathbf{b_0}(\mathbf{x})\right]^T (\mathbf{B}^{-1}(\mathbf{x}))^T \left[\tfrac{d\mathbf{b_0}}{d\mathbf{x}} - I\right]^T \mathbf{P}\left[\mathbf{x} - \mathbf{b_0}(\mathbf{x})\right] + \\
&+ \left[\mathbf{x} - \mathbf{b_0}(\mathbf{x})\right]^T \mathbf{P}\left[\tfrac{d\mathbf{b_0}}{d\mathbf{x}} - I\right] \mathbf{B}^{-1}(\mathbf{x})\left[\mathbf{x} - \mathbf{b_0}(\mathbf{x})\right] \\
&= \left[\mathbf{x} - \mathbf{b_0}(\mathbf{x})\right]^T \Big((\mathbf{B}^{-1}(\mathbf{x}))^T \left[\tfrac{d\mathbf{b_0}}{d\mathbf{x}} - I\right]^T \mathbf{P} + \\
&+ \mathbf{P}\left[\tfrac{d\mathbf{b_0}}{d\mathbf{x}} - I\right] \mathbf{B}^{-1}(\mathbf{x})\Big)\left[\mathbf{x} - \mathbf{b_0}(\mathbf{x})\right] < 0, \ \forall \mathbf{x} \neq 0
\end{aligned}$$

So $\dot{V}(\mathbf{x}) < 0 \ \ \forall \mathbf{x} \neq 0,$ if

$$[\mathbf{B}^{-1}(\mathbf{x})]^T \left[\frac{d\mathbf{b_0}(\mathbf{x})}{d\mathbf{x}} - I\right]^T \mathbf{P} + \mathbf{P}\left[\frac{d\mathbf{b_0}(\mathbf{x})}{d\mathbf{x}} - I\right] \mathbf{B}^{-1}(\mathbf{x}) < 0, \ \ \forall \mathbf{x} \geq 0$$

□.

Theorem 2 *The equilibrium of a free fuzzy system described by the general form of Takagi-Sugeno's model (see equation (4)), is aymptotically stable in the large if*

$$\begin{cases} \frac{\partial b_0}{\partial x} < 1 \\ \frac{\partial b_0(\mathbf{x})}{\partial x^{(l-1)}} < b_{l-1}(\mathbf{x}) \qquad \forall l = \{2, \ldots, n\} \end{cases}$$

Proof. For simplicity, the demonstration will be carried out for a second order system. Assume that

$$\mathbf{M}(\mathbf{x}) = \left[I - \frac{d\mathbf{b_0}}{d\mathbf{x}}\right] \mathbf{B}^{-1}(\mathbf{x})$$

where $\mathbf{M}(\mathbf{x})$ has the form

$$\mathbf{M}(\mathbf{x}) = \begin{bmatrix} \frac{1}{b_2(\mathbf{x})}\frac{\partial \mathbf{b_0}(\mathbf{x})}{\partial \dot{x}} & 1 - \frac{\partial \mathbf{b_0}(\mathbf{x})}{\partial x} + \frac{b_1(\mathbf{x})}{b_2(\mathbf{x})}\frac{\partial \mathbf{b_0}(\mathbf{x})}{\partial \dot{x}} \\ -\frac{1}{b_2(\mathbf{x})} & -\frac{b_1(\mathbf{x})}{b_2(\mathbf{x})} \end{bmatrix} = \begin{bmatrix} a_{11} & a_{12} \\ a_{21} & a_{22} \end{bmatrix}$$

Applying Theorem 1,

$$\mathbf{M}^T(\mathbf{x})\mathbf{P} + \mathbf{P}\mathbf{M}(\mathbf{x}) < 0$$

this means, for instance, that

$$\begin{bmatrix} a_{11} & a_{21} \\ a_{12} & a_{22} \end{bmatrix} \begin{bmatrix} p_{11} & p_{12} \\ p_{12} & p_{22} \end{bmatrix} + \begin{bmatrix} p_{11} & p_{12} \\ p_{12} & p_{22} \end{bmatrix} \begin{bmatrix} a_{11} & a_{12} \\ a_{21} & a_{22} \end{bmatrix} = \begin{bmatrix} -1 & 0 \\ 0 & -1 \end{bmatrix}$$

and so

$$\begin{cases} 2(a_{11}p_{11} + a_{21}p_{12}) = -1 \\ a_{11}p_{12} + a_{21}p_{22} + a_{12}p_{11} + a_{22}p_{12} = 0 \\ 2(a_{12}p_{12} + a_{22}p_{22}) = -1 \end{cases}$$

resulting in the following **P** matrix

$$\mathbf{P} = \frac{\begin{bmatrix} -(a_{11}a_{22} - a_{21}a_{12} + a_{21}^2 + a_{22}^2) & a_{12}a_{22} + a_{21}a_{11} \\ a_{12}a_{22} + a_{21}a_{11} & -(a_{11}a_{22} - a_{21}a_{12} + a_{11}^2 + a_{12}^2) \end{bmatrix}}{2(a_{11} + a_{22})(a_{11}a_{22} - a_{21}a_{12})}$$

which is positive definite if $|\mathbf{P}| > 0$ and $p_{11} > 0$:

1) $|\mathbf{P}| = \frac{(a_{11}+a_{22})^2+(a_{12}-a_{21})^2}{4(a_{11}+a_{22})^2(a_{11}a_{22}-a_{12}a_{21})}$
   is greater than zero when $(a_{11}a_{22} - a_{12}a_{21}) > 0$, i.e. $|\mathbf{M}(\mathbf{x})| > 0$

2) $p_{11} = -\frac{a_{11}a_{22}-a_{12}a_{21}+a_{21}{}^2+a_{22}{}^2}{2(a_{11}+a_{22})(a_{11}a_{22}-a_{21}a_{12})}$
   which is greater than zero when $a_{11} + a_{22} < 0$, because $|\mathbf{M}(\mathbf{x})| > 0$

By substituting the values of a_{11}, a_{21}, a_{12} and a_{22} in the last two conditions, we obtain

1) $|\mathbf{M}(\mathbf{x})| > 0$ this means,

$$\begin{vmatrix} \frac{1}{b_2(\mathbf{x})}\frac{\partial \mathbf{b}_0(\mathbf{x})}{\partial \dot{x}} & 1 - \frac{\partial \mathbf{b}_0(\mathbf{x})}{\partial x} + \frac{b_1(\mathbf{x})}{b_2(\mathbf{x})}\frac{\partial \mathbf{b}_0(\mathbf{x})}{\partial \dot{x}} \\ -\frac{1}{b_2(\mathbf{x})} & -\frac{b_1(\mathbf{x})}{b_2(\mathbf{x})} \end{vmatrix} > 0$$

   And by solving the above deteminant, we obtain that $\frac{\partial \mathbf{b}_0(\mathbf{x})}{\partial x} < 1$

2) $a_{11} + a_{22} = \frac{1}{b_2(\mathbf{x})}\frac{\partial \mathbf{b}_0(\mathbf{x})}{\partial \dot{x}} - \frac{b_1(\mathbf{x})}{b_2(\mathbf{x})} < 0$ this means, $\frac{\partial \mathbf{b}_0(\mathbf{x})}{\partial \dot{x}} < b_1(\mathbf{x})$

□.

5 Examples

The advantage of the previous theorems is the ease of checking the stability. In this section we present two examples. Example 1 analyzes the stability of a first order system. It applies Theorem 2 for a case in which the system is unstable although all the linear subsystems are not. Example 2 analyzes the stability of a second order fuzzy system.

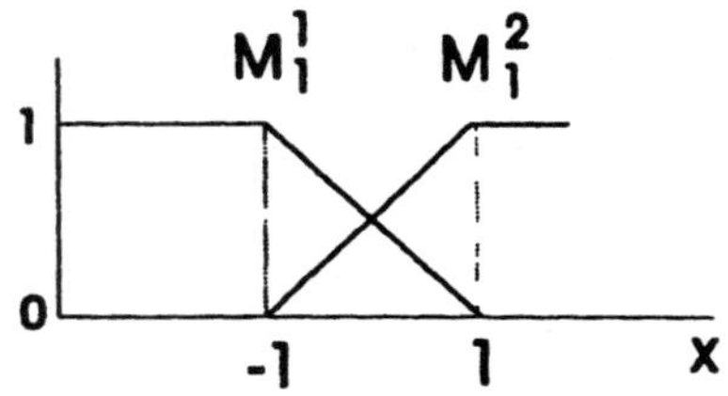

Fig. 3: Membership functions for the example

Example 1 Consider the fuzzy system described by:

$$\begin{array}{ll} R^1 & : IF \ (x \ is \ M_1^1) \ THEN \ \dot{x} + x = -2 \\ R^2 & : IF \ (x \ is \ M_1^2) \ THEN \ \dot{x} + x = 2 \end{array}$$

with fuzzy sets as shown in Figure 3.

In this case we have that $\mathbf{B}^1 = \mathbf{B}^2 = 1$ and that $\mathbf{b_0}^1 = -2$, $\mathbf{b_0}^2 = 2$. As we can see, the two linear sub-systems are stable, because $\mathbf{B_1}^1$ and $\mathbf{B_1}^2$ are positive definite. If we eliminate the independent term $\mathbf{b_0}^{i_1}$ in both rules, we could apply equation (3). By doing so, the system should be stable, because there exists a common positive definite matrix $\mathbf{P}$ such that $\forall i_1$, $(\mathbf{A}^{i_1})^T\mathbf{P} + \mathbf{P}\mathbf{A}^{i_1} < 0$. If we choose $\mathbf{P} = \lambda = 1$,

$$(\mathbf{A}^{i_1})^T\mathbf{P} + \mathbf{P}\mathbf{A}^{i_1} = \left[(-\mathbf{B}^{i_1})^{-1}\right]^T\mathbf{P} + \mathbf{P}(-\mathbf{B}^{i_1})^{-1} = -2 < 0$$

Nevertheless, the complete fuzzy system (with independent term) is unstable. In fact, by applying Theorem 2, it is not possible to determine the stability of the fuzzy system, as we will see. The parameters of the fuzzy system can be obtained as

$$\begin{array}{lll} \mathbf{B(x)} & = \dfrac{\sum_{i_1=1}^{2} w^{i_1}(\mathbf{x})\mathbf{B}^{i_1}(\mathbf{x})}{\sum_{i_1=1}^{2} w^{i_1}(\mathbf{x})} & = 1 \\ \mathbf{b_0(x)} & = \dfrac{\sum_{i_1=1}^{2} w^{i_1}(\mathbf{x})\mathbf{b_0}^{i_1}(\mathbf{x})}{\sum_{i_1=1}^{2} w^{i_1}(\mathbf{x})} & = 2x \end{array}$$

where $\mathbf{b_0(0)} = \mathbf{0}$ as required. In order to verify the conditions of theorem 2:

$$\frac{\partial \mathbf{b_0(x)}}{\partial x} < 1$$

but from the example

$$\frac{\partial \mathbf{b_0(x)}}{\partial x} = 2$$

so, it can not be concluded that the system is stable.

Example 2 Now a second order system is considered. Let there be a fuzzy system with the following rules:

$$\begin{array}{ll} R^{11} & : IF\ (x\ is\ M_1^1)\ AND\ (\dot{x}\ is\ M_2^1)\ THEN\ \ddot{x}+2\dot{x}+x=5 \\ R^{12} & : IF\ (x\ is\ M_1^2)\ AND\ (\dot{x}\ is\ M_2^1)\ THEN\ \ddot{x}+4\dot{x}+x=3 \\ R^{21} & : IF\ (x\ is\ M_1^1)\ AND\ (\dot{x}\ is\ M_2^2)\ THEN\ \ddot{x}+0.5\dot{x}+x=-2 \\ R^{22} & : IF\ (x\ is\ M_1^2)\ AND\ (\dot{x}\ is\ M_2^2)\ THEN\ \ddot{x}+\dot{x}+x=-6 \end{array}$$

where x and $\dot{x}$ have membership functions as in figure 3. It is easy to check that all the subsystems are stable. Taking into account the shape of the membership functions,

$$\begin{aligned} b_0(\mathbf{x}) &= \frac{\sum_{i_1=1}^{2}\sum_{i_2=1}^{2} w^{i_1 i_2}(\mathbf{x}) b_0^{i_1 i_2}}{\sum_{i_1=1}^{2}\sum_{i_2=1}^{2} w^{i_1 i_2}(\mathbf{x})} = \\ &= \tfrac{1}{4}\,[5(x-1)(\dot{x}-1) - 3(x+1)(\dot{x}-1)- \\ &+ 2(x-1)(\dot{x}+1) - 6(x+1)(\dot{x}+1)] \end{aligned}$$

with $b_0(0) = 0$. Following the same procedure we obtain that

$$\begin{aligned} b_1(\mathbf{x}) &= \tfrac{1}{4}\,[2(x-1)(\dot{x}-1) - 4(x+1)(\dot{x}-1)- \\ &- 0.5(x-1)(\dot{x}+1) + (x+1)(\dot{x}+1)] \end{aligned}$$

The next conditions must be satisfied:

1) for $\dot{x} = -1$:

$$b_0(\mathbf{x}) = 4 - x$$

$$\frac{\partial b_0}{\partial x} = -1 < 1$$

for $\dot{x} = 1$:

$$b_0(\mathbf{x}) = -2x - 4$$

$$\frac{\partial b_0}{\partial x} = -2 < 1$$

Between both points, the fuzzy system function is an interpolation, so the condition is also satisfied.

2) for $x = -1$:

$$b_0(\mathbf{x}) = \frac{3 - 7\dot{x}}{2}$$

$$b_1(\mathbf{x}) = \frac{5 - 3\dot{x}}{4}$$

$$\frac{\partial b_0}{\partial \dot{x}} = -3.5 < \frac{5 - 3\dot{x}}{4}$$

because $\dot{x} < \frac{19}{3}$.

for $x = 1$:

$$b_0(\mathbf{x}) = \frac{-3 - 9\dot{x}}{2}$$

$$b_1(\mathbf{x}) = \frac{5 - 3\dot{x}}{2}$$

$$\frac{\partial b_0}{\partial \dot{x}} = -4.5 < \frac{5 - 3\dot{x}}{2}$$

because $\dot{x} < \frac{14}{3}$.

Between both points, the fuzzy system function is an interpolation, so the condition is also satisfied.

So we can conclude that the fuzzy system is asymptotically stable.

6 Conclusions

In this chapter further theorems towards the stability analysis of Takagi-Sugeno fuzzy model have been presented. The suggested theorems extend the stability theorems of Tanaka-Sugeno to be more general, because now the linearization points are not always the origin. The presented stability conditions may be easily checked, as we have shown in the examples.

Acknowledgement

We acknowledge the funding oh the Spanish government through the CICYT project TAP96-0600 EVS (Virtual Platform for Autonomous Distributed Systems Engineering) and of the European Community through the TMR (Training and Mobility of Researchers program) project FMRX CT96-0070 MobiNet (Mobile Robotics Technology for Health Care Services).

References

[1] T. Takagi and M. Sugeno. Fuzzy identification of systems and its applications to modeling and control. *IEEE Transactions on Systems, Man and Cybernetics*, SMC–15(1):116–132, February 1985.

[2] B. Kosko. *Fuzzy Engineering.* Prentice Hall, 1997.

[3] F. Sheikholeslam and V. Tahani. Stability analysis of fuzzy systems and design of fuzzy controllers for linear plants. In *Seventh IFSA World Congress*, pages 404–409, 1997.

[4] K. Tanaka and M. Sugeno. Stability analysis and design of fuzzy control systems. *Fuzzy Sets and Systems*, 45:135–156, 1992.

[5] K. Tanaka and M. Sano. A robust stabilization problem of fuzzy control systems and its application to backing up control of a truck-trailer. *IEEE Transactions on Fuzzy Systems*, 2(2):119–134, May 1994.

[6] K. Tanaka, T. Ikeda, and H. O. Wang. Robust stabilization of a class of uncertain nonlinear systems via fuzzy control: Quadratic stabilizability, H_∞ control theory, and linear matrix inequalities. *IEEE Transactions on Fuzzy Systems*, 4(1):1–13, February 1996.

[7] H. O. Wang, K. Tanaka, and M. F. Griffin. An approach to fuzzy control of nonlinear systems: Stability and design issues. *IEEE Transactions on Fuzzy Systems*, 4(1):14–23, February 1996.

[8] A. Jiménez and F. Matía. A fast piecewise-linear implementation of fuzzy controllers. In *Proc. of NAFIPS'94, San Antonio, TX (EEUU)*, 1994.

9

FUZZY BEHAVIOR-BASED CONTROL WITH LOCAL LEARNING

K. IZUMI* and K.WATANABE**
**Department of Mechanical Engineering*
Faculty of Science and Engineering
Saga University, 1-Honjomachi, Saga 840-8502, Japan
*** Department of Advanced Systems Control Engineering*
Graduate School of Science and Engineering,
Saga University, 1-Honjomachi, Saga 840-8502, Japan

1. Introduction

A behavior-based control system, which can intelligently construct a controlled system for a mobile robot or manipulator, is now being attractive in the field of robotics [1]~[3]. One very interesting approach is known as a subsumption architecture [4]. In this method, the controlled system is separated into each elemental behavior, and then several results from elemental behaviors are competed or cooperated.

We developed a fuzzy behavior-based control system [5] that combines the concept of subsumption architecture and fuzzy reasoning technique, and then apply it for a mobile robot with two independent driving wheels. The present behavior-based control system is fundamentally similar to the system due to the subsumption architecture, but has a competition or co-operation unit consisting of a saturation function to generate a suitable consequent result. Thus, the proposed behavior-based control system is realized as a soft-computing.

The simplified fuzzy reasoning is assigned to each elemental behavior consisting of a single input-output relation. The mean value and inverse value of standard deviation for Gaussian membership function, and a constant value in the conclusion part are learned [6]~[8] by a genetic algorithm (GA). Thus, the total parameters associated with a fuzzy controller are encoded within one chromosome, like a Pittsburgh approach. However, it

S.G. Tzafestas (ed.), Advances in Intelligent Systems, 93–103.

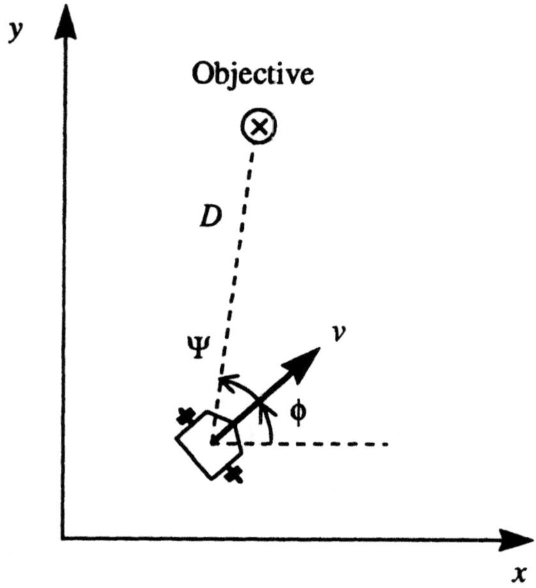

Figure 1. Behavior model for a mobile robot

should be noted that the results using the single input-output reasoning are not robust against the change of environment, so that a further method is required for making the fuzzy behavior-based control system robust against a change of environment.

In this paper, we discuss the local learning approach for the fuzzy behavior-based control system in order to make the system robust against an environment change. The effectiveness of the present learning method is illustrated with a simulation example for the acquisition of an objective point.

2. Problem Statement

Consider a terminal control problem in which a mobile robot with two independent driving wheels travels to a goal point as shown in **Fig. 1**. Here, D is the distance between the goal point and the mobile robot, Ψ is the relative angle between the azimuth of the robot and the goal point, v is the forward speed of the mobile robot, $\dot{\phi}$ is the forward azimuth rate of the mobile robot. These states are assumed to be obtained as the sensor information. To solve this problem, we construct a fuzzy behavior-based control system which has an objective behavior group and a free behavior group as shown in **Fig. 2**.

In Fig. 2, u_D^* denotes the force to change the goal distance, u_Ψ^* is the

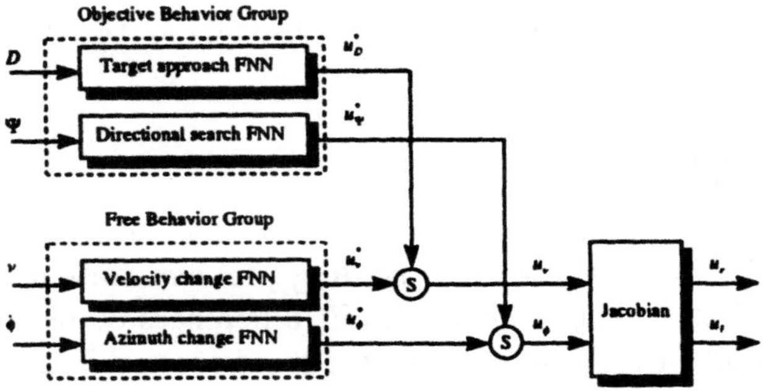

Figure 2. A fuzzy behavior-based control system

torque to change the relative angle to the goal point, u_v^* is the force to change the velocity of the robot, and u_ϕ^* is the torque to change the azimuth of the robot. The symbol S denotes a suppression unit that implies the cooperation and competition for the output results of two behavioral elements. If two reasoning results are expressed by a and b, then the suppression logic can be written by

$$c = (1-s)a + sb, \tag{1}$$
$$s = |\mathrm{sat}(a) + \mathrm{sat}(b)|/2, \tag{2}$$

where c denotes the result of competition or cooperation and sat$(\cdot)$ is a saturation function

$$\mathrm{sat}(x) = \begin{cases} \mathrm{sgn}(x) & \text{for } |x| > \epsilon \\ x/\epsilon & \text{for } |x| \le \epsilon \end{cases} \tag{3}$$

in which ϵ is a small positive constant. In this problem, the case in which the output results of two behavior groups have same sign is defined as the cooperation, whereas the case in which those results have opposite sign is defined as the competition. This logic differs from that in conventional subsumption. Also, the Jacobian matrix J of the mobile robot satisfies

$$\begin{bmatrix} u_r \\ u_l \end{bmatrix} = J^T \begin{bmatrix} u_v \\ u_\phi \end{bmatrix}, \tag{4}$$
$$J^T = \frac{r}{2}\begin{bmatrix} 1 & 1/l \\ 1 & -1/l \end{bmatrix}, \tag{5}$$

where u_l and u_r denote the driving torques for the left and right wheels, u_v is the force to change the speed of the robot, u_ϕ denotes the torque to change the azimuth of the robot, r is the radius of the wheel, l denotes the distance between one wheel and the center of gravity of the robot. In the following, we use a GA to acquire the knowledge on the control (i.e., fuzzy control rules that achieve the purpose).

3. Local Learning

In order to obtain global knowledge, the robot has to learn through a lot of experience. For example, in the preceding problem statement, the experience of the robot heavily depends on the differences of an initial condition and a goal point. A local learning method is discussed in here by using the structure of the fuzzy behavior-based control system, where the learning is assumed to be performed at each behavior group.

3.1. LEARNING OF OBJECTIVE BEHAVIOR GROUP

A goal domain is defined around the goal point (x_g, y_g). Learning or training positions are freely selected within the goal domain, where the number of learning patterns is to be l_g. The state variables are set to be anyone on a learning position. If the goal domain is the circle with the center (x_g, y_g) and radius $\bar{r}$, then the information used for the learning are to be v, ϕ, $\dot{\phi}$, x_g, y_g and $\bar{r}$. The fitness of GA is defined by

$$f_o = \sum_{i=1}^{l_g} D^i_{\text{end}} \tag{6}$$

where f_o denotes the fitness for the learning of the objective behavior group; D^i_{end} denotes the distance between the goal point and the final point of the robot at the simulation using the i-th learning information. The learning of the objective behavior group is performed without using the free behavior group. Thus, the fuzzy rules of the objective behavior group is obtained through this learning.

3.2. LEARNING OF FREE BEHAVIOR GROUP

The learning patterns of the free behavior group is similar to that of the objective behavior group. Let the number of learning patterns be l_f. The fitness of GA is then given by

$$f_f = \sum_{i=1}^{l_f} d^i_{\text{end}} \tag{7}$$

where f_f denotes the fitness for the learning of the free behavior group; d^i_{end} denotes the minimum distance between the goal domain and the final point of the robot at the simulation using the i-th learning pattern. The

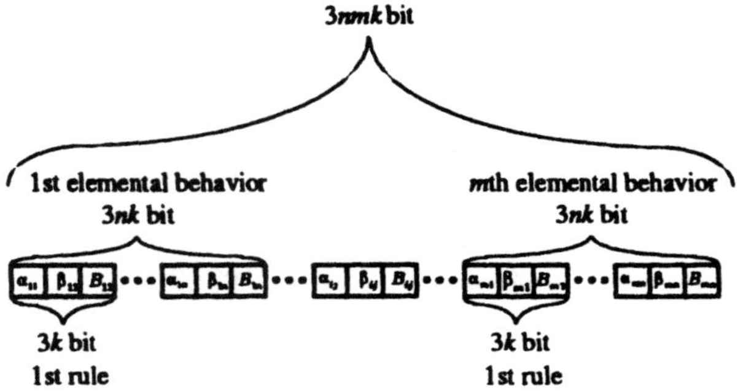

Figure 3. Individual structure

learning of the free behavior group is performed, together with the use of the outputs of the objective behavior group that has been already trained by the above procedure.

Consequently, all of fuzzy rules are obtained through each local learning.

4. GA Representation for Fuzzy Behavior-Based Control System

It is assumed that the fuzzy reasoning for each behavioral element consists of a simplified reasoning with single input and output and that a Gaussian type function is used as the membership function. In order to learn the fuzzy control rules with GA, the phenotype of one fuzzy rule is composed of the center value of the membership function α_{ij}, the reciprocal value of the standard deviation β_{ij} and the constant value of conclusion part B_{ij}, as shown in **Fig. 3**. The genotype is coded with 8 bits for each parameter. Here, the subscript i denotes the index of behavioral element and j denotes the fuzzy rule number. The transformation from the genotype to the phenotype is given by

$$x = x_{\min} + (x_{\max} - x_{\min}) \frac{\text{gcode}}{2^8 - 1} \tag{8}$$

where x is the center value of membership function or the reciprocal value of standard deviation or the constant value of conclusion, $x_{\min}$ and $x_{\max}$ denote the minimum and maximum values for each parameter, and "gcode" denotes the value obtained by transforming the genotype into the integer with a gray code.

If the number of labels is n, then an individual needs $n \times 8 \times 3$ bits for one behavioral element, because of the use of simplified fuzzy reasoning. Accordingly, if the total rules of the control system are expressed by one individual, like a Pittsburgh approach, then the individual length needs $96n$ bits for constructing four behavioral elements.

TABLE 1. Min and max values of central value

	Min	Max
v	-0.5	0.5
D	0	1
ϕ and Ψ	$-\pi$	π

TABLE 2. Min and max values of standard deviation

	Min	Max
v	1/14	1/4
D	1/14	1/4
ϕ and Ψ	$\pi/7$	$\pi/2$

5. Simulations

5.1. COMMON PARAMETER SETTING

The maximum and minimum values of the center value and the reciprocal value of the standard deviation for the membership function in the antecedent part are shown in **Table 1** and **Table 2**, and those of the constant value in the consequent part are also shown in **Table 3**. In addition, the model of a mobile robot studied in the reference [9] is used, where the radius r for the wheel is assumed to be 0.05 [m], and the half distance between driving wheels l is 0.165 [m].

The population size is set to be 60. The selection method is a tournament type with three individuals. The crossover method is a uniform type with its probability 0.6. The mutation probability is fixed to be 1/480. The number of generations is fixed to be 30. The boundary layer value ϵ, which is used

TABLE 3. Min and max values of consequent part

	Min	Max
v and D	-0.5	0.5
ϕ and Ψ	-0.5	0.5

TABLE 4. Learning pattern for the objective behavior group

	$v(0)$	$\phi(0)$	$\dot{\phi}(0)$	$x(0)$	$y(0)$	$\bar{r}$
pattern 1	0.25	$\pi/2$	0.0	5.0	4.0	1.0
pattern 2	0.25	$\pi/2$	0.0	4.0	5.0	1.0
pattern 3	0.25	$\pi/2$	0.0	3.0	5.0	1.0
pattern 4	0.25	$\pi/2$	0.0	4.0	3.0	1.0

TABLE 5. Learning factor for the free behavior group

	$v(0)$	$\phi(0)$	$\dot{\phi}(0)$	$x(0)$	$y(0)$	$\bar{r}$
pattern 1	0.0	0.0	0.0	0.0	0.0	1.0
pattern 2	0.0	$\pi/2$	0.0	0.0	0.0	1.0
pattern 3	0.0	$-\pi/2$	0.0	0.0	0.0	1.0
pattern 4	0.0	$-\pi$	0.0	0.0	0.0	1.0

in the saturation function to make a competition or cooperation of the reasoning result, is set to be 0.01. Finally, the goal point is (4.0, 4.0).

5.2. LOCAL LEARNING PATTERNS

In the learning of the objective behavior group, the learning patterns are selected 4 as shown in **Table 4**. The goal domain is the circle with the center (4.0, 4.0) and radius 1.0. On the other hand, in the learning of the free behavior group, the learning patterns are selected 4 as shown in **Table 5**.

5.3. SIMULATION RESULTS

In order to check the performance of the fuzzy rules obtained by the above local trainings, some simulations are conducted by unlearned initial conditions. The results are shown in **Fig. 4**. The robot arrives at the goal point for all initial conditions.

Next, the same fuzzy rules are applied for the environment with a different goal point, in which the position is assumed to be (8, 8). The result is depicted in **Fig. 5**. The robot also successfully arrives at the goal point.

Finally, the obtained fuzzy rules are given in **Fig. 6~Fig. 9** and **Table 6**.

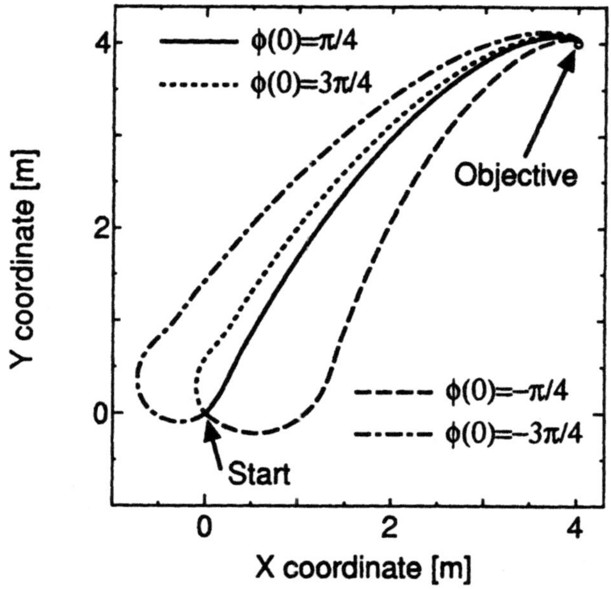

Figure 4. Resulting mobile robot path with different initial conditions

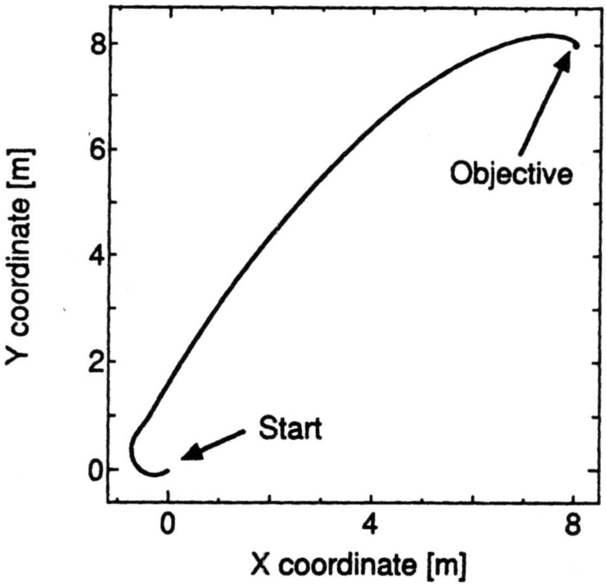

Figure 5. Resulting mobile robot path with different initial conditions

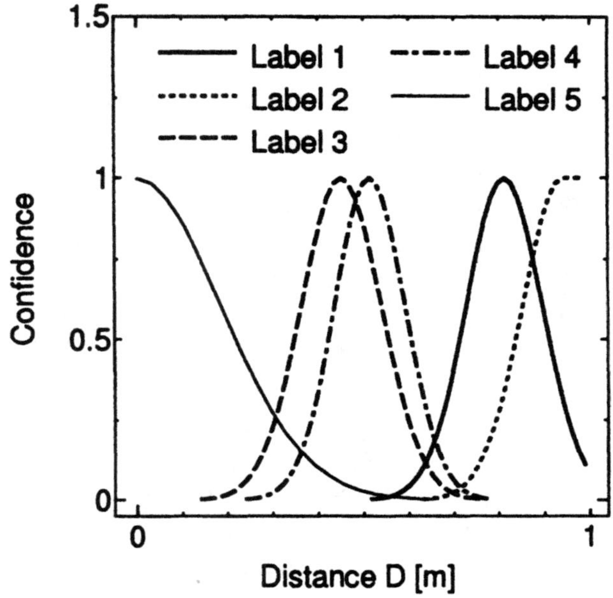

Figure 6. Membership functions obtained for distance v

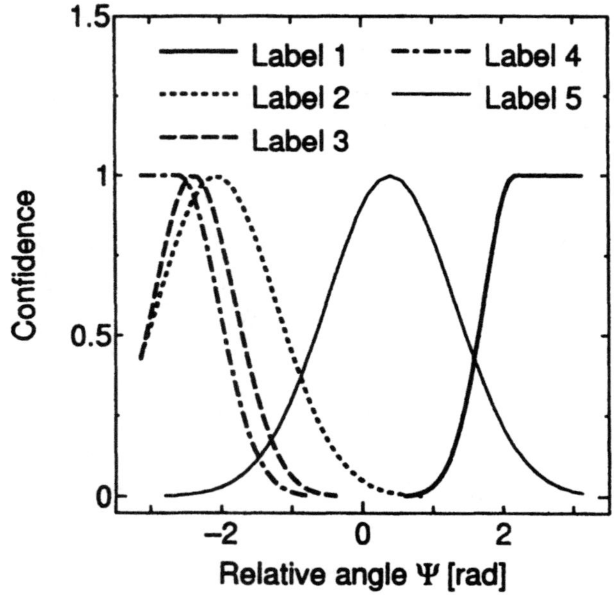

Figure 7. Membership functions obtained for relative angle Ψ

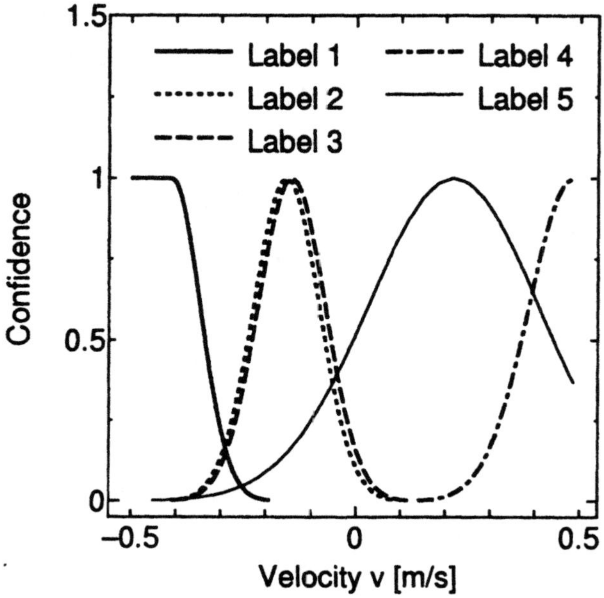

Figure 8. Membership functions obtained for velocity v

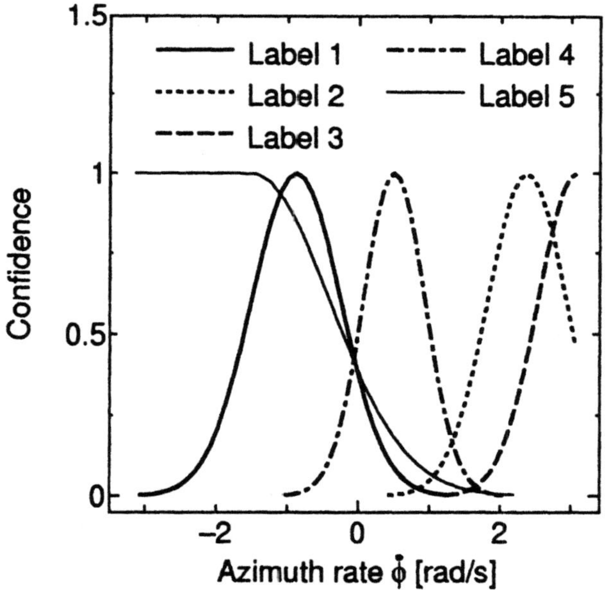

Figure 9. Membership functions obtained for forward azimuth rate $\dot{\phi}$

TABLE 6. Constant values of consequent part

	Label 1	Label 2	Label 3	Label 4	Label 5
D	0.061	0.469	0.073	−0.033	0.127
Ψ	0.045	−0.453	0.014	−0.5	0.065
v	−0.237	−0.159	0.296	−0.280	0.457
$\dot{\phi}$	−0.143	0.237	0.143	−0.029	0.186

6. Conclusions

The local learning for the fuzzy behavior-based control system has been applied to the acquisition problem of a goal point for a mobile robot. The efficacy of this method was shown by some simulations.

References

1. S. YamadaCLearning for Reactive Planning, *Journal of the Robotics Society of Japan*, Vol.13, No.1, pp.38-43, 1995 (in Japanese).
2. Y. KunoCBehaviors of Behavior-Based Robots, *Journal of the Robotics Society of Japan*CVol.11, No.8, pp.1178-1184, 1993 (in Japanese).
3. N. Yoshida, Robot Motion Control Using Subsumption Architecture, *Journal of the Robotics Society of Japan*CVol.11, No.8, pp.1118-1123, 1993 (in Japanese).
4. R.A. Brooks, A Robust Layered Control Systems for a Mobile Robot, *IEEE Robotics and Automation*, Vol. 2, No. 1, pp. 14-23, 1986.
5. K. Izumi and K. Watanabe, Fuzzy Behavior-Based Tracking Control for a Mobile Robot," *Procs. of 2nd Asian Control Conference*, Vol. I, pp. 685-688, 1997.
6. K. Watanabe and K. Izumi, "Construction of Fuzzy Behavior-Based Control Systems for a Mobile Robot," *Procs. of the Third Int. Symposium on Artificial Life and Robotics (AROB 3rd '98)*, Vol. 2, pp.518-523, 1998.
7. K. Izumi and K. Watanabe, "Fuzzy Behavior-Based Tracking Control for an Omnidirctional Mobile Robot," *Procs. of the Third Int. Symposium on Artificial Life and Robotics (AROB 3rd '98)*, Vol. 2, pp.524-527, 1998.
8. K. Izumi, K. Watanabe and T. Miyazaki, "Fuzzy Behavior-Based Control for a Miniature Mobile Robot," *Procs. of the 2nd Int. Conf. on Knowledge-based Intelligent Electronic Systems (KES '98)*, Vol. 3, pp. 483-490, 1998.
9. J. Tang, A. Nomiyama and K.Watanabe, "Stochastic Fuzzy Control Law for Path Tracking in Mobile Robot," *Procs. of 1996 Japan-U.S.A. Symposium on Flexible Automation*, Vol.1, pp.615-622, 1996.

10

A STABLE AND ROBUST FUZZY CONTROLLER FOR THE POSITION CONTROL OF ROBOTS

Z. DOULGERI and D. BISKAS
Department of Electrical and Computer Engineering
Aristotle University of Thessaloniki
54006 Thessaloniki, Greece

1. INTRODUCTION

Many successful applications of fuzzy control systems have been reported in recent years. One of the basic issues that only recently has been addressed concerns the stability analysis of such systems. Stability conditions for non-linear systems comprising a fuzzy plant model and a fuzzy logic controller have been derived [1] [2] and extended for the case of parameter uncertainties [3]. Practically parameter uncertainties are inevitable; for instance some parameters may vary during the operation of the system or the plant model used in designing the fuzzy logic controller may not be exact. In this chapter an efficient approach for designing a stable and robust fuzzy controller is presented and applied to the position control of a single link robot. This approach considers the system as a fuzzy plant model with parameter uncertainties. First the non-linear plant is represented by Takaki – Sugeno type fuzzy model [4]. In this type of fuzzy model, local dynamics in different state space regions are represented by linear models. The overall model of the system is achieved by fuzzy «blending» of these linear models. The control design is carried out based on the fuzzy model and the feedback linearization technique. The resulting overall controller is again a fuzzy blending of each individual controller. A stability condition for the resulting fuzzy system is given [3]. The approach is illustrated through an application example of controlling the position of a single link robot.

S.G. Tzafestas (ed.), Advances in Intelligent Systems, 105–114.

2. FUZZY PLANT MODEL WITH PARAMETER UNCERTAINTIES

Consider the non-linear system described by $\ddot{x} = f(x, \dot{x}, u)$, where x is the position and $\dot{x}$ the velocity. This system can be represented exactly by a fuzzy plant model with a number of IF – THEN rules, which locally represent input – output relations of the system:

Rule 1: IF x is F_1 THEN $\dot{\mathbf{x}} = (A_1 + \Delta A_1)\mathbf{x} + (B_1 + \Delta B_1)u$

Rule 2: IF x is F_2 THEN $\dot{\mathbf{x}} = (A_2 + \Delta A_2)\mathbf{x} + (B_2 + \Delta B_2)u$

. . . .

Rule n: IF x is F_n THEN $\dot{\mathbf{x}} = (A_n + \Delta A_n)\mathbf{x} + (B_n + \Delta B_n)u$

where $\mathbf{x} = [x_1 \quad x_2]^T$, with $x_1 = x$ and $x_2 = \dot{x}$, F_i are fuzzy sets and the matrices ΔA_1, ΔA_2, ..., ΔA_n, ΔB_1, ΔB_2, ..., ΔB_n represent the parametric uncertainties.

The overall fuzzy model of the system is achieved by fuzzy blending of the linear system models, so that the final output of the fuzzy system is inferred by:

$$\dot{\mathbf{x}} = \sum_{i=1}^{n} w_i [(A_i + \Delta A_i)\mathbf{x} + (B_i + \Delta B_i)u] \quad \text{and} \quad \sum_{i=1}^{n} w_i = 1, \ \forall i \tag{1}$$

where $w_i = \dfrac{\mu_i(x)}{\sum_{i=1}^{n} \mu_i(x)}$ are the normalized grades of membership of x and $\mu_i(x)$ are membership functions.

3. FUZZY CONTROLLER AND STABILITY OF THE CLOSED-LOOP SYSTEM

The fuzzy controller to be designed is described by the following rules:

Rule 1: IF x is F_1 THEN $u = G_1\mathbf{x}$

Rule 2: IF x is F_2 THEN $u = G_2\mathbf{x}$

. . . .

Rule n: IF x is F_n THEN $u = G_n\mathbf{x}$

where $G_i \in \Re^{1\times 2}$ is the feedback gain to be designed. The inferred controller output is given by

$$u = \sum_{i=1}^{n} w_i G_i' \mathbf{x} \tag{2}$$

From (1) and (2) the closed-loop system is given by

$$\dot{\mathbf{x}} = A_0\mathbf{x} + \sum_{i=1}^{n} w_i (H_i - A_0 + \Delta H_i)\mathbf{x}$$

where A_0 is the designed closed-loop system matrix governing the system behavior under no parameter uncertainty and

$H_i = A_i + BG_i$

$\Delta H_i = \Delta A_i + \Delta BG_i$

where B and ΔB are defined by $B=\sum_{i=1}^{n} w_i B_i$ and $\Delta B=\sum_{i=1}^{n} w_i \Delta B_i$.

The closed-loop system is reduced to a linear one if $\Delta H_i = 0$ and the gain of each subsystem G_i is designed by solving the following equations:

$$H_i = A_i + BG_i = A_0$$

The design method of this proposed fuzzy logic controller is based on feedback linearization of each sub-system and is very simple.

The stability of the closed-loop system can be stated according to the following theorem [3]:

Theorem: An uncertain fuzzy control system is guaranteed to be stable, if the following algebraic Riccati equation (ARE) has a positive definite solution $P \in \Re^{n\times n}$:

$$A_0^T P + PA_0 + \frac{PWP}{\varepsilon} + \frac{\varepsilon \max_i \|\Delta H_i\|_{max}^2 I}{\lambda_{min}(W)} = -Q$$

where $Q,\ W \in \Re^{n\times n}$ are positive definite matrices, ε is a positive real scalar, $\lambda_{min}(W)$ denotes the smallest eigenvalue of W, $\max_i \|\Delta H_i\|_{max} \geq \|\Delta H_i\|$ for all i, $\|\bullet\|$ denotes the l_2 induced norm and $I \in \Re^{n\times n}$ is an identity matrix. The ARE given in *Theorem* can be solved by the Schur method.

Proof of Theorem: The proof can be obtained by considering the following lemma:

Lemma: Given any two vectors, y and $z \in \Re^{n\times 1}$, the following inequality holds:

$$y^T z + z^T y \leq \frac{y^T W y}{\varepsilon} + \varepsilon z^T W^{-1} z \leq \frac{y^T W y}{\varepsilon} + \frac{\varepsilon z^T z}{\lambda_{min}(W)}$$

Let a Lyapunov 's function be

$$V = \frac{1}{2} x^T P x$$

Then

$$\dot{V} = \frac{1}{2}\sum_{i=1}^{n} w_i x^T \left[A_0^T P + PA_0 + (H_i + \Delta H_i - A_0)^T P + P(H_i + \Delta H_i - A_0)\right] x \leq$$

$$\leq \frac{1}{2}\sum_{i=1}^{n} w_i x^T \left(A_0^T P + PA_0 + \frac{PWP}{\varepsilon} + \frac{\varepsilon \|H_i + \Delta H_i - A_0\|^2 I_{n\times n}}{\lambda_{min}(W)} \right) x$$

If the non-linear feedback gains are designed as described before,

$$\dot{V} \leq \frac{1}{2} x^T \left(A_0^T P + PA_0 + \frac{PWP}{\varepsilon} + \frac{\varepsilon \max_i \|\Delta H_i\|_{max}^2 I_{n\times n}}{\lambda_{min}(W)} \right) x =$$

$$= -\frac{1}{2}\sum_{i=1}^{n} w_i x^T Q x = -\frac{1}{2} x^T Q x \leq 0$$

3.1. FUZZY LOGIC CONTROL TO A SINGLE LINK ROBOT

Consider a single link robot, which is shown in figure 1.

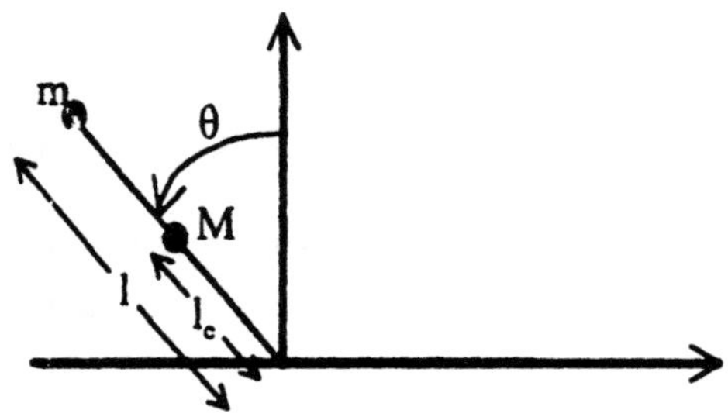

Figure 1: Single Link Robot

Let M be the mass of the link and m the mass of the load grasped by the robot, l denotes the length of the link and l_c denotes the location of the center of mass of the link.
The equation of motion is given by

$$T = h\ddot{\theta} + b\dot{\theta} - a\sin\theta$$

where θ and $\dot{\theta}$ are the angular displacement from the vertical axis (in rad), b is the viscous friction coefficient of the joint, T is the torque given to the robot by the motor, h is the generalized manipulator inertia given by the equation

$$h = J + ml^2$$ with J being the link 's moment of inertia

and a is the gravity coefficient given by

$$a = (Ml_c + ml)g$$ where g is the gravity acceleration.

The values of J, l_c, b and m are not exactly known, but it is assumed that their real values lie within a lower and an upper bound as follows:

$0.1 \le J \le 0.3$, $0.1 \le l_c \le 0.3$, $0.001 \le b \le 0.01$, $0 \le m \le 1$.

In order to apply the Takaki – Sugeno theory to this model, we first linearize the non-linear system and then design a Fuzzy Logic Controller.

3.2. LINEARIZATION ABOUT $(\theta_0, 0, 0)$

Consider that we want to linearize the equation of motion generally about a specific point θ_0. The state space equations are

$$\dot{x}_1 = x_2$$

$$\dot{x}_2 = \frac{a}{h}\sin x_1 - \frac{b}{h}x_2 + \frac{1}{h}u$$

The linearization about θ_0 gives:

$$\dot{x}_1 = x_2$$
$$\dot{x}_2 = \frac{a}{h\theta_0}\sin\theta_0 \cdot x_1 - \frac{b}{h}\cdot x_2 + \frac{1}{h}\cdot u$$

or $\dot{x} = Ax + Bu$

where

$$A = \begin{bmatrix} 0 & 1 \\ \frac{a}{h\theta_0}\sin\theta_0 & -\frac{b}{h} \end{bmatrix} \text{ and } B = \begin{bmatrix} 0 \\ \frac{1}{h} \end{bmatrix}$$

If we linearize about 0, $\pm\frac{\pi}{2}$ and π, the fuzzy plant model is

Rule 1: IF x_1 is about 0 THEN $\dot{x} = (A_1 + \Delta A_1)x + (B_1 + \Delta B_1)u$

Rule 2: IF x_1 is about $\frac{\pi}{2}$ THEN $\dot{x} = (A_2 + \Delta A_2)x + (B_2 + \Delta B_2)u$

Rule 3: IF x_1 is about $-\frac{\pi}{2}$ THEN $\dot{x} = (A_3 + \Delta A_3)x + (B_3 + \Delta B_3)u$

Rule 4: IF x_1 is about π THEN $\dot{x} = (A_4 + \Delta A_4)x + (B_4 + \Delta B_4)u$

where

$$A_1 = \begin{bmatrix} 0 & 1 \\ \frac{a}{h} & -\frac{b}{h} \end{bmatrix},\ A_2 = \begin{bmatrix} 0 & 1 \\ \frac{2a}{\pi h} & -\frac{b}{h} \end{bmatrix},\ A_3 = \begin{bmatrix} 0 & 1 \\ \frac{2a}{\pi h} & -\frac{b}{h} \end{bmatrix},\ A_4 = \begin{bmatrix} 0 & 1 \\ 0 & -\frac{b}{h} \end{bmatrix}$$

$$B_1 = B_2 = B_3 = B_4 = \begin{bmatrix} 0 \\ \frac{1}{h} \end{bmatrix}$$

If we consider the real robot parameters to be M=2Kgr m=1Kgr, l=0.5m, $l_c = 0.187\text{m}$, $J = 0.127$, $b = 0.008$ and take into account the bounds for the uncertain parameters which are given before then the range in which the parameter uncertainty matrices lie are as follows:

$$\Delta A_1 \in \left(\begin{bmatrix} 0 & 1 \\ -\frac{19g}{2\pi} & -0.1 \end{bmatrix}, \begin{bmatrix} 0 & 1 \\ \frac{-0.167g}{2\pi} & -0.00167 \end{bmatrix} \right)$$

$$\Delta A_2 = \Delta A_3 \in \left(\begin{bmatrix} 0 & 1 \\ 0 & -0.1 \end{bmatrix}, \begin{bmatrix} 0 & 1 \\ 0 & -0.00167 \end{bmatrix} \right)$$

$$\Delta A_4 \in \left(\begin{bmatrix} 0 & 1 \\ -\frac{19g}{3\pi} & -0.1 \end{bmatrix}, \begin{bmatrix} 0 & 1 \\ -\frac{0.167g}{3\pi} & -0.00167 \end{bmatrix} \right)$$

$$\Delta B_1 = \Delta B_2 = \Delta B_3 = \Delta B_4 \in \left(\begin{bmatrix} 0 \\ 1.67 \end{bmatrix}, \begin{bmatrix} 0 \\ 10 \end{bmatrix} \right)$$

According to the theory, the inferred system states of the model are governed by

$$\dot{x} = \sum_{i=1}^{4} w_i\left[(A_i + \Delta A_i)x + (B_i + \Delta B_i)u\right]$$

where the weights w_i are given at the table below (table 1), as shown at the figure 2. The usual specifications for the performance of the robot position controller include a critical damped response and a zero position steady state error. In this case we also require a settling time of 0.5sec. Furthermore, in this application example the desired equilibrium state is the desired position x_d, which is generally not zero. Thus the controller of each subsystem is of the form:

Rule 1: IF x_1 is about 0 THEN $u = G_1\mathbf{x} + G_{f1}x_d$

Rule 2: IF x_1 is about $\frac{\pi}{2}$ THEN $u = G_2\mathbf{x} + G_{f2}x_d$

Rule 3: IF x_1 is about $-\frac{\pi}{2}$ THEN $u = G_3\mathbf{x} + G_{f3}x_d$

Rule 4: IF x_1 is about π THEN $u = G_4\mathbf{x} + G_{f4}x_d$

where G_i and G_{fi} are the gain matrices. The inferred controller output is given by

$$u = \sum_{i=1}^{4} w_i(G_i\mathbf{x} + G_{fi}x_d)$$

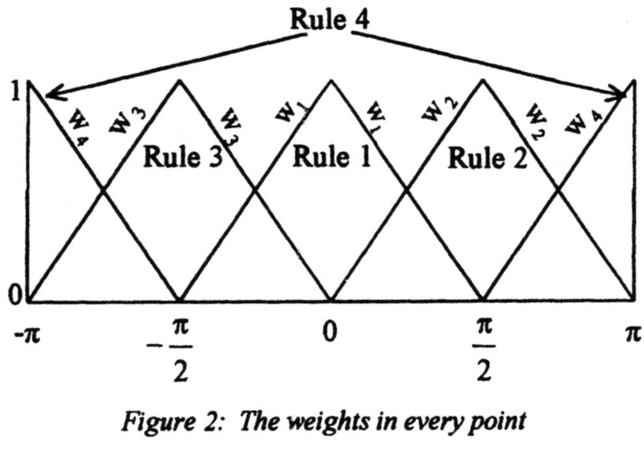

Figure 2: The weights in every point

$0 - \frac{\pi}{2}$	$w_1 = \frac{\frac{\pi}{2} - x_1}{\frac{\pi}{2}}$	$w_2 = 1 - w_1$	$w_3 = 0$	$w_4 = 0$
$\frac{\pi}{2} - \pi$	$w_1 = 0$	$w_2 = 1 - w_4$	$w_3 = 0$	$w_4 = \frac{x_1 - \frac{\pi}{2}}{\frac{\pi}{2}}$
$(-\pi) - \left(-\frac{\pi}{2}\right)$	$w_1 = 0$	$w_2 = 0$	$w_3 = 1 - w_4$	$w_4 = \left\lvert \frac{\frac{\pi}{2} + x_1}{\frac{\pi}{2}} \right\rvert$
$\left(-\frac{\pi}{2}\right) - 0$	$w_1 = \frac{\frac{\pi}{2} + x_1}{\frac{\pi}{2}}$	$w_2 = 0$	$w_3 = 1 - w_1$	$w_4 = 0$

Table 1: The weights values

For the calculation of the control gains we assume the following estimates for the uncertain parameters: $\hat{l}_c = 0.25\text{m}$, $\hat{J} = 0.2\text{m}$, $\hat{b} = 0.005$. According to the above specifications $A_0 = \begin{bmatrix} 0 & 1 \\ -64 & -16 \end{bmatrix}$ and the values for the control gains for each subsystem are found to be:

$$G_1 = [-54.6 \quad -11.195]$$
$$G_2 = [-51 \quad -11.195]$$
$$G_3 = [-51 \quad -11.195]$$
$$G_4 = [-44.8 \quad -11.195]$$
$$G_{f1} = G_{f2} = G_{f3} = G_{f4} = -44.8$$

Regarding the stability of the system the limits of the matrices ΔA_i and ΔB_i were used to find that $\max\|\Delta H_i\|_{max}^2 = 52.5314$ and supposing that:

$$W = Q = I_2$$
$$\varepsilon = 1$$
$$\lambda_{min}(W) = 1$$

then the ARE is solved for P and results in:

$$P = \begin{bmatrix} 17.4902 & -2.003 \\ -2.003 & 1.3642 \end{bmatrix}$$

with eigenvalues:

$$\lambda_1 = 17.7353 > 0$$
$$\lambda_2 = 1.1191 > 0$$

So, the system is stable. Simulation results are shown in figure 3.

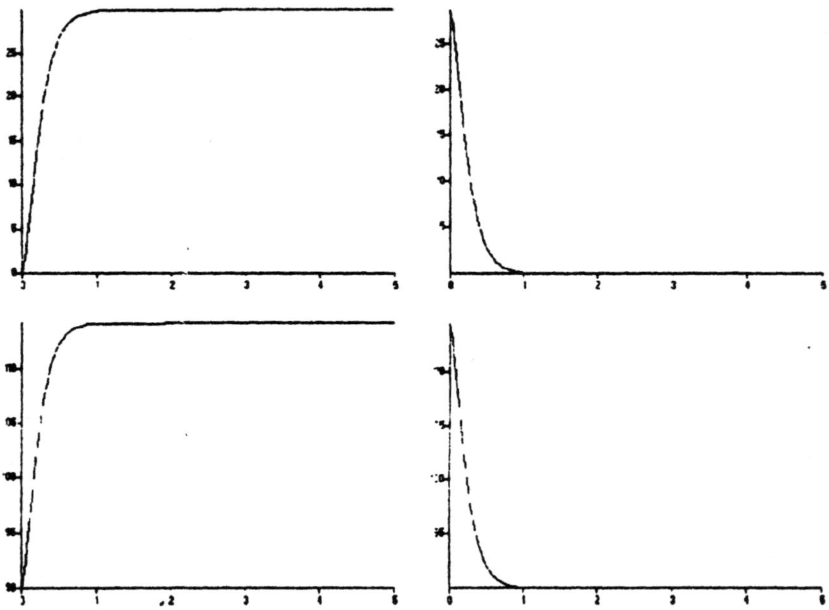

Figure 3: The motion of robot. a) From 0 to 30 deg b) From 30 to 0 deg c) From 90 to 120 deg d) From 120 to 90 deg

We can see that the dynamic performance of the system is according to the specifications irrespective of the desired position. However, steady state error is eliminated only at linearization points. The existing steady state errors in other desired positions are in the range of two degrees and it is maximum (2.2deg) at the mid-point between two linearization points. Therefore, a refinement of the division of the state space region is expected to reduce the steady state error. So, next we linearize about the following additional position points: $45°, 135°, 225°, 315°$ and repeat the design procedure for the new fuzzy control system. Simulation results are shown in figure 4, where for the non-linearization points, steady state position error is less than $0.5°$.

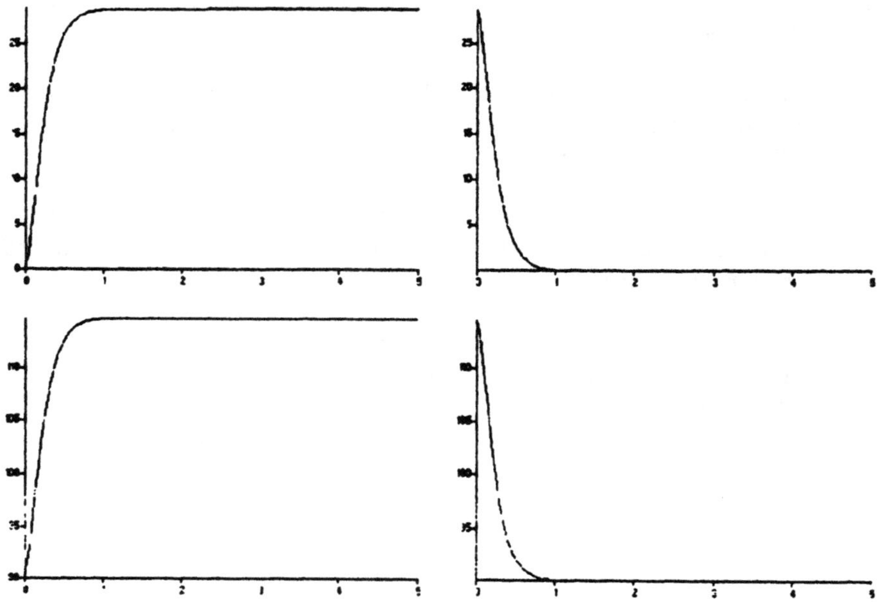

Figure 4: The motion of robot. a) From 0 to 30 deg b) From 30 to 0 deg c) From 90 to 120 deg d) From 120 to 90

4. CONCLUSIONS

Feedback linearization has been employed to design a fuzzy position control system. The stability and robustness of the closed loop system have been analyzed and robustness issues have been discussed. The design procedure is easily applicable to uncertain nonlinear systems in general. It is here applied to the position control of a single link robot. Simulation results have shown that a very good dynamic performance of the fuzzy position controller can be achieved and that the robot can successfully reach the desired position when there is an adequate division of the state space region.

REFERENCES

[1] Tanaka, K., and Sugeno, M.: «Stability analysis and design of fuzzy control systems». Fuzzy Sets and Systems, 1992, 45, pp. 135-156

[2] Wang, H. O., Tanaka K., and Griffin, M. F.: «An approach to fuzzy control of non-linear systems: Stability and the design issues», IEEE Trans. Fuzzy Syst., 1996, 4, (2), pp. 14-23

[3] F. H. F. Leung, H. K. Lam, P. K. S. Tam, «Design of stable and robust fuzzy controller for an inverted pendulum using feedback linearization», submitted to the Electronic Letters, 1988

[4] Takaki, T., and Sugeno, M.: «Fuzzy identification of systems and its applications to modeling and control». IEEE Trans. Sys., Man., Cybern., 1985, 15, (2), pp. 116-132

11

STABILITY ANALYSIS AND FUZZY SELF-TUNING CONTROL OF A NONLINEAR PROCESS WITH LIMIT POINTS

D. I. SAGIAS, E.N.SARAFIS, C.I.SIETTOS
and G.V. BAFAS
Department of Chemical Engineering
National Technical University of Athens
GR - 15780, Athens, Greece

1. INTRODUCTION

What is required for the qualitative and quantitative study of many practical engineering systems is the construction of a set of time-depended non-linear partial differential equations (PDE's). Generally speaking, the state equations resulting from mass, energy and moment balances are of the form

$$\frac{\partial x}{\partial t} = L(x, z, p) \tag{1}$$

where **L** is a partial differential operator defined over an Euclidean space **V**, **z** is the one, two or three-dimensional Euclidean space vector, **x** is the distributed, over **z**, state vector and **p** is the vector of structural parameters. A striking characteristic on the solution branch (**x**(**p**), **p**) of such systems is the occurrence of limit points (or turning points) which represent the boundary between stability and instability of the system. It thus follows that the solution branch may contain physically unstable (or unobservable) regions.

This chapter deals with the stability analysis of a laminar flow tubular reactor which is a typical time depended non-linear distributed parameter system. Furthermore a fuzzy self- tuning controller is proposed for the control problem. In many cases, adaptation comprises a necessity for the control loop. One reason for this, is the requirement of satisfactory performance over a wide range of the operating branch; process time varying characteristics, is another motivation for using adaptive control.

Due to the built in complexity, the analysis of the system, requires the implementation of ad hoc computational schemes. Here the solution of equation (1) all over the operating branch is obtained by applying finite element basis functions and Galerkin's method of weighted residuals along with an arc-length continuation method for finding the limit points.

The performance of a self tuning fuzzy controller over a wide range of step disturbances and operating regions, is compared against a conventional model identification adaptive controller (MIAC). The two schemes are compared with several methods including, time integral performance criteria such as Integral of Absolute Error (IAE) [1].

S.G. Tzafestas (ed.), Advances in Intelligent Systems, 115–126.

2. MATHEMATICAL MODELLING AND DISCRETIZITION

The case under study is shown in Figure 1 and concerns the control of a jacketed plug flow tubular reactor where a first order exothermic, irreversible reaction A->B takes place.

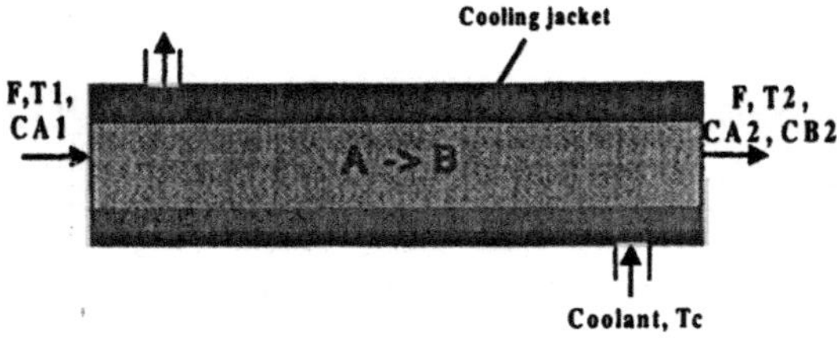

Figure1: The case study: a tubular reactor

Assuming constant temperature for the coolant, the state modelling equations are given by the following partial differential equations [1]:

$$\frac{\partial C_A}{\partial t}+v\frac{\partial C_A}{\partial z}=-k_o\exp\left(-\frac{E}{RT}\right)C_A \tag{2a}$$

$$c_p\rho A\frac{\partial T}{\partial t}+c_p\rho vA\frac{\partial T}{\partial z}=UA_t\left(T_C-T\right)+\left(-DH_R\right)k_o\exp\left(-\frac{E}{RT}\right)AC_A \tag{2b}$$

with the boundary conditions:

$$C_A(0,t)=C_o,\ T(0,t)=T_o,\ C_A(\mathbf{z},0)=C_{\mathbf{PROFILE}},\ T(\mathbf{z},0)=T_{\mathbf{PROFILE}},\ 0<z<l, \tag{2c}$$

where $C_A(z,t)$, $T(z,t)$ are the concentration and temperature respectively profile inside the reactor, ρ is the mass density, v represents the fluid flow velocity, c_P for specific heat capacity, $-DH_R$ denotes the heat of reaction, E is the activation energy, k_o is the Arrhenius factor, R is the thermodynamic constant, l is the length of reactor, U is the overall heat transfer coefficient, A_t is the heat transfer unit area, A is the cross section, Tc is the coolant temperature.

Clearly, the set of equations (2a) and (2b) consist a time depended non-linear distributed parameter system with time delay; the governing equations are discretized in space and in time by a combination of Galerkin's method of weighted residuals[2], finite element basis functions and the differential algebraic system solver (DASSL) [3].

The in issue domain is subdivided into a finite number of elements; next are assigned the mid points or nodes of each element.

The solution $C_A(z, t)$ and $T(z, t)$ is being approximated numerically by a set of piecewise quadratic polynomials (basis functions):

$$C_A(z,t)=\sum_{i=1}^{N} C_{A_i}\ \varphi_i(z), \tag{3a}$$

$$T(z,t) = \sum_{i=1}^{N} T_i \, \varphi_i(z), \tag{3b}$$

where N is the number of nodes, φ_i (z) are the basis functions which are taken to be unity at the ith node and zero at all other nodes, C_{A_i} and T_i are the approximations of the solution at the ith node.

Galerkin's method seeks for the solution of eq. (1), setting the weighted residuals equal to zero:

$$R_i = \int_V \left(\left(\frac{\partial x}{\partial t} \right) - L(x,z,t) \right) \varphi_i \, dz, \; i=1, \ldots, N \tag{4}$$

The weighting factors are taken to be the basis functions which are used for the construction of the approximate solution (eq. 3).

Substituting equation (3a) and (3b) in equation (4) the set of non-linear partial differential equations reduces to a large system of coupled first order ordinary differential equations of the form

$$\mathbf{B} \; \dot{x} \; = \mathbf{g} \; (\mathbf{x}, \mathbf{p}), \tag{5}$$

where $\mathbf{x} = [x_1, x_2, \ldots, x_N]$ is the set of unknowns at the nodes and $\mathbf{x}_i = [C_{ai}, T_i]$, $\mathbf{B}$ is a nxn matrix with elements given by $b_{ij} = \int_V \varphi^i \varphi^j dz$, $\mathbf{g}$ is a nx1 time depended vector which corresponds to the set of Galerkin's weighted residuals calculated at steady state i.e. setting $\frac{\partial x}{\partial t} = 0$ in equation (1).

To this end, the discretized problem given by equation (5) is solved in time using the DASSL.

3. STABILITY ANALYSIS

Depending on the values of the set of parameters $\mathbf{p}$, the solution path of the tubular reactor contains limit points. A equilibrium point (**x**, p) is said to be a limit point (or regular turning point), when it satisfies the conditions [4]: a) the Jacobian $\mathbf{L}(\mathbf{x}, p)$ with respect to x i.e. $\mathbf{L}_x(\mathbf{x}, p)$ is singular b) the system $\mathbf{L}_x(\mathbf{x}, p) = \boldsymbol{\varphi} \, \mathbf{L}_p(\mathbf{x}, p)$ has no solution, c) the Jacobian $\mathbf{L}_x(\mathbf{x}, p)$ at (**x**, p) has only one zero eigenvalue.

Newton's iteration method for solving the linearized steady state equation set,

$$\mathbf{R}_x \, \delta\mathbf{x} = -\mathbf{R} \tag{6}$$

fails to converge as the solution approaches limit points, since the Jacobian $\mathbf{R}_x$ comes very close to singularity. Instead using Newton's method, the solution branch through the limit points is tracked solving the linearized equation set given by eq. (6) by an arc-length parameter continuation method [5].

Stability of solution, change across a limit point and is determined by monitoring its response subject to small disturbances. If the disturbance grows with time then the solution is unsteady; otherwise the solution is steady.
Let $\delta\mathbf{x}$ be a disturbance with infinitesmal amplitude, then linearizing $\mathbf{L}(\mathbf{x}, p)$ around $\mathbf{x}_o$ one obtains:

$$\mathbf{L}_{\mathbf{x}}(\mathbf{x}_0, 0, p)\delta\mathbf{x} + \mathbf{L}_{\dot{\mathbf{x}}}(\mathbf{x}_0, 0, p)\,\delta\,\dot{\mathbf{x}} = \mathbf{0} \tag{7}$$

The stability of the above linearized equation is governed by the generalized eigenvalue problem

$$\mathbf{L}_{\mathbf{x}}(\mathbf{x}_0, 0, p)\, v_i = \lambda_i\,[-\mathbf{L}_{\dot{\mathbf{x}}}(\mathbf{x}_0, 0, p)]v_i, \tag{8}$$

where λ_i are the eigenvalues, v_i represent the modes of the solution derived by eq. (8), $\mathbf{L}_{\mathbf{x}}(\mathbf{x}_0, 0, p)$ is the Jacobian matrix as it is calculated by the weighted residuals at a given value p, using parameter continuation, $\mathbf{L}_{\dot{\mathbf{x}}}(\mathbf{x}_0, 0, p) \equiv \mathbf{B}$ is the overlap of the finite element basis functions.
The Jacobian which has the following form in block notation:

$$\mathbf{L}_{\mathbf{x}}(\mathbf{x}_0, 0, p) = \begin{bmatrix} \frac{\partial L_{C_A}}{\partial C_A} & \frac{\partial L_{C_A}}{\partial T} \\ \frac{\partial L_T}{\partial C_A} & \frac{\partial L_T}{\partial T} \end{bmatrix} \tag{9}$$

is a large, sparse and non-symmetric matrix of size 2Nx2N.
Hence, stability analysis is reduced to the solution of the above generalized eigenvalue problem. What is really required, is finding only the lower modes of the generalized eigenvalue problem i.e. eigenvectors corresponding to eigenvalues with algebraically largest real parts. The information gathered by finding eigenvalues with algebraically smallest real parts are in practice of minor significance. In this chapter, the determination of the most dangerous modes of equation (8) is attained using standard Arnoldi algorithm [6].

4. INDIRECT FUZZY ADAPTIVE CONTROL DESIGN

A major problem encountered in non-linear or/and time depended systems, is the degradation of the closed loop performance as the system shifts away from the initial operational settings. This drawback imposes the need of using adaptive controllers i.e. controllers which adjust their parameters optimally, according to some objective criterion. So far, many fuzzy schemes have been proposed for adaptive control implementation, including self-organizing control ([7], [8]), scaling factors adjustment([9], [10], [11]). However most of the schemes, are based on heuristic approaches, since fuzzy controllers suffice of the lack of a mathematical model and an analytical adaptive control algorithm is difficult, if not impossible, to derive. Conventional techniques, in turn, are well established, offering a variety of adaptive control algorithms. In this chapter, a model identification fuzzy adaptive controller (MIFAC) for scaling factors adjustment is proposed. The adaptive algorithm is based upon the thoroughly used recursive least squares (RLS) algorithm [12]. The structure of the proposed fuzzy adaptive control mechanism is shown in figure 2. It consists from a fuzzy logic based controller (FLC), a system identification unit, the controller synthesis unit and the desired closed loop performance.

The fuzzy logic controller is a two input- single output incremental controller with the following variables: e(t)=r(t)-y(t), ce(t)=e(t)-e(t-1), cu(t)=u(t)-u(t-1), where: r(t) is the set point at time t, y(t) is the process output at time t, e(t), ce(t) are the error and the change of error at time t, cu(t) is the change of controller output at time t. The fuzzification of the input -output variables, were obtained using seven fuzzy sets, with fixed symmetrical triangular shaped membership functions defined over the normalized universe of discourse [-1, 1]. The structure of the rule base is constructed heuristically and provides negative feedback control in order to maintain stability under any condition. The control rules are of the form:

IF e is A_i AND ce is B_j THEN cu is C_l (10)

System identification is carried out monitoring input - output data. In a general manner the model of the system at the neighborhood of a equilibrium can be posed as a linear discrete equation

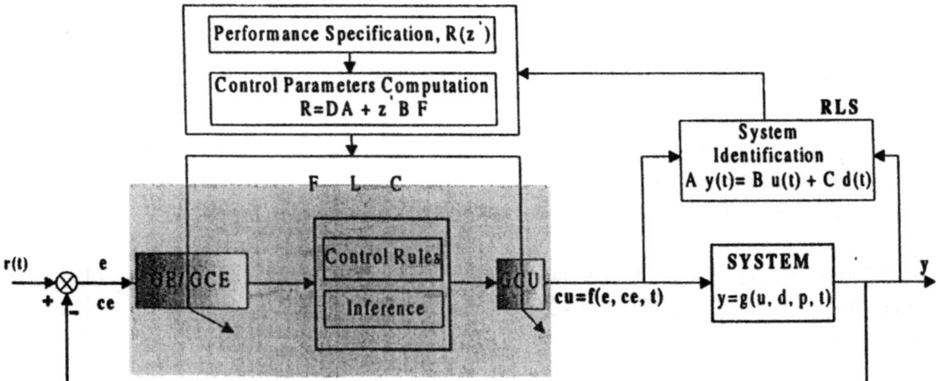

Figure 2: Basic Structure of the Fuzzy Adaptive Control Mechanism

$$y(t)+a_1\, y(t-1)+a_2\, y(t-2)+ \dots +a_n\, y(t-n) =$$
$$b_o u(t)+ \dots +b_m u(t-m) + c_o d(t) + \dots c_l\, d(n-l) \qquad (11)$$

where y(t) is the system output at time t, u(t) is the control action at time t, d(t) is the disturbance imposed in the system at time t.

The parameters of the system are estimated on-line using recursive least squares.

Assume that the error and change of error at time t are given by e'(t), ce'(t). The derivation of the contoller output is obtained utilizing max-dot fuzzy inference method and centroid defuzzification method [13]:

$$c\,u = \frac{\sum_{i,j} \mu_{ij} u_{ij}}{\sum_{i,j} \mu_{ij}} \tag{12}$$

where $\mu_{ij} = A_i(e(t))\, B_j\, (ce(t))$ (13)

In a general form the incremental control action cu can be represented as a non-linear function of the input variables e(t), ce(t):

$$cu = f(e', ce', t) = f\,(GE\ e, GCE\ ce, t) \tag{14}$$

For small perturbations δe, δce and δcu around equilibrium the above equation is approximated by the linearized equation

$$cu = \left[\frac{\partial f}{\partial e}\right]_{ce} \delta e + \left[\frac{\partial f}{\partial ce}\right]_{e} \delta ce \tag{15}$$

Substituting equation (12) in equation (14) one obtains finally the simplified discretized equation [14]:

$$cu(t) = GE\ e(t) + GCE\ ce(t) \tag{16a}$$

which gives the contro output at time t; GE, GCE are the scaling factors for the error and change of error respectively.
For the backward time instances (t-1) and (t-2), equation (16a) yields:

$$cu(t-1) = GCE_1\ e(t-1) + GCE_2\ ce(t-1) \tag{16b}$$
$$cu(t-2) = GCE_2\ e(t-2) + GCE_2\ ce(t-2) \tag{16c}$$
$$cu(t-3) = GCE_3\ e(t-3) + GCE_3\ ce(t-3) \tag{16d}$$

Summing equations (16b) - (16d) and banding a scaling factor GCU_i for each control output cu(t-i), the following difference equation is obtained

$$GCU_1\ cu(t-1) + GCU_2\ cu(t-2) + GCU_3\ cu(t-3) = (GE_1 + GCE_1)\ e(t-1) + (GE_2 + GCE_2 - GCE_1)\ e(t-2) + (GE_3 + GCE_3 - GCE_2)\ e(t-3) - GCE_3\ e(t-4) \tag{17}$$

Using the z-Transform, equation (17) can be represented in the Z-Domain as
$D\ cu(t-1) = F\ e(t-1)$ (18a)

where,

$D = GCU_1 + GCU_2\, z^{-1} + GCU_3\, z^{-2}$, (18b)

$F = (GE_1 + GCE_1) + (GE_2 + GCE_2 - GCE_1)\, z^{-1} +$
$(GE_3 + GCE_3 - GCE_2)\, z^{-2} + GCE_3\, z^{-3}$ (18c)

The self tuning control of the fuzzy controller states as follows

1. Acquisition of system output at time t.
2. System identification by a discrete equation of the form
   $A(z^{1})\, y(t) = B(z^{-1})\, cu(t)$.
   using recursive least squares estimation
3. Solve $R = D\,A + z^{-1}\,B\,F$ (19)
   in terms of D and F; R is the desired closed loop characteristic equation.
4. Update scaling factors GE, GCE, GCU
5. Implement control using the simplified equation
   $cu(t) = GCU_1\, cu(t-1) + GCU_2\, cu(t-2) + GCU_3\, cu(t-3)$ (20)
6. t=t+1; goto step 1.

5. SIMULATION: RESULTS AND DISCUSSION

5.1 EQUILIBRIA PATHS

Simulation results were obtained taken the following values of the parameters: k_o=3.34 10^8 min^{-1}, E/R =8600 K, C_{A1}=1.6 kmol/m^3, T_1=423 K, T_2=440 K, A=0.002 m^2, A_t=0.01 m, DH_R=-44000 kcal/kmol, T_c =293 K, U=25 Kcal $m^2 min^{-1} grad^{-1}$, v=2 m/min, c_p=25 kcal/ kmol K, ρ=47 kmol/m^3, l=5 m, con2=U A_t /(c_p ρ A).

The steady state solution branches (**x**(con2), con2), for fixed values of fluid velocity are shown in figure 1. As it is shown, the equilibrium solution of the system (2a) and (2b) exhibits some very interesting non-linearities. Particularly, there is a value of the fluid velocity, say v* for which the steady state solution of L(**x**(con2), con2)=0 has two for v>v*, one for v=v* or no limit points for v<v*. The computed value of v* was v*=1.7335 m/s.

Across limit points an odd number of eigenvalues of the Jacobian changed sign indicating that they crossed the imaginary axis in the complex plane; this implies that stability of solution changes.

Assume that the fluid velocity is taken to be v>v*; hence the solution branch contains two limit points (fig. 1a). When con2>con2u or con2<con2l the system is stable. Between con2l<con2<con2u the solution branch exhibits a hysteresis behavior: as con2 decreases following the solution branch towards con2l from the equilibrium state D to the equilibrium state C, the system remains stable until con2 equals con2l. At con2=con2l a limit point occurs; imposing a small decrement in con2, the system loses stability and the system jumps discontinuously to the state A, where it gains stability back. Increasing con2 towards con2u, the solution recovers its observability tracking the branch from state A to state B.

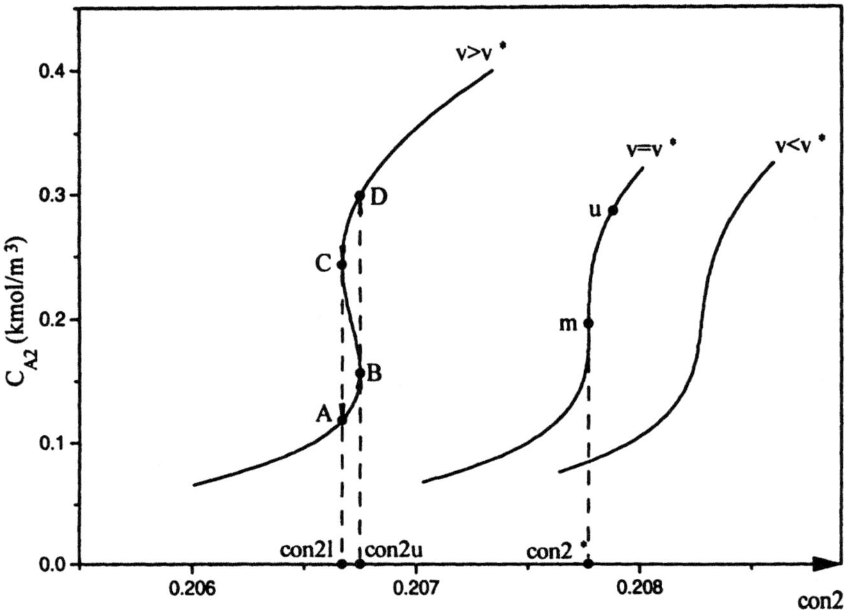

Figure 3: Steady state paths (C_{A2}(con2), con2) for v*<v=1.8, v=v*=1.7335, v*>v=1.7

Further increment of con2 beyond limit point B, results to a discontinuous jump of the system to state D. The hysteresis behavior renders the solution branch from B to C physically unstable, thus unobservable and uncontrollable. The system becomes observable all over the solution branch by changing the fluid velocity to a value v<v*.

5.2 THE CONTROL PROBLEM: MIFAC IMPLEMENTATION AND PERFORMANCE EVALUATION

The control objective here, is to maintain the control variable, which is the composition of the reacting mixture at the output of the reactor, within the desired operational settings and particularly to keep the A reactant concentration C_{A2} at the outlet, below its nominal steady state value, eliminating concentration disturbances at the inlet. The manipulated variable is taken to be the coolant temperature T_C.

The performance of the proposed fuzzy adaptive controller, is compared against a conventional model identification adaptive controller. For the performance evaluation of the two adaptive controllers, step change disturbances in input concentration, ranging from -5% up to -15% are imposed into the process. Simulation results are presented for two different operating points: either for the upper point u where con2=1.78, or for the middle point m where con2=con2*.

It is assumed that the dynamics of the process are poorly known; therefore the control loop (fig. 2) involves on-line process identification for both fuzzy and conventional controllers. The process is modelled by the difference equation:

$$(1+a_1z^{-1}+a_2z^{-2}+a_3z^{-3})\, y(t)=(b_1z^{-1}+b_2z^{-2}+b_3z^{-3})u(t) \qquad (21)$$

where the structure of the conventional controller is taken as

$$(1+f_1z^{-1}+f_2z^{-2}+f_3z^{-3})\, y(t)=- (g_o+g_1z^{-1}+g_2z^{-2}+g_3z^{-3})\, u(t) \qquad (22)$$

The process parameters are obtained using the recursive least squares estimation, while the determination of the controller coefficients is achieved using the pole assignment identity[ ]. The desired closed loop pole set in eq. (19) is specified by the polynomial $R=1+ r_1 z^{-1}+r_2 z^{-2}+ r_3 z^{-3} +r_4 z^{-4} +r_5 z^{-5} +r_6 z^{-6} +r_7 z^{-7}$, where r_1=-4.192, r_2=8.111, r_3=-9.734, r_4=8.028, r_5=-4.602, r_6=1.686, r_7=-0.295.

The solution of the identification problem for the upper point u, gives: a_1=-2.338, a_2=1.927, a_3=-0.580, $b_1=7.9498\cdot10^{-5}$, $b_2=3.560\cdot10^{-4}$, $b_3=6.561\cdot10^{-5}$, f_1=-1.066, f_2=0.752, f_3= -0.508, g_0=-14.75, g_1=35.110, g_2=-27.435, g_3=9.979 while for the middle point m, gives: a_1=-2.327, a_2=1.938, a_3=-0.600, $b_1=7.787\cdot10^{-5}$, $b_2=2.61\cdot10^{-4}$, $b_3=8.756\cdot10^{-5}$, f_1=-0.991, f_2=0.788, f_3=-0.510, g_0=24.894, g_1=-60.998, g_2=49.770, g_3=-13.508.

The gains of the fuzzy controller are derived using the proposed algorithm of section 4. For the upper point u, the algorithm gives: GCU_1=-4.255, GCU_2=4.455, GCU_3=-2.285, GE_1=0.0129, GE_2=0.0351, GE_3=-0.3149, GCE_1=0.5789, GCE_2=1.5784, GCE_2=-0.3347, while for the middle point the algorithm gives: GCU_1=-4.220, GCU_2=4.450, GCU_3=-2.280, GE_1=0.013, GE_2=0.0355, GE_3=-0.31874, GCE_1=0.5855, GCE_2=1.5971, GCE_3=-0.3391.

Figure 4 depicts the responses of the controlled variable using the fuzzy or the conventional adaptive schemes on the upper operating point and figure 5 depicts the responses of the controlled variable on the middle operating point. As it shown, the performance for all the range of step disturbances of the proposed fuzzy adaptive controller is better than that of the conventional scheme, for both the upper and middle operating points. The proposed controller exhibits exhibits a much robust response, with up to 90% faster settling time (Fig. 5c), overshooting less up to 8% the controlled variable (fig. 5b).

In addition to these simulation results, fig. 4d and fig. 5d depicts in summary the Integral of Absolute Error (IAE) dynamic performance criterion for the manipulated variable on the upper and middle point respectively. Based on fig. 4d and fig. 5d, the proposed controller exhibits up to 50% lower IAE (for -15% step disturbance) achieving therefore significant energy savings.

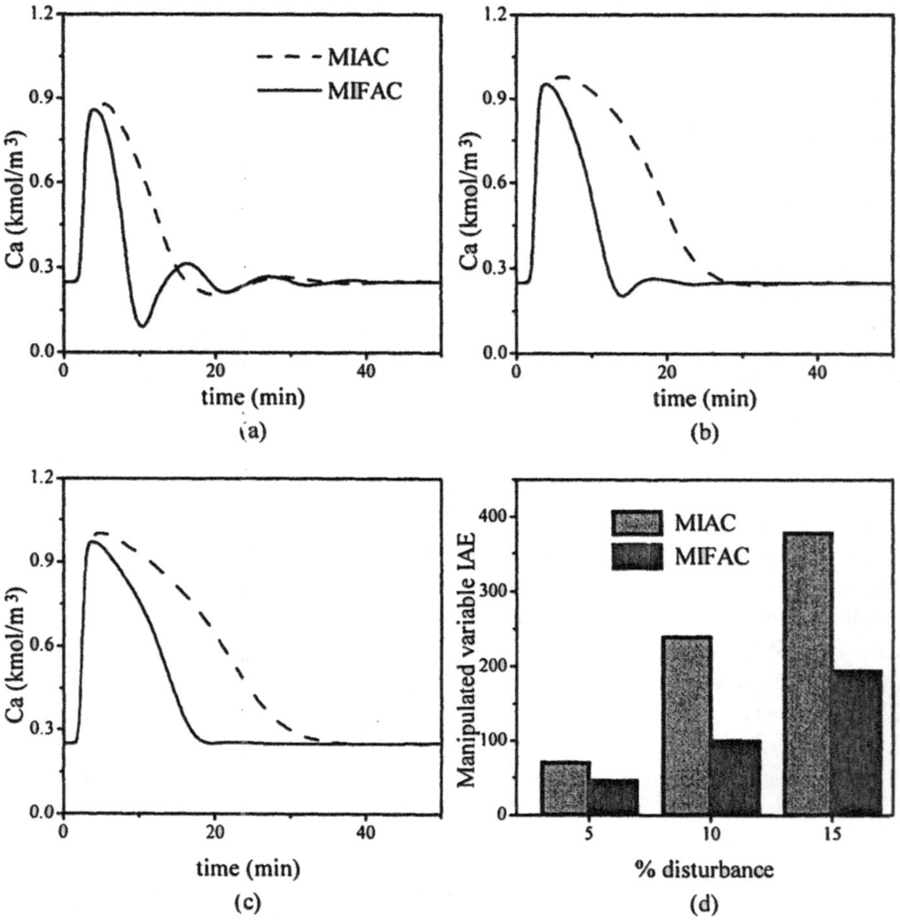

Figure 4. System response on the upper operating point: ——MIFAC, ········MIAC. (a) -5% step disturbance in input concentration, (b) -10% step disturbance in input concentration, (c) -15% step disturbance in input concentration (d) Performance comparison via the IAE criterion.

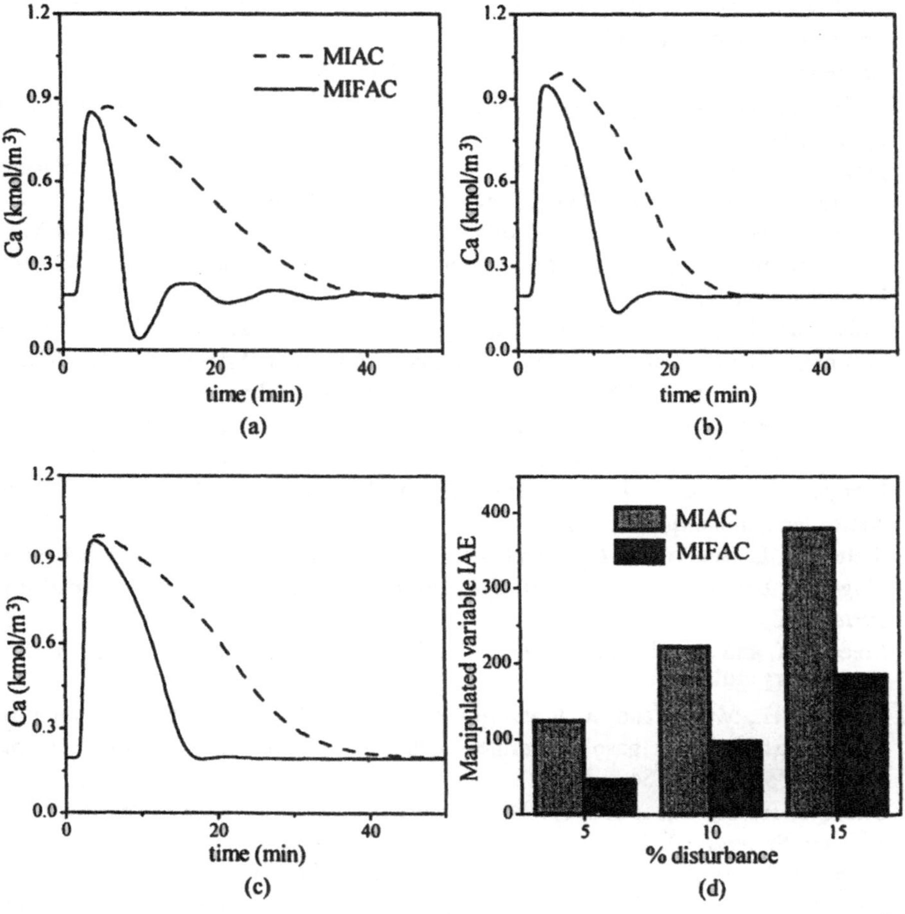

Figure 5. System response on the middle operating point: —— MIFAC, ······· MIAC. (a) -5% step disturbance in input concentration, (b) -10% step disturbance in input concentration, (c) -15% step disturbance in input concentration. (d) Performance comparison via the IAE criterion.

REFERENCES

1. Stephanopoulos, G. (1984). *Chemical process control: an introduction to theory and practice*. Prentice Hall, NY.
2. Boudouvis, A. G. and L. E. Scriven (1986). States of parameter-dependent, nonlinear, distributed systems. *International Journal of Modelling and Simulation*, **6**, 102-105.
3. Petzold, L. R. et al. (1989). *Numerical solution of initial-value problems in differential-algebraic equations*. Elsevier Science Publishing Co., Inc.
4. Keller, H. B. (1977). Numerical solution of bifurcation and non-linear eigenvalue problems. *In Applications of Bifurcation Theory* (P. H. Rabinowitz ed.), Academic Press, New York.
5. Abbot, P. J. (1978). An efficient algorithm for the determination of certain bifurcation points. *Journal of Computational and Applied Mathematics*, **4**, 19-26.
6. Siettos, C. I. et. al. (1996). Stability analysis of magnetohydrostatic equilibrium by the finite element method and Arnoldi and Lanczos eigensolvers. *Advances in Engineering Software*, **27**, 145-153.
7. Procyk, T. J. and E. H. Mamdani (1979). A linguistic self-organizing process controller. *Automatica*, **15**, 15-30.
8. Siettos, C. I., Boudouvis, A. G. and Bafas, G. V. (1998). Fuzzy adaptive control of a plug flow tubular reactor. *5th IFAC Symposium on Dynamics and Control of Process Systems*, Corfu, Greece.
9. Maeda, M. and Murakami, S. (1992). A self-tuning fuzzy controller. *Fuzzy Sets and Systems*, **51**, 29-40.
10. Bare, W. H., Mulholland, R. J. and Sofer, S. S. (1990). Design of a self-tuning rule base controller for a gasoline refinery catalytic reformer. *IEEE Trans. Automat. Control*, **35**(2), 156-164.
11. Chou, C. H. and Lu, H. C. (1994). A heuristic self-tuning fuzzy controller. *Fuzzy Sets and Systems*, **61**, 249-264.
12. Isermann, R. (1989). *Digital control system*, Vol. II. Springer, Germany.
13. Zimmermann, H. J., 1991, Fuzzy set theory and its applications, Kluwer, Boston, MA.
14. Mizumoto, M. (1995). Realization of PID controls by fuzzy control methods. *Fuzzy Sets and Systems*, **70**, 171-182.

PART III

GENETIC ALGORITHM-BASED SYSTEM ANALYSIS AND DESIGN

12

APPLICATION OF THE GENETIC ALGORITHM TO A CELLULAR DYNAMIC CHANNEL ALLOCATION MODEL

H. G. SANDALIDIS*,**, P. P. STAVROULAKIS*
and J. RODRIGUEZ-TELLEZ**
* *Telecommunication Systems Institute of Crete (T.S.I)*
Technical University of Crete, Chania, Greece
** *University of Bradford, Bradford, U.K.*

1. INTRODUCTION

Genetic algorithms (GAs) are adaptive heuristic methods that may be used to solve optimisation problems by generating recursively a new population of solutions from an initial one. They have been applied to several combinatorial problems such as the travelling salesman problem with satisfactory efficiency [1]. *Combinatorial* is a problem which is related with the optimisation of a discrete cost or energy function that occurs from a specific problem modelling.

Channel assignment is the process of assigning appropriate channels to the individual members of a cellular network. The problem is crucial for the normal operation of a mobile system due to the limitation of frequency spectrum and the existence of interference constraints. Channel assignment is generally classified into fixed and dynamic. In *fixed channel assignment (FCA)*, channels are statically allocated to each cell whereas in *dynamic channel assignment* channels are assigned dynamically. DCA gives the opportunity every cell to use any of the available channels but requires more complex operations than FCA [2].

The use of GAs to various FCA models has been extensively studied in [3], [4] and [5]. A genetic algorithm based dynamic channel assignment for uniform traffic distribution has been suggested in [6]. This chapter extends the application of GAs to include further assumptions than [6]. Section II gives the basic characteristics of GAs and section III describes the model. Simulation results are provided in section IV.

S.G. Tzafestas (ed.), Advances in Intelligent Systems, 129–137.

2. GAS FEATURES

GAs are search algorithms based on the mechanics of natural selection, genetics and evolution. They work with a *population of solutions* that are known as *chromosomes* or *individuals* or *strings*. Strings consist of *genes* that are usually binary numbers. A simple GA is shown in fig. 1

```
t=0
initialise population P(t)
evaluate P(t)
repeat
   select P(t+1)  from P(t)
   alter P(t+1) using crossover and mutation
   evaluate P(t+1)
   t=t+1
until termination criterion
Print best solution
```

Fig. 1: Simple genetic algorithm

At first, an initial population is provided either at random or by using problem specific information. Then, the fitness of each chromosome in the population is measured according to an optimisation criterion and the fitter individuals are selected. Some of them undergo transformations to produce offspring for the next generation. The main transformations are crossover and mutation. *Crossover* creates two children by combining material from the initial chromosomes (*parents*) whereas *mutation* alters one or more genes. After that, the new population is ready for its next evaluation. The process is repeated and when a termination criterion is reached, the best chromosome is selected [1], [5], [6].

A GA for a particular problem must have the following components:

- A *genetic representation* of solutions
- An *evaluation or fitness function* that plays the role of the optimisation criterion
- *Genetic operators*
- *Values for various parameters* that GA uses (population size, probabilities of genetic operators, etc.)
- A *termination criterion* [1]

3. GA-DCA MODEL

DCA model extends the initial assumptions provided in [6],[7] and [8] with the introduction of the cosite and adjacent channel constraints. The main characteristics are the following:

i. The model is composed of hexagonal cells that form a parallelogram.

ii. For every incoming call a channel is selected with the restraints of no cochannel cosite and adjacent channel constraints. *Cochannel interference* does not permit the same channel to be assigned in cells that are located in a distance less than a minimum (*reuse distance*). *Cosite constraint* does not permit adjacent channels to be allocated in the same base station and *adjacent channel* in the neighbouring ones. The above conditions cannot be violated and are named as *hard conditions*

iii. *Channel rearrangement*, that is the ability of changing channels without call interruption, is taken into account only to the cell where a new call refers.

iv. Other conditions that improve DCA performance but may be violated (*soft conditions*) are:

   1. *Packing condition*. The aim of this condition is to exploit better the number of channels by selecting channels already used in other cells, without violating hard conditions. In case of more possible solutions, channels used in the nearest cells are taken into account.
   2. Cells that are located in a distance equal to the reuse distance, belong to the same *reuse scheme* and may use the same channels unconditionally. This assumption is known as *resonance condition*.
   3. The number of channels that are assigned in the cell where a new arrival occurs is tried to be equal with the number of channels that the cell has to serve.
   4. Channels are assigned by limiting the rearranging process in the cell where a call arrives. This is achieved by assigning the same channels assigned before in the allocation table.

By considering the following definitions:

- *CE:* the number of cells in the model.
- *CH:* the number of channels for the cellular system.
- *k:* cell involved in a new arrival of a call.
- $A_{i,j}$: the *ij*-th element of the allocation table *A*, $A_{i,j} = 1$ if channel *j* is assigned to cell *i*, $A_{i,j} = 0$ otherwise ($i = 1,...,CE, j = 1,...,CH$).
- V_k: Output vector for cell *k*, with dimension *CH* (problem variable). $V_{k,j} = 1$ if channel *j* is assigned to cell *k*, otherwise $V_{k,j} = 0$.
- *interf(i,k):* function whose value is 1 if there is cochannel interference between cells i and k, otherwise 0.
- $CSC_{i,j}$: the ij-th element of matrix *CSC*, $CSC_{i,j}=1$ if channel *j* cannot be assigned to cell *i* because of the cosite constraint, $CSC_{i,j}=0$ otherwise ($i=1,...,CE, j=1,...,CH$).
- $ADJ_{i,j}$: the ij-th element of matrix *ADJ*, $ADJ_{i,j}=1$ if channel *j* cannot be assigned to cell *i* because of the adjacent channel constraint, $ADJ_{i,j}=0$ otherwise ($i=1,...,CE, j=1,...,CH$).
- *dist(i,k):* the distance between cells *i* and *k*, normalised to the distance between the centres of two adjacent cells.
- *res(i,k):* function whose value is 1 if cells *i* and *k* belong to the same reuse scheme, otherwise 0.
- *traf(k):* the number of channels that cell *k* must serve.

an energy function is defined as follows:

$$E = \frac{A}{2}\cdot\sum_{j=1}^{CH}\sum_{\substack{i=1\\ i\neq k}}^{CE} V_{k,j}\cdot A_{i,j}\cdot interf(i,k)$$

$$+\frac{B}{2}\cdot\sum_{j=1}^{CH} V_{k,j}\cdot CSC_{k,j}$$

$$+\frac{C}{2}\cdot\sum_{j=1}^{CH} V_{k,j}\cdot ADJ_{k,j}$$

$$-\frac{D}{2}\cdot\sum_{j=1}^{CH}\sum_{\substack{i=1\\ i\neq k}}^{CE} V_{k,j}\cdot A_{i,j}\cdot\frac{(1-interf(i,k))}{dist(i,k)}$$

$$+\frac{F}{2}\cdot\sum_{j=1}^{CH}\sum_{\substack{i=1\\ i\neq k}}^{CE} V_{k,j}\cdot A_{i,j}\cdot(1-res(i,k))$$

$$+\frac{G}{2}\cdot\left(\sum_{j=1}^{CH} V_{k,j}-traf(k)\right)^2$$

$$-\frac{H}{2}\cdot\sum_{j=1}^{CH} V_{k,j}\cdot A_{k,j}$$

Each row in the above function expresses one of the 7 conditions (3 hard and 4 soft). Taking for example the first row one can see that the energy increases if channel j is assigned in cells i and k that are subject to cochannel interference ($interf(i,k) = 1$). Similarly the fourth term reduces the energy when channel j assigned to cell i is also selected by cell k and $interf(i,k) = 0$. For this term energy reduction is inversely proportional to $dist(i,k)$ (Packing condition). Coefficients A, B, C, D, F, G, H are positive constants that determine the significance of conditions by biasing the respective terms.

GA is applied to solve the above model with the following characteristics [6]:

- Representation: A chromosome represents a cell from the cellular system where a call is referred, and a binary gene corresponds to a channel. The number of bits in a chromosome is the number of channels that the cell may serve. If gene is 0 the channel is free and if 1 the channel is occupied.
- Evaluation function: The evaluation function that determines the fitness of the chromosomes is the energy function of the model.
- Genetic operators: Biased random selection together with two-point crossover and simple mutation are used. *Biased random selection* is implemented by a roulette wheel, where the size of each slot in the wheel is proportional to the individual's fitness. The probability P(i) of any individual to be selected from the population is defined as:

$$P(i) = \frac{f(i)}{\sum_j f(j)}$$

where f(i) is the fitness of the ith chromosome in the population. This method favours the selection of the fittest individuals. Two point crossover selects two random chromosomes:

$$(b_1\ b_2 \ldots b_{pos1-1}\ b_{pos1} \ldots b_{pos2}\ b_{pos2+1} \ldots b_m)$$
$$(c_1\ c_2 \ldots c_{pos1-1}\ c_{pos1} \ldots c_{pos2}\ c_{pos2+1} \ldots c_m)$$

and replace them by the pair of their offspring:

$$(b_1\ b_2 \ldots b_{pos1-1}\ c_{pos1} \ldots c_{pos2}\ b_{pos2+1} \ldots b_m)$$
$$(c_1\ c_2 \ldots c_{pos1-1}\ b_{pos1} \ldots b_{pos2}\ c_{pos2+1} \ldots c_m)$$

where *pos1*, *pos2* are random uniform numbers. Mutation simply alters the value of a selected bit from 0 to 1 or vice versa.

- GA's parameters: A population size of 50 chromosomes is used. The population is initialised randomly. The probability of crossover (p_c) was set to 0.75 whereas the probability of mutation (p_m) to 0.05. These values are in line with the common GAs' parameters and were chosen after many experimental trials
- Termination criterion The whole search is terminated after a maximum number of iterations where there are not presented significant changes in the energy value of the best chromosome between successive generations. For our case 100 generations were used. The best individual is the solution to the DCA problem and corresponds to the problem variable V_k referred above. If the solution violates any of the hard conditions the call is blocked.

4. SIMULATION RESULTS

The proposed GA-DCA was compared with other channel assignment schemes with criterion the probability of blocking in call arrivals. These schemes are the FCA, LODA(Locally Optimised Dynamic Assignment), BCO(Borrowing with Channel Ordering), BDCL(Borrowing with Directional Channel-Locking) implemented and evaluated in [8] and a Hopfield Neural Network suggested in [7]. FCA is studied theoretically in [8] with Erlang-B distribution whereas all the other results are numerical. The performance of the above strategies was examined in a model that is in line with the one described in this chapter under the effect of cochannel interference constraint only. The model has 49 cells and 70 channels. Channel reuse distance was assumed to be three cell units, and simulation results were taken by the 9 central cells only. The number of incoming calls follows a Poisson distribution whereas call duration is a random variable with exponential distribution. The product of the mean arrival rate and the mean call duration expresses the load offered to the cellular network in erlangs. The *erlang* represents a radio channel being occupied continuously for a duration of one hour. Figure 2 shows the performance of the above techniques when all cells in the network have the same arrival rate (traffic distribution is uniform).

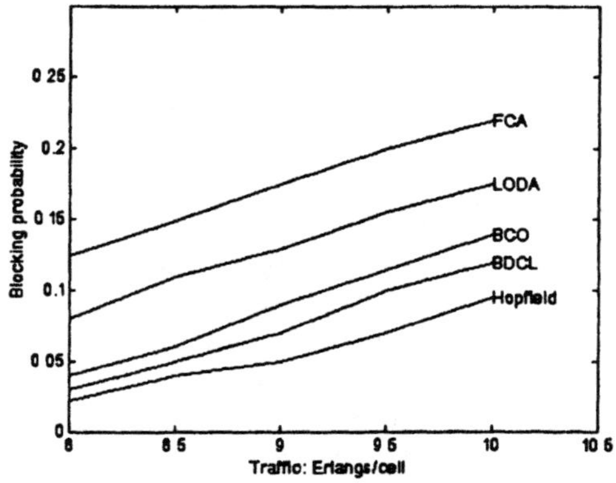

Fig. 2: Performance of various channel assignment techniques with uniform traffic

A simulation model for GA-DCA has been implemented in MATLAB programming language and follows all the above hypotheses. Figure 3 shows the behaviour of the proposed strategy by considering the effects of: a) cochannel interference, b) cochannel and cosite interferences and c) all the hard conditions (cochannel, cosite and adjacent channel). For cases b and c, adjacent channels are assumed to be the two nearest channels of an already used channel. The performance of GA-DCA for case a is slightly lower than the Hopfield neural network. In the other cases, the effects of more interference constraints reduce the resources of the system and decrease blocking probability as it is expected.

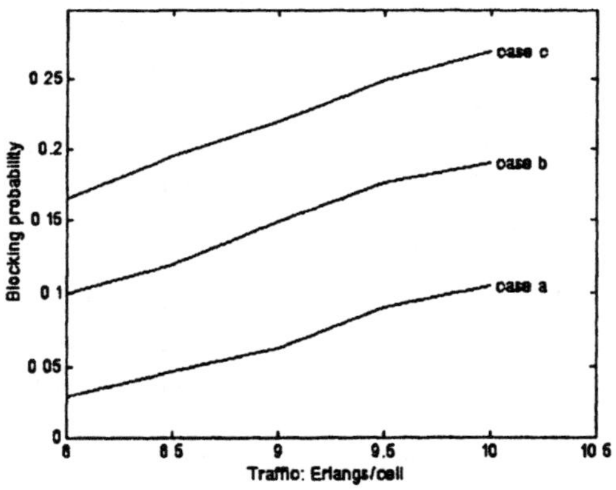

Fig. 3: Performance of GA-DCA for various interference constraints

For nonuniform traffic distribution the traffic patterns proposed in [8] were considered (Fig 4, 5). The numbers in the cells represent initial Poisson arrival rates (calls/hour). Simulation results are summarised in figures 6, 7. Hopfield neural network performance is not presented as it was slightly lower than BDCL. The traffic load for each cell is increased by a percent factor over the base load.

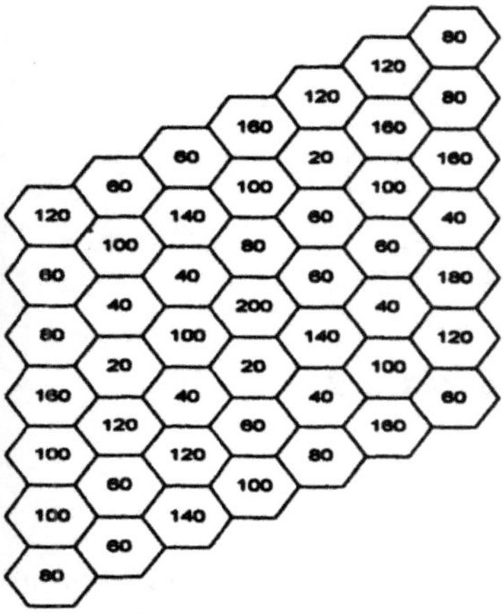

Fig. 4: Pattern 1 (calls/hour)

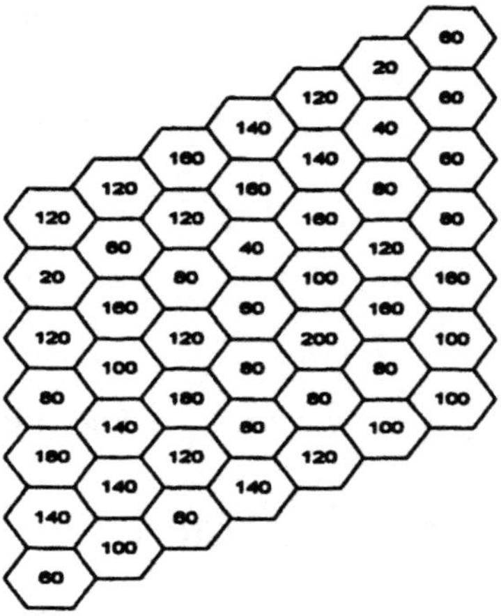

Fig. 5: Pattern 2 (calls/hour)

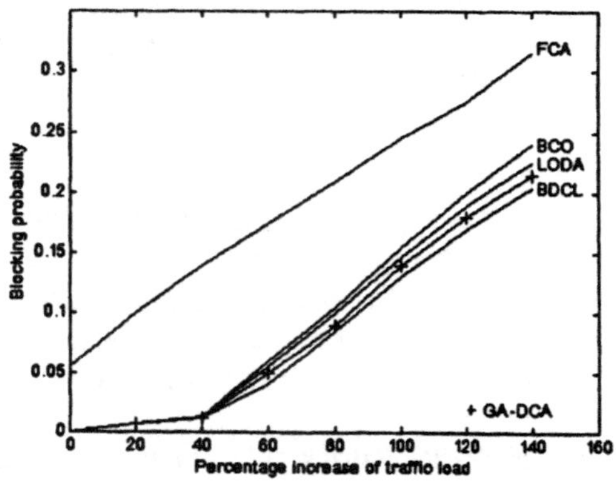

Fig. 6: Blocking probability curves for pattern 1

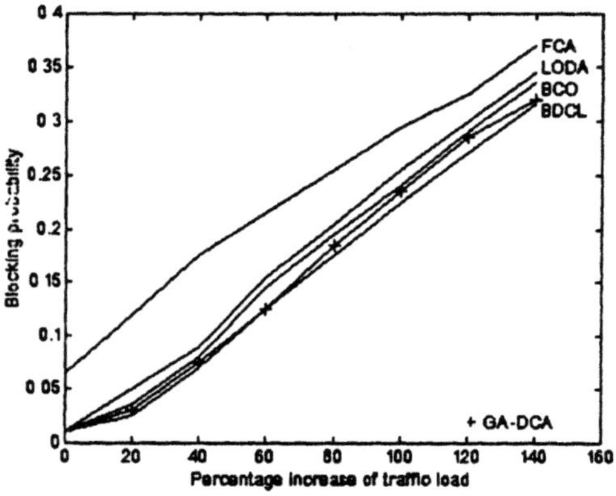

Fig. 7: Blocking probability curves for pattern 2

5. CONCLUSIONS

The performance of GA-DCA has been investigated through this chapter. The ability of GAs to search for a solution from a broad spectrum of possible solutions, rather than restrict the search to a narrow domain, led to the extraction of quite satisfactory results over other channel assignment methods. Moreover the convergence of GAs without problem specific parameter tuning, makes it advantageous compared to a Hopfield neural network which had superior performance than GA-DCA in uniform traffic distribution. The main drawback of the proposed technique is the possible maintenance of infeasible solutions to the population that occur mainly because of the crossover operator. Therefore the use of different GA representations to the problem or other heuristic techniques are topics of further research.

REFERENCES

[1] Zbigniew Michalewicz, *Genetic Algorithms + Data Structures = Evolution Programs,* Springer - Verlag, 3rd edition, 1996

[2] A. Mehrotra, *Cellular Radio. Analog and Digital Systems*, Artech House Inc., 1994.

[3] K. Lai and George G. Coghill, *Channel Assignment through Evolutionary Optimization,* IEEE Transactions on Vehicular Technology, Vol. 45, No. 1, pp. 91-96, February 1996

[4] Jae-Soo Kim, Sahng H Park, Patrick W. Dowd and Nasser M. Nasrabadi, *Comparison of Two optimization Techniques for Channel Assignment in Cellular Radio Network,* IEEE International Conference on Communications, Vol.3, pp. 1850-1854, June 1995

[5] W. Crompton, S. Hurley and N.M Stephens, *A Parallel Genetic Algorithm for Frequency Assignment Problems,* Proceedings of the IMACS/IEEE Conference on Signal Processing, Robotics and Neural Networks, pp. 81-84, 1994

[6] Harilaos G. Sandalidis, Peter P. Stavroulakis and J. Rodriguez-Tellez, *Implementation of Genetic Algorithms to a Channel Assignment Problem in*

Cellular Communications, 6th International Conference on Advances in Communications and Control (COMCON 6), Corfu, Greece, June 1997

[7] Enrico Del Re, Romano Fantacci and Luca Ronga, *A Dynamic Channel Allocation Technique Based on Hopfield Neural Networks,* IEEE Transactions on Vehicular Technology, Vol. 45, No. 1, pp. 26-32, February 1996

[8] Ming Zhang and Tak-Shing P. Yum, *Comparisons of Channel Assignment Strategies in Cellular Mobile Telephone Systems,* IEEE Transactions on Vehicular Technology, Vol. 38, No. 4, pp. 211-215, November 1989

13

APPLYING GENETIC ALGORITHMS TO WORKSHOP CELL DECOMPOSITION

P. DE LIT, A. DELCHAMBRE and E. FALKENHAUER
Department of Applied Mechanics
University of Brussels (ULB)
Brussels, Belgium

1. Introduction

Competitivity being more than ever a primary goal for the industry, productivity's and product's quality improvement ask for a global approach. We talk about "production systems".

In the 1960s, J. L. Burbidge [1] developed a systematic planning approach on the concept according to which parts with similar features could be manufactured together with standardized processes. Today large facilities regroup small independent units acting themselves as little factories. Creation of these units is based on the concept of Group Technology (GT) — a theory of management based on the principle that *similar things should be done similarly*.

Layout problems arise with the study of a new production system, or during reorganization due to introduction of new ressoucres or product design modification. This layout aims to reduce carriage and handling, which leads to simplifications of the management, lessening of lead times, and indirectly, to an increasing product's quality and a quicker response to market's fluctuations.

A dogma of the GT advocates the division of a unit in little groups or cells. Products needing similar operations and a common set of ressources are grouped into families, the ressources being regrouped into production subsystems. This GT has revealed itself as a key in production control and optimization, as well as in material transport.

One major problem to tackle in GT is the partition of the set of ressources into as independent as possible cells. As it is most of the time illusory in industrial applications to get totally independent cells, several approaches have been proposed to group machines, for example Production Flow Analysis [2], use of Similarity Coefficient Methods [3], [4], or graph partitioning methods [5]. We considered the problem as pertaining to the latter category, and consisting in minimizing product's traffic between cells. This decomposition of a production system into cells is known to be a NP-hard *grouping problem*.

S.G. Tzafestas (ed.), Advances in Intelligent Systems, 139–150.

This chapter is organized as follows. We briefly mention work related to ours in section 2. Section 3 is devoted to the description of the problem to be solved. We then describe in section 4 the philosophy and priciples of Grouping Genetic Algorithms, a class of algorithms well suited to Grouping problems. Implementation details are explained in section 5. Results of our algorithm will be given at section 6 together with our conclusions.

2. Related works

We here mention some work accomplished in the domain of workcell decomposition that influenced ours. The problem can be seen as the partition of an undirected weighted graph, the nodes representing the machines and the weighted edges the traffic between these ressources (this notion of traffic will be explained at section 3). Each subgraph will represent a group. The aim is to find the minimal cut, restraining the size of each subgraph to N_{max} nodes.

2.1 Haralakis' heuristic

G. Haralakis *et al.* [5] have proposed a simple heuristic to minimize the intercell traffic,which is divided into two phases: an aggregation and a local refinement. At the beginning of the aggregation each machine is in a cell. The possible aggregations are the one not exceeding the maximum cell size. The two cells between which there exists the highest traffic are grouped. After this aggregation, a refining phase tries to convert the intercell traffic into intracell traffic. Each machine is considered as a separate entity, and its traffic with each cell is computed. A given machine is attributed to the group it has the most important interaction with. Note that most of the time a machine will be reattributed to its cell, but some changes may occur.
This algorithm is simple and rather quick (his complexity is $O(x^3)$), but does not always yields the optimal solution. The smallest problem which the algorithm is deceived by is illustrated at Figure 1. The optimal solution yields groups {1,3,4} and {2} and an intercell traffic of 7.

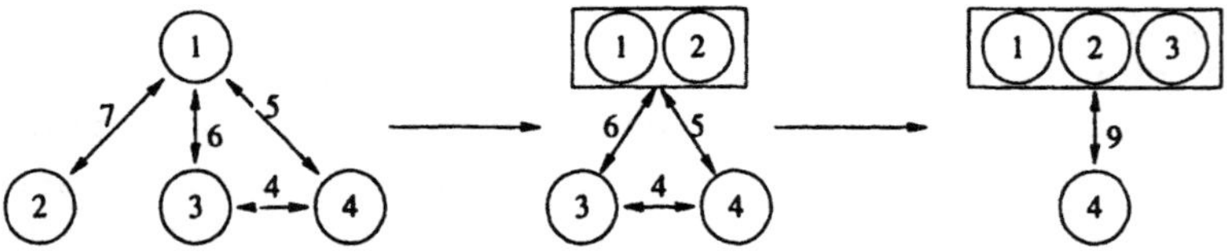

Figure 1 : Minimal deceiver for Haralakis' heuristic ($N_{max} = 3$).

2.2 Nodes swapping

Another popular heuristic for graph partitioning is Kernighan and Lin's heuristic ([6]), which can be adapted to multiple groups and variable group sizes [7]. This procedure starts from a given partition and first tries to find the best possible swap between the groups (swap may concern subsets of the two partitions). Once the best swap has been performed, the process restarts. This heuristic yields good results (for example, it will solve the deceptive problem illustrated at Figure 1), but is inoperant if the improvement of a partition needs a swap between more than two groups. On the following example, showed at Figure 2, the heuristic will not be able to lead to the optimal solution, no swap improving the proposed decomposition[1]. The optimal solution asks for a swap of three machines together (1 from G_0 to G_2, 4 from G_2 to G_1, and 7 from G_1 to G_0).

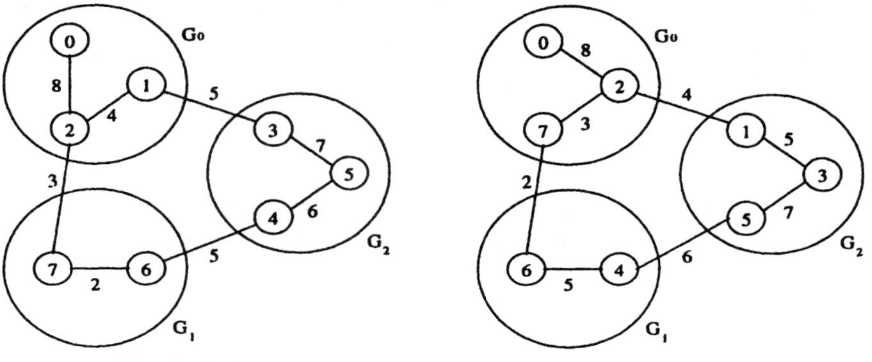

Figure 2 : Two cells swap deceptive problem.

2.3 Simulated annealing

Simulated annealing [8] was applied to workshop cell decomposition problems ([9]), defining the neighbourhood of a partition *C* of the set of machines *M* by the set of partitions derived from *C* by switch of two machines in two different cells, creation of a new cell by extracting a machine from a given one, or reattribution of a machine from a cell to another.

This simulated annealing will more rarely fall in local extrema than heuristics, but has the major drawback to be extremely slow. However the approach vehicles the interesting idea to accept bad attributions or swiches of machines to escape from local optima.

3. Mathematical aspects of the problem to solve

Let $P=\{(p_0,u_0),\ldots,(p_i,u_i),\ldots(p_{n-1},u_{n-1})\}$ be a set of *n* couples (products, product weight[2]) and $M=\{m_0,\ldots,m_j,\ldots,m_{m-1}\}$ a set of *m* machines.

Let $\{r_k(p_k)\}$ be the manufacturing sequence of product p_k, constituted by machines $m_j^k \in M$, and $C=\{C_0,\ldots,C_{\omega-1}\}$ a partition of the set of machines, each C_i representing a cell. Let us define for p_k, x_{lm}^k as the number of times m_j is immediately preceeded by m_i in $\{r_k\}$, with $C_l, C_m \in C$ and $m \neq l$. We call *traffic* between the two partitions :

$$T_{lm} = \sum_{k=1}^{n} u_k (x_{lm}^k + x_{ml}^k).$$

---

[1] This first decompositon was obtained using Harhalakis' aggregation procedure.

[2] By weight we mean fraction of a production volume or cost factor.

The problem is to find the partition $C^* = \{C_0,\ldots,C_{\omega-1}\}$ minimizing

$$\sum_{i=0}^{\omega-2}\sum_{j=i+1}^{\omega-1} T_{ij} \quad \text{with } \mathrm{card}(C_k \subset C^*) \leq N_{\max},$$

where N_{max} is the maximum size of a cell.

The traffic between two partitions can also be defined as follows. Let y_{ij}^k be for p_k the number of times m_j is immediat successor of m_i in the manufacturing sequence of p_k. We call traffic between two machines m_i and m_j

$$t_{ij} = \sum_{k=0}^{n-1} u_k (y_{ij}^k + y_{ji}^k).$$

These traffics between machines are independent from partition C and can be represented as an $m \times m$ matrix. The traffic between cells for a given partiton C^* is

$$T_{ij} = \sum_{m_k \in C_i} \sum_{m_l \in C_l} t_{kl}, \quad k \neq l.$$

4. The Grouping Genetic Algorithm

4.1 *The Grouping Problems*

The Grouping Problems constitute a large family of problems, many of them naturally arising in practice, which consist in partitioning a set U of items into a collection of mutually disjoint subsets U_i of U, i.e., such that:

$$\begin{cases} \bigcup_i U_i = U \\ U_i \cup U_j = \varnothing \quad \forall i \neq j. \end{cases}$$

In most of these problems, not all possible groupings are allowed: a solution of the problem must comply with various hard constraints, the solution being otherwise invalid. That is, usually an item cannot be grouped with all possible subsets of the remaining items.

The objective of the grouping is to optimise a cost function defined over the set of all valid groupings. This cost function depends on the composition of the groups, that is, where one item taken isolatedly has little or no meaning.

4.2 *The method*

Introduced by J. Holland [10], the Genetic Algorithm (GA) is an optimization technique inspired by the process of evolution of living organisms. The basic idea is to maintain a population of chromosomes, each chromosome being the encoding (a description or genotype) of a solution (or phenotype) to the problem being solved. The worth of each chromosome is measured by its fitness, which is often simply the value of the objective function in the point of the search space defined by the (decoded) chromosome (in a maximization problem).

Starting with an initial population generated mostly at random, the GA proceeds in quite the same manner as Nature in evolving ever better solutions: chromosomes with high fitness are crossed over, producing progeny that replaces chromosomes with low fitness. A low rate of mutation, a small random modification of a chromosome, is applied to prevent a premature convergence to a local optimum. A good introduction to GAs is given in [11].
E. Falkenauer [12] pointed out the weaknesses of standard GA when applied to grouping problems and introduced the Grouping Genetic Algorithm (GGA) [13], a GA heavily modified to match the structure of grouping problems. The GGA differs from the classic GA in two important aspects. First, a special encoding scheme is used in order to make the relevant structures of grouping problems become genes in chromosomes. Second, given the encoding, special genetic operators are used, suitable for the chromosomes.

4.3 The encoding

The GGA uses an specific encoding scheme: the standard chromosome is augmented with a group part, encoding the groups on a one gene for one group basis. More concretely, let us consider a chromosome of a standard GA. Numbering the items from 0 through 5, the item part of the chromosome can be explicitly written

```
012345
ADBCEB :...
```

meaning the item 0 is in the group labelled (named) *A*, 1 in the group *D*, 2 and 5 in *B*, 3 in *C*, and 4 in *E*. The group part of the chromosome represents only the groups. Thus

```
...: BECDA
```

expresses the fact that there are five groups in the solution. Of course, what names are used for each of the bins is irrelevant in our grouping problem: only the contents of each group counts in this problem. We thus come to the *raison d'être* of the item part. Indeed, by a lookup there, we can establish what the names stand for. Namely,

$$A=\{0\},\ B=\{2,5\},\ C=\{3\},\ D=\{1\} \text{ and } E=\{4\}.$$

In fact, the chromosome could also be written

$$\{0\}\{2,5\}\{3\}\{1\}\{4\}.$$

The important point is that the genetic operators will work with the *group part* of the chromosomes, the standard item part of the chromosomes merely serving to identify which items actually form which group[3].

[3] Note that this implies that the operators will have to handle chromosomes of variable length, and that the order of the groups in the chromosome is irrelevant in the GGA.

4.4 The Crossover

Given the fact that the hard constraints and the cost function vary among different grouping problems, the ways groups can be combined without producing invalid or too bad individuals are not the same for all those problems. Thus the crossover used will not be the same for all of them. However, it will fit the following pattern:

1. Randomly select two crossing sites, delimiting a crossing section, in each of the two parents.
2. Inject the contents of the crossing section of the first parent at the first crossing site of the second parent. This means injecting some of the groups from the first parent into the second.
3. Eliminate all items now occurring twice from the groups they were members of in the second parent, so that the "old" membership of these items gives way to the membership specified by the "new" injected groups. Consequently, some of the "old" groups coming from the second parent are altered: they do not contain all the same items anymore, since some of those items had to be eliminated[4].
4. If necessary, adapt the resulting groups, according to the hard constraints and the cost function to optimise. At this stage, local problem-dependent heuristics can be applied.
5. Apply points 2 to 4 inverting the roles of the two parents in order to generate the second child.

As can easily be seen, the idea behind the GGA crossover is to promote promising groups by inheritance. We describe in section 5.3.1 the adaptation of the crossover operator to our grouping problem.

4.5 The mutation

A mutation operator for grouping problems must work with groups rather than items. As for the crossover, the implementation details of the operator depend on the particular grouping problem on hand. Our mutation operator is described at section 5.3.2.

4.6 The inversion

The inversion operator serves to shorten good schemata in order to facilitate their transmission from parents to offspring, thus ensuring an increased rate of sampling of the above-average ones ([10]). In a Grouping GA, it is applied to the group part of the chromosome.

---

[4] The injection operation has some of the flavour of the cut & splice recombination operator of Goldberg's messy GA (e.g. [14]). The fundamental difference lies with the fact that the GGA's crossover always ends up with each item present exactly once, through a possible modification of genes coming from the second parent.

Thus for instance, the chromosome

$$ADBCEB:BECDA$$

could be inverted into

$$ADBCEB:CEBDA.$$

The example illustrates the utility of this operator: should B and D be promising genes (i.e., well-performing groups), the probability of transmitting both of them during the next crossover is improved after the inversion, since they are now closer together, i.e., safer against disruption.

5. Implementation

5.1 Validity of an individual

We fixed two conditions for an individual to be valid: the size of the cells may not exceed N_{max}, and the subgraphs associated to the different cells must be connex[5].

5.2 Cost function

The intracell traffic for a cell C_i is

$$T_i = \sum_{k \in C_i} \sum_{l \in C_i} \frac{t_{lk}}{2}.$$

As the total traffic between machines stays constant (T_{total}), the intercell traffic is given by

$$T_{inter} = T_{total} - T_{intra}.$$

So minimizing the intercell traffic is equivalent to maximizing the total intracell traffic, given for q cells by

$$\max(T_{intra}) = \max\left(\sum_{i=0}^{q-1} T_i\right).$$

This cost function is well adapted to our problem, the evolution of the traffic only depending on affected cells during a perturbation.

5.3 Genetic operators

5.3.1 Crossover

Crossover is applied as described in section 0, with the difference that affected groups are emptied. Reinjection of the machines uses the following heuristic. The traffic be-

[5] Note that this constraint influences the number of groups proposed in the optimal solution (we could otherwise propose to group machine without traffic between them), but has no influence on the quality of the solution according to our cost function.

tween a chosen[6] non attributed machine and the existing cells is computed, and this machine is attributed to a group with a probability *pro rata* of this traffic. Note that the complexity of this heuristic is $O(x^2)$. We create a new cell if none can accept a chosen machine thrice in a row. Figure 3 illustrates our words. The traffics between machine 4 and groups 1 and 2 amount to 24 and 36 respectively. The machine will be injected in group 1 and with a probability of 0.4 for the former and 0.6 for the latter. The same reasoning goes for machine 3. This one having no connections with the existing cells, a new one is created to greet it. This ranking and selection of the greeting cells aims to help the GGA to leave local optima it could get stuck in.

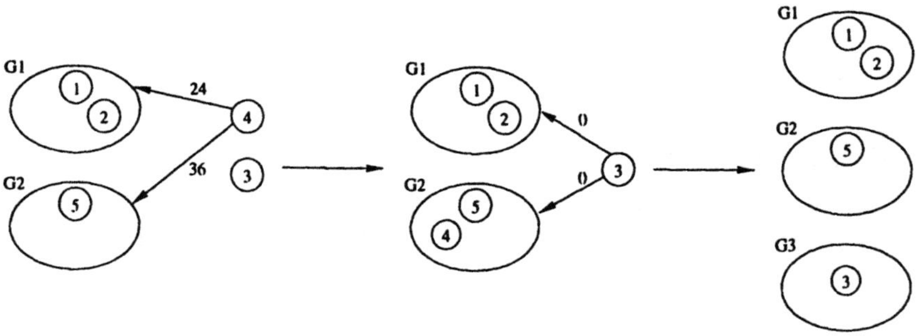

Figure 3: Machine reattribution heuristic.

After having applied the heuristic described above, if the GGA seems to be stuck in a local optimum, we search for the best possible swap between two machines that reduces the intercell traffic (the complexity of this search is $O(x^2)$). If there is none, a swap worsening the solution will sometimes be applied, to help the GGA to leave that local optimum. Note that this probabilistic swap is crucial to leave a local optimum in case of problems requiring cyclic swaps like the one illustrated at Figure 2. Without this heuristic, the algorithm get stuck in local optimas for about 1000 generations for rather small instances of the problem (about 300 machines[7]).

5.3.2 Mutation

The mutation operator is only applied if the crossover does not generate a new individual in the population. The mutation removes one tenth of the objects among the groups and reinjects them according to the heuristic we described in section 5.3.1.

5.4 Generating the first population

The first individual of the population is generated by Haralakis' aggregation procedure, the others using heuristic described at section 5.3.1. These aggregations are followed by a swapping heuristic: if a group has external connections, the best swap between it and all others is made. The swap will occur once per group (so a given group will not undergo two successive swaps). This first population generation procedure insures us to

[6] by "to choose", we mean "to draw lots".

[7] These instances are in fact several independent elementary problems presented at Figure 2.

find the global optimum at initialization if a perfect decomposition is possible. The reader could object that the generation of an individual which could sometimes be highly better fit than the others will lead to a broad speading of schemata belonging to a suboptimal solution among the population. Tests on deceptive problems indicate that the proposed genetic operators make the GGA "forget" the effect of Haralakis' aggregation at population initialization.

6. Experimental results

For all simulations, performed on a 266 Mhz Pentium, size of the population was 30 individuals. Results presented here are mean and standard deviation values of 20 runs.

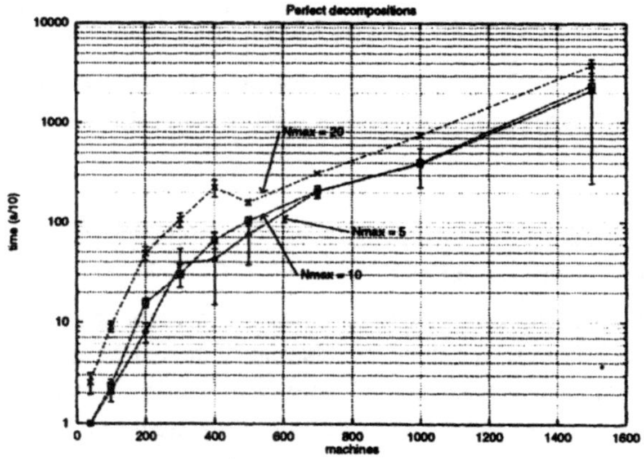

Figure 4 : Results for perfect decompositon without Haralakis' heuristic (time to optimum).

6.1 Perfect decompositions

Those simple problems, for which Haralakis' heuristic always converges to the optimum, allow the evaluation of the algorithm on unimodal functions. To avoid finding the optimal solution at evaluation, we suppressed Haralakis' procedure and machine swaps at initialization. The evolution of the computing time and generations to the optimum according to the size of the problem are given at Figure 4 and Figure 5.

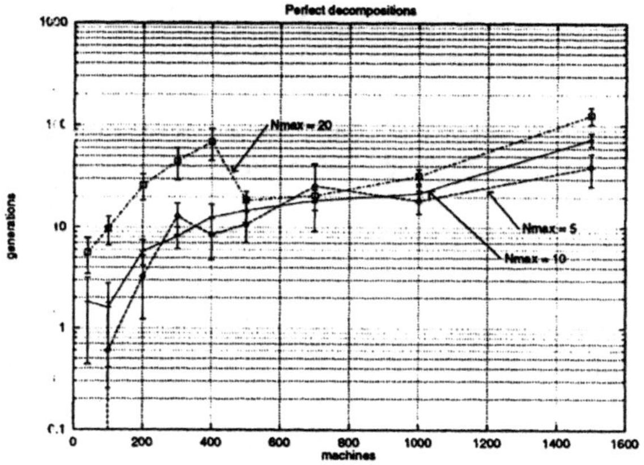

Figure 5 : Results for perfect decompositions without Haralakis' heuristic (generations to optimum).

6.2 *Deceptive problems*

6.2.1 Haralakis' minimal deceiver

We studied Haralahis' minimal deceiver to see the effects of the swapping heuristic on the search of the optimum. We disabled Haralakis' heuristic at initialization, as it leads to the optimal solution when associated with initialization's swap heuristic. Results are reported at Figure 6. One can see that that the swap heuristic enabled after 10 generations without improvement increases the speed to reach the optimal solution.

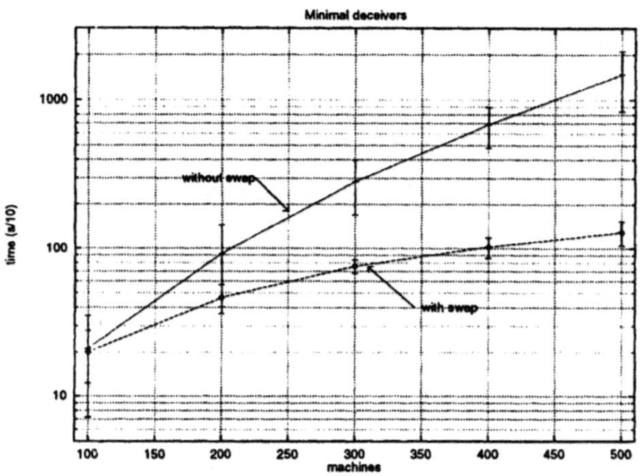

Figure 6 : Minimal deceivers with and without swap heuristic ($N_{max} = 3$).

6.2.2 Complete problem

The graph associated to the elementary deceptive problem we studied is given at Figure 7. Optimal solution, for N_{max}= 5 yields groups {0,3,4,5,6} and {1,2}. Note that the application of Haralakis' followed by the swap heuristic is inoperant in this case, because the algorithm will swap machines 2 and 3. The instances are composed of independent elemntary deceivers. The results at Figure 8 show that the GGA is able to deal with important instances of the problem in.reasonnable amount of time.

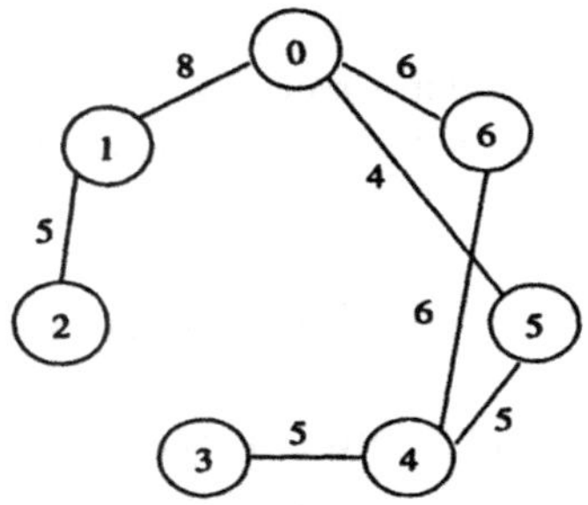

Figure 7 : Deceptive problem.

Note that the important dispersion on the results is due to the fact that the swap heuristic is triggered during the generational evolution when the GGA has not improved the solution for 30 generations. This heuristic being time consuming, it notably increases computation time when it is enabled.

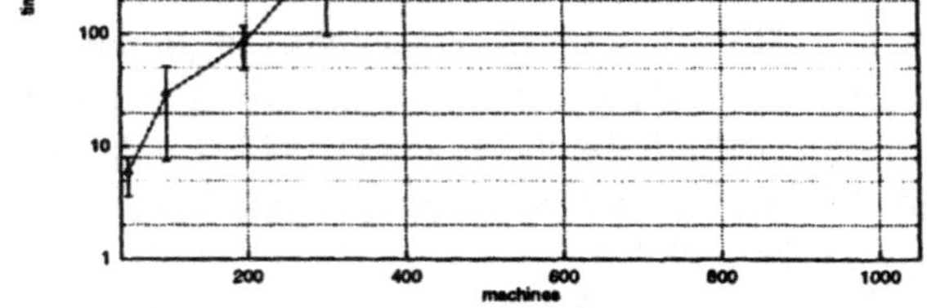

Figure 8: Results for the complete problem. (N_{max}= 5).

7. Conclusions

We addressed in this chapter the problem of workshop cell decomposition, which is an important aspect of Group Technology (GT). We proposed to solve it using a matrix formalism an unoriented weighted graph can be associated to, the problem to tackle being to find the partition which gives the minimal cut for a given maximal size of each subgraph. Since this problem is *NP*-hard, enumerative methods will crash down for important instances of it. We thus proposed a Grouping Genetic Algorithm which does not present this major drawback, making it applicable to industrial cases, and offering the advantage not to be stucked in local optima like heuristics.

8. Aknowledgement

This paper is based on results of the project "Outils d'aide à la conception interactive des produits et de leur ligne d'assemblage". This project is made in collaboration with the Université Catholique de Louvain (UCL), the Faculté Polytechnique de Mons (FPMs), and the Belgian Research Center for the metalworking industry (CRIF). We particularly thank the "Région Wallonne" which has funded this project.

References

[1] J. L. Burbidge, *The Introduction of Group Technology*, John Wiley & Sons, New York., 1975.

[2] J. L Burbidge, "Production flow analysis", *Production Engineer*, pp. 139-152, Apr./May 1971.

[3] J. McAuley, "Machine grouping for efficient production", *Production Engeneer*, pp. 53-57, Feb 1972.

[4] H. Seifoddini and P. M. Wolfe, "Application of the similarity coefficient method in group technology", *IIE Transactions*, vol. 18, no. 3, pp. 271-277, 1986.

[5] G. Harhalakis, R. Nagi, and J. M. Proth, "An efficient heuristic in manufacturing cell formation for group technology applications", *International Journal of Production Research*, vol. 28, no. 1, pp. 185-198, 1990.

[6] B. W. Kernighan and S. Lin, "An efficient heuristic procedure for partitioning graphs", *The Bell System Technical Journal*, vol. 49, pp. 291-307.

[7] R. G. Askin and C. R. Standridge, *Modeling and Analysis of Manufacturing Systems*, John Wiley & Sons, 1993.

[8] S. Kirkpatrick, C. D. Gelatti, and M. P. Vecchi, "Optimization by simulated annealing", *Science*, vol. 220, pp. 671-680, 1983.

[9] B. Pommerenke, "Choix optimal des cellules de production par un algorithme génétique de groupement", *Travail de fin d'études présenté en vue de l'obtention du grade d'Ingénieur Civil Physicien*, Université Libre de Bruxelles, 1995.

[10] J. H. Holland, *Adaptation in Natural and Artificial Systems*, University of Michigan Press, Ann Arbor, 1975.

[11] D. E. Goldberg, "Genetic Algorithms in Search, Optimization and Machine Learning", Addison--Wesley Publishing Company Inc., 1989.

[12] E. Falkenauer, "A hybrid grouping genetic algorithm for bin packing", *Journal of Heuristics*, vol. 2, no. 1, pp. 5-30, 1996.

[13] E. Falkenauer, *Genetic Algorithms and Grouping Problems*, John Wiley & Sons Inc., 1998.

[14] D. E. Goldberg, D. Kalyanmoy and K. Bradley, "Don't worry, be messy", *Proceedings of the Fourth International Conference on Genetic Algorithms, University of California*, R. K. Belew and L. B. Booker eds., Morgan Kaufmann Publishers, San Mateo, CA, Jul 1991.

14

A GENETIC ALGORITHM FOR EFFICIENT VIDEO CONTENT REPRESENTATION

A. D. DOULAMIS, Y.S.AVRITHIS, N.D.DOULAMIS
and S.D. KOLLIAS
Department of Electrical and Computer Engineering
National Technical University of Athens (NTUA)
Zografou Campus, Zografou, 15773, Athens, Greece

1. Introduction

The rapid development of video and multimedia applications has enabled users to handle large amounts of visual information. At the same time, new requirements have emerged for more intelligent access to such databases, i.e., content-based indexing, retrieval and video browsing. The traditional text-based approach to accessing image or video databases has the drawback that it is difficult to characterize the rich content of images or video based only on text information [1]. For this reason, a new standardization phase is currently in progress by the Moving Picture Expert Group (MPEG) in order to develop an integrated framework for a multimedia content description interface (MPEG-7).

Many techniques have been developed in this research area and many image/video retrieval systems have been built. The active research effort has been reflected in many special issues of leading journals dedicated to this topic [2], [3]. Some of these approaches are now in the first stage of commercial exploitation, such as the VisualSEEK and QBIC [4] prototypes. Color image retrieval has been examined in [5] based on a hidden Markov model, and extraction of detailed image regions for indexing and retrieval has been proposed in [6]. Object modeling and segmentation for indexing in video databases has been reported in [7] while single frame extraction based on the frame properties has been proposed in [8] to perform the queries. A new progressive resolution motion indexing has been presented in [9] using 3-D wavelet decomposition of video sequences as well as rigid polygonal shapes. Finally, an approach for automatic video segmentation and content-based retrieval based on a temporally windowed principal component analysis of a sub-sampled version of a video sequence has been reported in [10].

However, these systems cannot be easily extended to video databases since it is practically impossible to perform queries on every video frame. Furthermore, due to the strong temporal correlation of video frames, examination of each frame is very

S.G. Tzafestas (ed.), Advances in Intelligent Systems, 151–162.

inefficient. To make retrieval in video databases more efficient, a pre-indexing stage should be introduced which extracts a small but meaningful information of the video content. Then, video queries can be directly applied to this small amount of information. In this chapter, we propose an efficient video content representation using optimal extraction of a limited number of key frames and scenes of video sequences. This approach not only provides a more efficient way for video indexing, but also results in reducing storage requirements and thus permits easy management of multimedia databases. Then, video queries are performed on this small but meaningful collection of frames instead of the entire video stream.

The first stage of the proposed algorithm includes a scene cut detection mechanism. Then, video processing and image analysis techniques are applied to each video frame for extracting color, motion and texture information. Color information is extracted by applying a hierarchical color segmentation algorithm to each video frame. Consequently, apart from the color histogram of each frame additional features are collected concerning the number of color segments, and their location, size and shape. Motion information is also extracted in a similar way by using a motion estimation and segmentation algorithm.

All the above features are gathered in order to form a multidimensional feature vector for each video frame. The representation of each frame by a feature vector, apart from reducing storage requirements, transforms the image domain to another domain, more efficient for key frame selection. Since similar frames can be characterized by different color or motion segments, due to imperfections of the segmentation algorithms, a fuzzy representation of feature vectors is adopted in order to provide more robust searching capabilities. In particular, we classify color as well as motion and texture segments into pre-determined classes forming a multidimensional histogram and a degree of membership is allocated to each category so that the possibility of erroneous comparisons is eliminated.

Optimal selection of representative scenes is performed by minimizing a distortion criterion. This is accomplished by clustering similar scenes and selecting a limited number of cluster representatives. The generalized Lloyd-Max algorithm has been used for this purpose as described in [11]. The next step is to select the key frames within each one of the selected scenes. This is achieved by minimizing a correlation criterion, so that the selected frames are not similar to each other. This approach gives better results than the one proposed in [12], where frame selection was based on detection of feature vectors that reside in extreme locations of the feature vector trajectory. Since similar frames may be characterized by different segments, the latter approach was rather sensitive and heavily dependent on the adopted segmentation algorithm.

Unfortunately, the complexity of an exhaustive search for the minimum value of a correlation measure is such that a direct implementation would be practically unfeasible. For this reason, a genetic algorithm approach [13] is adopted in this chapter. Possible solutions of the optimization problem, i.e., sets of frames, are represented by chromosomes whose genetic material consists of frame numbers (indices). An initial population of chromosomes is then generated by selecting sets of frames whose feature vectors reside in extreme locations of the feature vector trajectory. The objective function used to estimate the fitness values of all chromosomes, is defined as the sum of squares of cross-correlations between all combinations of feature vectors, for all frame numbers that belong to the genetic material of the respective chromosome. Following a proportionate scheme for parent selection, a set of new chromosomes (offspring) is produced by mating the parent chromosomes and applying uniform crossover and mutation operations.

This chapter is organized as follows: Section 2 briefly describes the feature extraction module, including the scene cut detection as well as the color / motion segmentation procedure. Section 3 refers to the feature vector formulation, while in sections 4 and 5 the scene and frame selection mechanisms are presented respectively. Experimental results illustrating the performance of the proposed scheme are presented in section 6, while conclusions are given in section 7 of this chapter.

2. Feature Extraction

The feature extraction procedure is performed in a way similar to [14] and is briefly discussed in the sequel.

Scene Cut Detection. The first stage of the feature extraction procedure includes a scene cut detection technique, in order to locate the main shots of a video stream. Since visual content is typically stored in MPEG compressed format, it is preferable to perform the feature extraction directly in the compressed domain. As a result, in our approach scene cut detection is achieved by computing the sum of the block motion estimation error over each frame and detect frames for which this sum exceeds a certain threshold [14].

Color / Motion Segmentation. Color and motion segmentation provides a powerful representation of each video frame, more oriented to the human perception. In general, the number, size and location of objects as well as their color, motion, or texture characteristics give more meaningful information for an image than raw pixels. Thus, a color and motion segmentation technique is applied to each video frame. Block resolution has been adopted both for reducing the required computational time and for exploiting information that already exists in the MPEG coding standard. To avoid oversegmentation problems we have proposed a hierarchical block-based segmentation algorithm described in [14]. Moreover, object tracking is supported by taking into account motion compensated segmentation results of previous frames [12]. Apart from information provided by color or motion segmentation other features are included in the feature vector, such as information provided by color and motion histograms or appropriate ac coefficients of the DCT transform.

3. Feature Vector Formulation

All of the above frame features are gathered in order to form a multidimensional feature vector which is used for collection of information content for each frame. Properties of color or motion segments cannot be used directly as elements of feature vectors, since its size will differ between frames. To overcome this problem, we classify color as well as motion segments into pre-determined classes, forming a multidimensional histogram. To eliminate the possibility of classifying two similar segments to different classes, causing erroneous comparisons, a degree of membership is allocated to each class, resulting in a fuzzy classification [15].

This kind of classification is illustrated in Figure 1 for the simple case of a single one-dimensional feature x, normalized between 0 and 1 (e.g., normalized segment size).

A fuzzy partition of the feature space [0,1] into Q=5 classes is defined by using Q membership functions $\mu_n(x) \in [0,1]$, $n = 1,\ldots,Q$. Triangular membership functions with 50% overlap between successive partitions are used in Figure 1, but their exact shape and overlap percentage can be greatly varied. Using this partition scheme for feature x, a fuzzy histogram can be constructed from a large number of feature samples, corresponding to different image segments. Moreover, this histogram can be meaningful even when the total number of segments is small, since similar features always produce similar classification results.

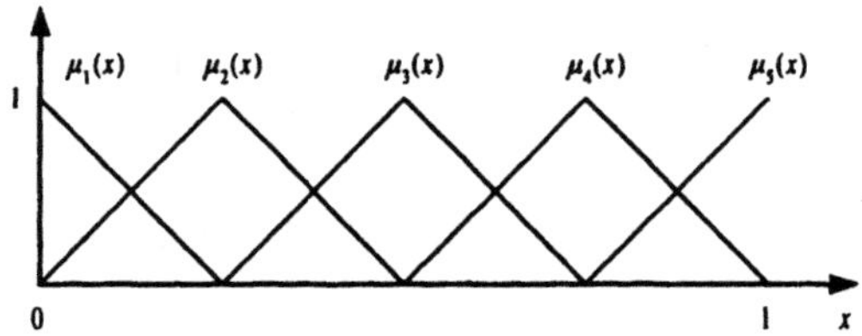

Figure 1: One-dimensional fuzzy classification.

In the more general case of multiple segment properties (such as size, color and motion), multidimensional classification is applied. Let $P(S_i)$, $\mathbf{c}(S_i)$, $\mathbf{v}(S_i)$, and $\mathbf{l}(S_i)$ denote the size, color, motion vector and location of the i-th segment S_i. The $L\times 1$ vector

$$\mathbf{x}^{(i)} = [P(S_i)\,\mathbf{c}(S_i)^T\;\mathbf{v}(S_i)^T\;\mathbf{l}(S_i)^T]^T \tag{1}$$

then fully describes the properties of segment S_i using a total of L segment features. Since the length of vectors **c**, **v** and **l** is 3×1, 2×1 and 2×1 respectively, L will be equal to 8 in the above formulation. However, it can be greater if we also include segment shape or texture information. Each feature space is then partitioned into Q regions and a partition index $n_j \in \{1,2,\ldots,Q\}$ is assigned to the j-th feature element, $x_j^{(i)}$, of $\mathbf{x}^{(i)}$. The degree of membership of segment S_i into the L-dimensional class $\mathbf{n} = [n_1 \ldots n_L]^T$ is defined as

$$M_i(\mathbf{n}) = \prod_{j=1}^{L} \mu_{n_j}(x_j^{(i)}) \quad \in [0,1] \tag{2}$$

where $\mu_{n_j}(x_j^{(i)})$ is the degree of membership of feature $x_j^{(i)}$ in partition n_j. The sum, over all segments, of the corresponding degrees of membership results in a fuzzy classification of a whole frame in class $\mathbf{n} = [n_1 \ldots n_L]^T$:

$$F(\mathbf{n}) = \sum_{i=1}^{K} M_i(\mathbf{n}) = \sum_{i=1}^{K} \left\{ \prod_{j=1}^{L} \mu_{n_j}(x_j^{(i)}) \right\} \tag{3}$$

where K is the total number of segments of the frame. The above summation actually corresponds to a multidimensional histogram, using segments S_i (or equivalently features $\mathbf{x}^{(i)}$) as samples. Finally, the frame feature vector is formed by gathering values $F(\mathbf{n})$ for all categories **n**, i.e., for all combinations of indices $n_1,\ldots,n_L \in \{1,2,\ldots,Q\}$, resulting in a total of M=Q^L feature elements.

Global frame characteristics, obtained through global frame analysis, are also included in the feature vector. In particular, the color histogram of each frame is calculated using YUV coordinates for color description and the average texture complexity is estimated using the high frequency DCT coefficients of each block derived from the MPEG stream. Finally, a scene feature vector, which characterizes a whole scene, is constructed by calculating the mean value of feature vectors over the whole duration of a scene.

4. Scene Selection.

The first stage for an efficient video content representation is the extraction of a small but sufficient number of scenes that satisfactorily represent the video content. This is accomplished by clustering similar scene feature vectors (that is, vectors whose distance in the feature space is small) and selecting only a limited number of cluster representatives. For example, in TV news recordings, consecutive scenes of the same person would reduce to just one. The extraction of the most representative scenes can be used in applications such as automatic generation of low resolution video clip previews.

Let $\mathbf{s}_i \in \Re^M$, $i = 1,2,\ldots,N_S$ be the scene feature vector for the i-th scene, where N_S is the total number of scenes. Then $S = \{\mathbf{s}_i, i = 1,2,\ldots,N_S\}$ is the set of all scene feature vectors. Let also K_S be the number of scenes to be selected and $\mathbf{c}_i$, $i = 1,2,\ldots,K_S$ the feature vectors which best represent those scenes. For each $\mathbf{c}_i$, an influence set is formed which contains all scene feature vectors $\mathbf{s} \in S$ which are closer to $\mathbf{c}_i$:

$$Z_i = \{\mathbf{s} \in S : d(\mathbf{s},\mathbf{c}_i) < d(\mathbf{s},\mathbf{c}_j)\ \forall\ j \neq i\} \tag{4}$$

where $d(\cdot)$ denotes the distance between two vectors. A common choice for $d(\cdot)$ is the Euclidean norm. In effect, the set of all Z_i defines a partition of S into clusters of similar scenes which are represented by the feature vectors $\mathbf{c}_i$. Then the average distortion, defined as

$$D(\mathbf{c}_1,\mathbf{c}_2,\ldots,\mathbf{c}_{K_S}) = \sum_{i=1}^{K_S} \sum_{\mathbf{s} \in Z_i} d(\mathbf{s},\mathbf{c}_i) \tag{5}$$

is a performance measure of the representation of scene feature vectors by the cluster centers $\mathbf{c}_i$. The optimal vectors $\hat{\mathbf{c}}_i$ are thus calculated by minimizing D:

$$(\hat{\mathbf{c}}_1,\hat{\mathbf{c}}_2,\ldots,\hat{\mathbf{c}}_{K_S}) = \underset{c_1,c_2,\ldots,c_{K_S} \in \Re^M}{\arg\min}\ D(\mathbf{c}_1,\mathbf{c}_2,\ldots,\mathbf{c}_{K_S}) \tag{6}$$

Direct minimization of the previous equation is a tedious task since the unknown parameters are involved both in distances $d(\cdot)$ and influence zones. For this reason, minimization is performed in an iterative way using the generalized Lloyd or K-means algorithm [16]. Starting from arbitrary initial values $\mathbf{c}_i(0)$, $i = 1,2,\ldots,K_S$, the new centers are calculated through the following equations for $n \geq 0$:

$$Z_i(n) = \{\mathbf{s} \in S:\ d(\mathbf{s},\mathbf{c}_i(n)) < d(\mathbf{s},\mathbf{c}_j(n))\ \forall\ j \neq i\} \tag{7a}$$

$$\mathbf{c}_i(n+1) = cent(Z_i(n)) \tag{7b}$$

where $\mathbf{c}_i(n)$ denotes the i-th center at the n-th iteration, and $Z_i(n)$ its influence set. The center of $Z_i(n)$ is estimated by the function

$$cent(Z_i(n)) = \frac{1}{|Z_i(n)|} \sum_{\mathbf{s}_i \in Z_i(n)} \mathbf{s}_i \tag{8}$$

where $|Z_i(n)|$ is the cardinality of $Z_i(n)$. The algorithm converges to the solution $(\hat{\mathbf{c}}_1, \hat{\mathbf{c}}_2, \ldots, \hat{\mathbf{c}}_{K_S})$ after a small number of iterations. Finally, the K_S most representative scenes are extracted as the ones whose feature vectors are closest to $(\hat{\mathbf{c}}_1, \hat{\mathbf{c}}_2, \ldots, \hat{\mathbf{c}}_{K_S})$:

$$\hat{\mathbf{s}}_i = \arg\min_{\mathbf{s} \in S} d(\mathbf{s}, \hat{c}_i), \quad i = 1,2,\ldots,K_S \tag{9}$$

5. Genetic Algorithm for Frame Selection.

After extracting the most representative scenes, the next step is to select the key frames within each one of the selected scenes. This is achieved by minimizing a correlation criterion, so that the selected frames are not similar to each other. In particular, the key frames are selected as the ones with the minimum correlation among them. The selection could also be performed using the previous optimization technique. However, that approach does not exploit the temporal relation of feature vectors, which is significant for the frame selection procedure, as it is described in the sequel.

Let us denote by $\mathbf{f}_i \in \Re^M$, $i \in V = \{1, \ldots, N_F\}$ the feature vector of the i-th frame, where N_F is the total number of frames of a scene, and suppose that the K_F most characteristic ones should be selected. The correlation coefficient of the feature vectors $\mathbf{f}_i, \mathbf{f}_j$ is defined as $\rho_{ij} = C_{ij}/(\sigma_i \sigma_j)$ where $C_{ij} = (\mathbf{f}_i - \mathbf{m})^T (\mathbf{f}_j - \mathbf{m})$ is the covariance of the two vectors, $\mathbf{m} = \sum_{i=1}^{N_F} \mathbf{f}_i / N_F$ is the average feature vector of the scene and $\sigma_i^2 = C_{ii}$ is the variance of $\mathbf{f}_i$. In order to define a measure of correlation between K_F feature vectors, we first define the *index* vector $\mathbf{x} = (x_1, \ldots, x_{K_F}) \in W \subset V^{K_F}$ where

$$W = \{(x_1, \ldots, x_{K_F}) \in V^{K_F} : x_1 < \cdots < x_{K_F}\} \tag{10}$$

is the subset of V^{K_F} which contains all sorted index vectors $\mathbf{x}$. Thus, each index vector $\mathbf{x} = (x_1, \ldots, x_{K_F})$ corresponds to a set of frame numbers. The *correlation measure* of the feature vectors $\mathbf{f}_i$, $i = x_1, \ldots, x_{K_F}$ is then defined as

$$R(\mathbf{x}) = R(x_1, \ldots, x_{K_F}) = \left(\sum_{i=1}^{K_F - 1} \sum_{j=i+1}^{K_F} (\rho_{x_i, x_j})^2 \right)^{1/2} \tag{11}$$

Based on the above definitions, it is clear that searching for a set of K_F minimally correlated feature vectors is equivalent to searching for an index vector $\mathbf{x}$ that minimizes $R(\mathbf{x})$. Searching is limited in the subset W, since index vectors are used in order to construct sets of feature vectors, therefore any permutations of the elements of

$\mathbf{x}$ will result in the same sets. The set of the K_F least correlated feature vectors, corresponding to the K_F most characteristic frames, is thus represented by

$$\hat{\mathbf{x}} = (\hat{x}_1, \ldots, \hat{x}_{N_F}) = \arg\min_{\mathbf{x} \in W} R(\mathbf{x}) \tag{12}$$

Unfortunately, the complexity of an exhaustive search for the minimum value of $R(\mathbf{x})$ is such that a direct implementation would be practically unfeasible, since the multidimensional space W includes all possible sets (combinations) of frames. A dramatic reduction in complexity is achieved, however, through *logarithmic search,* which has been introduced in [11] and is performed in a way similar to the search for block motion estimation in video sequences. The algorithm is restricted to the special case $N_F = 2^G$ and its implementation includes the definition of an *initial step size* $S_0 = 2^{G-2} = N_F / 4$ and an *initial index* $\mathbf{x}_0 \in W$ as the element of W which is closest to the *middle point* $\tilde{\mathbf{x}}_0 = (\mu, \ldots, \mu)$, where $\mu = 2^{G-1} - 1$. Successive index vector estimates are then obtained by the recursive equations

$$\mathbf{x}_n = \arg\min_{\mathbf{x} \in N(\mathbf{x}_{n-1}, S_{n-1})} R(\mathbf{x}), \quad S_n = S_{n-1} / 2 \tag{13}$$

where the *neighborhood* $N(\mathbf{x}, S)$ of $\mathbf{x}$ is a small set of index vectors whose distance from $\mathbf{x}$ is S. The final result $\hat{\mathbf{x}} = \mathbf{x}_{G-2}$ is obtained by applying the above recursion for $n = 1, \ldots, G-2$ (until $S_n = 1$). The algorithm, whose implementation details are fully described in [11], provides a very fast convergence to a sub-optimal solution. However, since the search procedure is by definition confined to a very small, pre-defined subset of the search space W, there is always a significant possibility of converging to a local minimum of $R(\mathbf{x})$, resulting in poor performance.

For this reason, a genetic algorithm (GA) [13] approach is adopted in this chapter. This approach seems to be very efficient for the particular optimization problem, given the size and dimensionality of the search space and the multimodal nature of the objective function. Possible solutions of the optimization problem, i.e., sets of frames, are represented by chromosomes whose genetic material consists of frame numbers (indices). Chromosomes are thus represented by index vectors $\mathbf{x} = (x_1, \ldots, x_{K_F}) \in V^{K_F}$ following an integer number encoding scheme, that is, using integer numbers for the representation of genes $x_i \in V, i = 1, \ldots, K_F$.

An *initial population* of P chromosomes, $\mathbf{X}(0) = (\mathbf{x}_1, \ldots, \mathbf{x}_P)$ is then generated by selecting sets of frames whose feature vectors reside in extreme locations of the feature vector trajectory. This selection is accomplished by locating points where the magnitude of the second-order derivative of feature vector trajectory is locally maximized. Traditionally, initial populations are randomly generated, but the above approach exploits the temporal relation of feature vectors and increases the possibility of locating sets of feature vectors with small correlation within the first few GA cycles. Note that this approach has been used directly for key frame selection in [12], [14]. The initial population $\mathbf{X}(0)$ is used for the creation of new generation populations $\mathbf{X}(n)$, $n>0$. The creation of $\mathbf{X}(n)$ at generation (or GA cycle) n is performed of by applying a set of operations on population $\mathbf{X}(n\text{-}1)$, described below. This procedure is repeated in an iterative way, until $\mathbf{X}(n)$ converges to an optimal solution of the problem.

The correlation measure $R(\mathbf{x})$ is used as an objective function to estimate the performance of all chromosomes $\mathbf{x}_i, i=1,\ldots,P$ in a given population. However, a *fitness function* is used to map objective values to fitness values, following a *linear normalization scheme*. In particular, chromosomes $\mathbf{x}_i$ are ranked in ascending order of $R(\mathbf{x}_i)$, since the objective function is to be minimized. Let $r(\mathbf{x}_i) \in \{1,\ldots,P\}$ be the rank of chromosome $\mathbf{x}_i, i=1,\ldots,P$. Defining an arbitrary fitness value F_B for the best chromosome, the fitness of the i-th chromosome is given by the linear function

$$F(\mathbf{x}_i) = F_B - [r(\mathbf{x}_i) - 1]D, \quad i=1,\ldots,P \tag{14}$$

where D is a decrement rate. Thus, the average objective value of the population is mapped into the average fitness [17]. After fitness values, $F(\mathbf{x}_i), i=1,\ldots,P$, have been calculated for all members of the current population, *parent selection* is then applied so that a fitter chromosome gives a higher number of offspring and thus has a higher chance of survival in the next generation. A *proportionate scheme*, implemented by the *roulette wheel selection* procedure [18], is used for parent selection, ensuring that each chromosome has a growth rate proportional to its fitness value.

A set of new chromosomes (offspring) is then produced by mating the selected parent chromosomes and applying a *crossover operator*. The genetic material of the parents is combined in a random way in order to produce the genetic material of the offspring. Figure 2 depicts an example of the crossover operator with four crossover points used for exchanging genes. A generalized *uniform crossover* scheme is employed in the context of this chapter, by considering each parent gene to be a potential crossover point. *Mutation* is then applied to the newly created chromosomes, introducing random gene variations that are useful for restoring lost genetic material, or for producing new material that corresponds to new search areas. In particular, each offspring gene x_i is replaced by randomly generated one $x_i' \in V = \{1,\ldots,N_F\}$, if a probability test is passed. A small mutation probability p_m ensures that only a small gene proportion is altered in each generation.

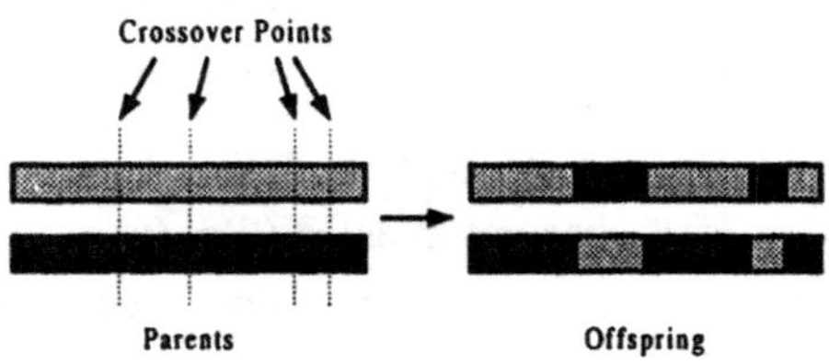

Figure 2: Example of the crossover operator.

Once new chromosomes have been generated for a given population $\mathbf{X}(n)$, $n \geq 0$, the next generation population, $\mathbf{X}(n+1)$, is formed by inserting those new chromosomes into $\mathbf{X}(n)$ and deleting an appropriate number of older chromosomes, so that each population consists of P members. The exact number, C, of old chromosomes to be replaced by new ones defines the *replacement strategy* of the GA and greatly affects its convergence rate. All of the above description refers to simple GA cycle. Several cycles need to take place, that is, several generations $\mathbf{X}(n), n>0$ need to be produced until the population converges to an optimal solution. For this reason, the procedures of fitness evaluation, parent selection, crossover and mutation are repeated until a termination criterion is reached. Usually the GA terminates when the best chromosome fitness remains constant for a large number of generations, indicating that further optimization is unlikely.

The above algorithm, as well as the logarithmic search algorithm, is based on the assumption that frames which are close to each other (in time) should have similar properties, and therefore indices which are close to each other (in W) should have similar correlation measures. However, the technique performs equally well even in the case of random feature vectors, as shown by experiments.

6. Experimental Results

The proposed algorithms were integrated into a system that was tested using several video sequences from video databases. The results obtained from a TV news reporting sequence of total duration 10 minutes (about 15000 frames) are illustrated in the following Figures. The sequence was first partitioned in 52 scenes, using the scene cut detection procedure described in Section 2. Then, the frame and scene feature vectors were extracted using the aforementioned methodology. In particular, an average feature vector was formulated, for each scene, based on the multidimensional feature vectors of the frames composed the scene. The generalized Lloyd algorithm was used for the selection of the most representative scenes.

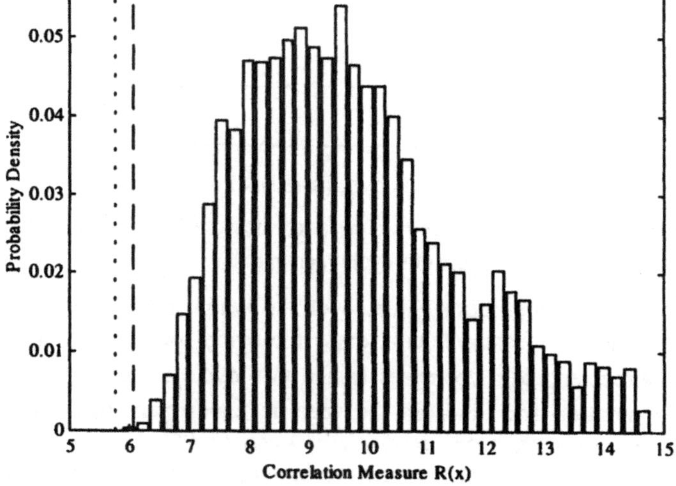

Figure 3. Probability density function of the correlation measure $R(\mathbf{x})$. The vertical dashed line shows the minimum value located by the logarithmic search algorithm, while the dotted line the one located by the genetic algorithm.

The frame selection procedure was then applied to the most representative scenes for extraction of the key video frames. Two methods have been used for selection of K_F=6 key frames out of a total of 293 frames (about 12 sec) of a specific scene: the logarithmic search procedure described in [11], and the genetic algorithm proposed in this chapter. Figure 3 indicates the probability density function of the correlation measure $R(\mathbf{x})$. This function is actually estimated by a histogram obtained through Monte-Carlo simulation, using a large number of random sample vectors $\mathbf{x} = (x_1, \ldots, x_{K_F}) \in V^{K_F}$. In the above Figure, the minimum values obtained by the logarithmic and the genetic search algorithms are also depicted (dashed and dotted line respectively). It is firstly observed that both algorithms return minimum values very close to the actual global minimum of $R(\mathbf{x})$. This, of course, is an approximate result, since the actual minimum value cannot be calculated. Secondly, it is clearly shown that the genetic algorithm provides more accurate results since the logarithmic search procedure can be "trapped" in a local minimum.

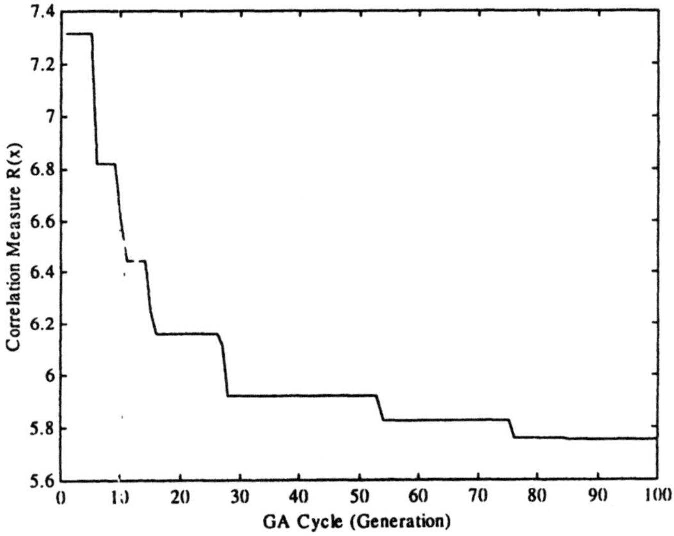

Figure 4: Genetic algorithm convergence: correlation measure $R(\mathbf{x})$ (objective function) versus the GA cycle (generation).

Figure 4 shows the minimum value, over the whole population, of the objective function (or correlation measure $R(\mathbf{x})$) versus the cycle (or generation) of the genetic algorithm. As expected, $R(\mathbf{x})$ decreases as the GA cycle increases, until it reaches a minimum at generation 85. Since in the specific experiment half chromosomes are replaced by new ones at each generation (P=80 and C=P/2=40 have been used), there are cases where all generated offspring have lower fitness than their parents. In these cases the value of the $R(\mathbf{x})$ remains at the same level, hence the "stepwise" appearance of the curve in the above Figure. Note that the step "width" increases with the GA cycle, since it is directly related to the probability of further optimization.

Figure 5. Six selected key frames of a scene after applying the genetic algorithm.

The six selected video frames of the given scene are shown in the Figure 5. Although a very small percentage of frames is retained, it is obvious that one can perceive the content of the scene by just examining the 6 selected frames. Consequently, it is clear that the selected frames give a meaningful representation of the content of the 12-sec video sequence.

7. Conclusions

In this chapter, an efficient video content representation system is presented which permits automatic extraction of a limited number of key frames or scenes that provide sufficient information about the content of a video sequence. In particular, a small but meaningful amount of information is extracted from a video sequence, which is capable of providing a representation suitable for visualization, browsing and content-based retrieval in video databases. Queries are then performed in a more efficient way since only a small number of representative frames involved in the process. In our approach a genetic search algorithm have been adopted for the key frame selection.

The GA approach seems to be very efficient for the particular optimization problem, given the size and dimensionality of the search space and the multimodal nature of the objective function. This estimation is supported by experimental results, demonstrating fast convergence to optimal solutions. The performance of the technique could be further improved by considering parallelization methods such as global, migration or diffusion [19]. Several other improvements are also possible for the proposed system, such as integration of color and motion segmentation results, more robust object tracking algorithm, more intelligent object extraction (e.g., extraction of human faces [6]), and interweaving of audio and video information. These topics are currently under investigation.

References

[1] Y. Rui, T. Huang and S.-F. Chang, "Digital Image / Video Library and MPEG-7: Standardization and Research Issues," *Proc. of ICASSP*, Seattle, USA, May 1998.

[2] Special issue on content-based image retrieval systems, *IEEE Computer Magazine*, Vol. 28, No. 9, 1995. Guest Editors: Venkat N. Gudivada and Jijay V. Raghavan.
[3] Special issue on visual information management, *Communications of ACM*, Dec. 1997. Guest Editor:Ramesh Jain.
[4] M. Flickner, H. Sawhney, W. Niblack, J. Ashley, Q. Huang, B. Dom, M. Gorkani, J. Hafner, D. Lee, D. Petkovic, D. Steele and P. Yanker, "Query by Image and Video content: the QBIC System," *IEEE Computer Magazine*, pp. 23-32, Sept. 1995.
[5] H.-C. Lin, L.-L. Wang and S.-N. Yang, "Color Image Retrieval Based on Hidden Markov Models," *IEEE Trans. Image Processing*, Vol. 6, No. 2, pp. 332-339, Feb. 1997.
[6] D. Androutsos, K. N. Plataniotis, and A. N. Venetsanopoulos, "Extraction of Detailed Image Regions for Content-Based Image Retrieval," *Proc. of ICASSP*, Seattle, USA, May 1998.
[7] M. Gelgon and P. Bouthemy, "A Hierarchical Motion-Based Segmentation and Tracking Technique for Video Storyboard-Like Representation and Content-Based Indexing," *Proc.of WIAMIS*, June 1997, Belgium.
[8] Y. Ariki and Y. Saito, "Extraction of TV News Articles Based on Scene Cut Detection using DCT Clustering," *Proc. of ICIP*, Sept. 1996, Switzerland.
[9] J. Nam and A. Tewfik, "Progressive Resolution Motion Indexing of Video Object," *Proc. of ICASSP*, Seattle, USA, May 1998.
[10] K. J. Han and A. H. Tewfik, "Eigen-Image Video Segmentation and Indexing," *Proc. of IEEE ICIP*, pp. 538-541, Santa Barbara, USA, Oct. 1997.
[11] N. Doulamis, A. Doulamis, Y. Avrithis and S. Kollias, "Video Content Representation Using Optimal Extraction of Frames and Scenes," *Proc.* of *ICIP*, Chicago, USA, Oct. 1998.
[12] Y. Avrithis, N. Doulamis, A. Doulamis and S. Kollias, "Efficient Content Representation in MPEG Video Databases" *Proc. of CVPR*, Santa Barbara, USA, June 1998.
[13] D.E. Goldberg, Genetic Algorithm in Search, Optimization and Machine Learning, Addison Wesley, 1989.
[14] A. Doulamis, Y. Avrithis, N. Doulamis and S. Kollias, "Indexing and Retrieval of the Most Characteristic Frames/Scenes in Video Databases," *Proc.of WIAMIS*, June 1997, Belgium.
[15] B. Kosko, *Neural Networks and Fuzzy Systems: A Dynamical Systems Approach to Machine Intelligence*, Prentice Hall, 1992.
[16] A. Gersho and R. M. Gray, *Vector Quantization and Signal Compression*, Kluwer Academic Publishers, 1993.
[17] K. S. Tang, K. F. Man, S. Kwong and Q. He, "Genetic Algorithms and Their Applications," *IEEE Signal Processing Magazine*, pp. 22-37, Nov. 1996.
[18] H. Holland, *Adaptation in Natural and Artificial Systems*, Ann Arbor: The University of Michigan Press, 1975.
[19] L. Davis, *Handbook of Genetic Algorithms*, Van Nostrand Reinhold, 1991.

15

FAST SYNTHETIC GA COMBINED WITH BP -- APPLICATION TO SHORT-TERM ECONOMIC DISPATCH OF HYDROTHERMAL POWER SYSTEMS

X. MENG*, H. ZHANG*, X. YANG and W. TAN**
* *Department of Automatic Control, Box 406*
Northeastern University, Liaoning 110006 P.R. China
** *Changchun Water Conservancy*
and Electric Power Institute
Changchun, Jilin 130012 P.R. China

1. INTRODUCTION

A need for optimality exists in the highly nonlinear and computationally difficult Power System Analysis environment. Artificial intelligent, unlike strict mathematical methods, has the apparent ability to adapt to nonline-arities and discontinuities found commonly in physical systems.

The linearizations and assumptions made in short-term economic dispatch of hydrothermal power systems(SEDHPS) problems present a classical example. The classical approach to solving this problem uses approximations is not desirable. Fast synthetic genetic algorithm (FSGA) combined with BP algorithm can alleviate this undesirable trait by reproducing the optimization techniques found in nature. It does not require the strict continuity of classical search techniques, instead it allows non-linearity and discontinuities to appear in the solution space. This chapter has made simulation studies on a three unit test system, the simulation results demonstrate that this method is effective and training time is short. This algorithm is a suitable method for solving most optimization problems. The results produced are acceptable, yet still imperfect.

2. FAST SYNTHETIC GENETIC ALGORITHM

Genetic algorithms (GAs) are probabilistic heuristic search processes based on natural genetic system. They are highly parallel and believed to be robust in searching global optimal solution of complex optimization problems. GAs are capable of solving wide range of complex optimization problems only using three simple genetic operations (selection / reproduction, crossover, and mutation) on coded solutions (strings / chromosomes) for the parameter set, not the parameters themselves in an iterative fashion. GAs consider several points in the search space simultaneously, which reduces the chance of convergence to a local optimum. But GAs have some main disadvantages:

S.G. Tzafestas (ed.), Advances in Intelligent Systems, 163–170.

(1) slow convergence speed;

(2) premature convergence or stalling;

(3) the larger number of populations and individuals for some complex optimization problems.

To overcome the disadvantages and to acquire better performances of GAs, some improved GAs have been proposed and published, some of them are improved by parameter selection and initialization, and the others by crossover operation or mutation operation. To use a better GA effectively, some different improved GAs have been compared and appraised by theory analysis and simulation tests, and we find that these improved GAs have certain advantages and disadvantages, respectively. To make full use of the advantages of each GA and to make up its disadvantages, we propose a fast synthetic genetic algorithm (FSGA), and analyze and simulate it, the results demonstrate that FSGA has faster convergence speed and higher computation precision, and the number of individuals and populations have been decreased, respectively, therefore it is a effective optimization algorithm. The basic principles of FSGA are as follows:

A. Selection / Reproduction Operation

Selection is a process in which potential strings (strings with higher fitness value) of the population are copied to a mating pool depending on their fitness function values. Suppose that there be N individuals in the population, decoding the N individuals obtains N fitness function values f_i ($i = 1, 2, \cdots, N$), copy the k-th string to mating pool as following formulation:

let $$p = rnd \times \sum_{i=}^{N} f_i \tag{1}$$

if $$\sum_{i=1}^{k-1} f_i < p, \sum_{i=1}^{k} f_i >= p \tag{2}$$

then copy the k-th string.

where rnd is a random value, $0 < rnd < 1$.

Arrange f values of the N individuals in order from large to small, copy preceding $N/2$ individuals to mating pool for waiting crossover operation. These selected $N/2$ individuals are operated by crossover operation and replaced to mating pool, replace the latter $N/2$ individuals, i.e., eliminate the latter $N/2$ individuals. The N individuals in the mating pool are rearranged in order of f values from large to small , employ elitist strategy, i.e., replacing the best individual in parent generation to the mating pool to replace the last individual in rearranged population to guarantee the elitist individual exist in population.

B. Crossover Operation

Crossover operation produces offspring by exchanging information between two potential strings selected randomly from the mating pool generated by selection process. This paper employs the technique combining single point crossover and /or two point crossover with crossover site being selected with unequal probability[1] as crossover operation, if the length of potential string of the chromosome is short, two point crossover operation is employed; but if the length of potential string of that is long, especially the number of bits of substring are many, single point crossover in each substring is employed and the technique of crossover sit being selected with unequal probability is introduced to make the crossover probability between the first and the second bit be double that other 2-m bits, the operation is determined depending on different problems.

C. Mutation Operation

Mutation means an occasional random alteration of a bit position. A random bit of a randomly selected string in the population is selected and the bit value is replaced by some other value, i.e., for a binary string a 0 is replaced by a 1 and vice versa. In this paper, mutation operation employs the technique that combines directed mutation[2] with genetic-catastrophic algorithm[3]. Directed mutation follows from the concept of induced mutation in biological systems. This operation does not involve probabilistic decision rules but the information acquired in the previous generations. Again it does not alter the probabilistic nature of the search technique. In certain environment this operation will deterministically introduce a new point in the population. The new point is directed (guided) by the solutions obtained earlier and therefore we call it directed mutation.

Let $B_t^{'}$ be the best fit string of the t -th generation and $f_t^{'}$ be the corresponding fitness value. Let $x_t^{'}=(x_{t1},x_{t2},\cdots,x_{tm})$ be the decoded value of the parameters. Now if in the τ -th ($\tau>$ t) generation the best fit string is $B_\tau^{'}$ with fitness value $f_\tau^{'}$ ($f_\tau^{'}>f_t^{'}$), then a new point $x^{'}$ is generated as

$$x^{'}=x_\tau^{'}+\alpha(1-\frac{t}{t_{max}})g(x_t^{'},x_\tau^{'}) \tag{3}$$

where $0<\alpha<1$ is a constant multiplier and τ represents the first generation after t -th generation with $f_\tau^{'}>f_t^{'}$. The point $x^{'}$ is then encoded as a string $B^{'}$ and is introduced in the population. $g(*)$ can be defined as an acceleration function.

If the function is differentiable everywhere, we can use

$$g(x_t^{'},x_\tau^{'})=f^{'}(x_\tau^{'}) \tag{4}$$

If the function $f(x_\tau^{'})$ is not differentiable everywhere, one can use the following acceleration function

$$g(x_t^{'},x_\tau^{'})=\frac{f_\tau^{'}-f_t^{'}}{x_\tau^{'}-x_t^{'}} \tag{5}$$

or use simple linear extrapolation

$$g(x_t^{'},x_\tau^{'})=x_\tau^{'}-x_t^{'} \tag{6}$$

i.e.

$$x^{'}=x_\tau^{'}+\alpha(1-\frac{t}{t_{max}})(x_\tau^{'}-x_t^{'}) \tag{7}$$

The extrapolated point $x^{'}$ lies on the line joining $x_\tau^{'}$ and $x_t^{'}$.

If the algorithm is premature convergent, one use genetic-catastrophic algorithm to increase suddenly the mutation probability, the mutation probability can be 0.9~0.99, it can be determined depending on different conditions. The combined method can obtain a very good effect.

3. BACK PROPAGATION (BP) ALGORITHM

Artificial Neural Networks (ANN) are gaining popularity in various field of engineering including electrical power systems due to their high computational rates and robustness as well as their modeling complex and nonlinear relationship through a training process with historical data. One of the ANN models extensively used for power system application is the multilayer perceptron model based on back propagation (BP) algorithm[4]. A three layer BP model, as shown in Fig. 1, containing one hidden layer has been selected in the present work for solving short - term economic load dispatch problem. When a input pattern was given to the network, it will be propagated from input layer to hidden layer, after being adjusted in hidden layer it will be propagated again to output layer, this is called forward propagation process. A typical neural node or neuron(say jth node) in hidden and output layers are represented by sigmoidal function having output

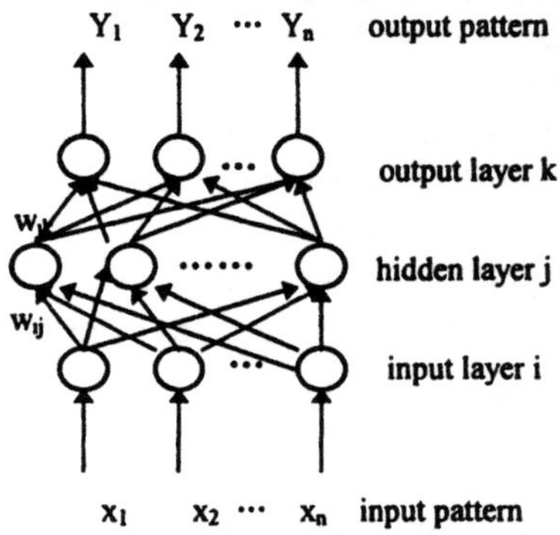

Fig. 1 a 3 layeren perceptron model of ANN

$$Y_j = \frac{1}{1+\exp(-s_j + e_j)} \tag{8}$$

where e_j is the threshold value or bias factor; s_j is the weighed sum of the iuput signals to jth node.

If total number of nodes in the input layer to jth node are N, and weight of interconnection from ith node having input x_i to jth node is w_{ij}, then

$$S_j = \sum_{i=1}^{N} w_{ij} x_j \tag{9}$$

If the output response is different from the expected output pattern and the error of the response dose not meet the demand, the error will turn to be propagated backward and the weight of interconnection will be amended layer by layer. The formulas are:

$$w_{ij}(k+1) = w_{ij}(k) + \Delta w_{ij}(k) \tag{10}$$

$$\Delta w_{ij}(k) = \eta \delta_{pi} Y_{pi} \tag{11}$$

where k is the repeating steps; η is the learning factor.

When Y_{pi} is the output of the output layer nodes,

$$\delta_{pi} = (1 - Y_{pi}) Y_{pi} (d_{pi} - Y_{pi}) \tag{12}$$

when Y_{pi} is the output of the hidden layer nodes

$$\delta_{pi} = Y_{pi}(1 - Y_{pi}) \sum_k \delta_{pk} w_{ki} \tag{13}$$

where d_{pi} is the expecting output of jth node of pth sample.

For a set of given training pattern, to use every training pattern to train the network and repeat the process of forward propagation and error back propagation until all training patterns satisfy the demand. Now, the Neural Network's training is achieved.

4. METHODS AND STEPS OF COMBINING GA WITH BP

Although BP algorithm can surely obtain the final convergence of the network learning course. Its marked weakness is that time of learning and training is too long and easy to converge on partial ultra-miniaturized value. Though some advanced BP algorithms are often put forward, their effect is not obvious. However, designing ANN with combined GA[5] can effectively solve above-mentioned problems. The typical forward neural network is shown in Fig. 1. Here, the node number of input layer is n_i, the node number of hidden layer is n_j, the node number of output layer is n_k, and the weighted sum of input layer node j is:

$$u_j(t) = \sum_i w_{ji} v_i(t) + \theta_j \tag{14}$$

the input weighted sum of output layer node k is:

$$u_k(t) = \sum_j w_{kj} v_j(t) + \theta_k \tag{15}$$

the output of each node is expressed as Sigmoid function:

$$f(x) = 1/(1+\exp(-x)) \tag{16}$$

The steps of training network optimization weight value using GA are as follows

(1) Chromosome coding and deciding on the length and architecture of chromosome

From Fig. 1, the weight value number of network is:

$$n_i \times n_j + n_i \times n_k + n_j \times n_k \tag{17}$$

each weight value is a stochastic number from -1.0 to +1.0. By now, the architecture and the length of chromosome have been defined.

(2) Determine fitness function

Fitness function is determined as follows:

$$f(i) = 1/E(i) \tag{18}$$

$$E(i) = \sum_p \sum_k (v_k - v_{tk})^2 \tag{19}$$

where i = 1, 2, Λ, n_i (the number of chromosome); k = 1, 2, Λ, n_k (the node number of output layer); p = 1, 2, Λ, n_t (the number of learning samples); v_{tk} is the teacher signal.

(3) Choice of control parameter

Depending on experience and test, the chromosome number of primary population is chosen as N = 10. On the basis of principle that the chromosomes with higher fitness have more chances to be chosen as paternal chromosome, k (which is equal to n_k) pairs of chromosomes will be chosen to be paternal chromosome and the genes of half chromosome's number are exchanged via cross over operation.

(4) Course of learning and training

a. Making the network weight value coded, ordering and constructing chromosomes. Randomly producing ten chromosome in order to form primary population, every weight value gets a smaller random number between -1.0 and +1.0;

b. Decoding every chromosome and obtaining the relevant weight value. According to formula (14)-(19) to calculate the error function E(i) of network and the fitness function f(i) of network;

c. Choosing a pair of chromosome C_1 and C_2 in accordance with probability P_{cross};

d. Executing the cross over operation on C_1 and C_2 , and obtaining chromosome C_1' and C_2';

e. Executing the mutation operation respectively to C_1' and C_2' according to probability P_{mut} and obtaining new son- chromosome C_1'' and C_2'';

f. Obtaining relevant weight value after decoding C_1'' and C_2'', and calculating the error function E of network and fitness value f of network according to formula (14)-(19);

g. Eliminating 2k chromosomes with smaller fitness among N+2*k chromosomes, thus a new generation population is formed;

h. If $E < \varepsilon$ (given error precision), choosing chromosomes which include the minimum error function E(i), namely the chromosome including the maximum fitness. Then decoding it and obtaining the network weight value, now learning and training end, otherwise turn to (c).

5. MATHEMATICS MODEL OF SEDHPS

What is called economic dispatch is rationally to distribute the loads in each power plant for making the total fuel (coal) cost minimize. The short-term economic dispatch refer to load dispatch by the day for a lead time ranging from one day to several days or a week out[6]. The load in a day use varying time interval model and the water consumption of each hydropower plant in a day is a given value.

The objective of the SEDHPS problem is to minimize the total fuel cost at thermal plants:

$$OBJ = \sum_{i=1}^{P} F_i(P_{Ti}) \tag{20}$$

Subject to the constraint of equality in real power and water consumption balance

$$\sum_{i=1}^{P} P_{Ti} + \sum_{j=1}^{q} P_{Hj} - P_L - P_{LD} = 0 \tag{21}$$

$$W_{j\Sigma} - \int_0^T W_j(P_{Hj})dt = 0 \qquad (j = 1,2,\Lambda ,q) \tag{22}$$

where $F_i(P_{Ti})$ is the fuel cost in term of its real power generation P_{Ti}.

P_{Ti} = output generation for unit i in thermalpower plants.

P = the total number of generators in all thermalpower plants.

P_{Hj} = output generation for unit j in hydropower plants.

q = the total number of generators in all hydropower plants.

P_L = the total network real power loss.

P_{LD} = the total real power demand.

$W_{j\Sigma}$ = the given water consumption of hydropower plant in considered time.

$W_j(P_{Hj})$ = the water consumption property of hydropower plant.

The thermal plant can be expressed as input-output models (cost function), where the input is the fuel cost and the output is the power output of each unit. In practice, the cost function could be represented by a quadratic function

$$F_i(P_i) = a_i P_i^2 + b_i P_i + c_i \tag{23}$$

where a_i, b_i, and c_i are fuel cost coefficients of ith generator and P_i is real power output. Incremental cost curve data are obtained by taking the derivative of the unit input-output equation resulting in the following equation for each generator:

$$\frac{dF_i(P_i)}{dP_i} = 2a_i P_i + b_i \tag{24}$$

Transmission losses are a function of the unit generations and are based on the system topology. Solving the SEDHPS equations for a specified system requires an iterative approach since all unit generation allocations are embedded in the equation for each unit. In practice, the loss penalty factors are usually obtained using on-line power flow software. This information is updated to ensure accuracy. They can also be calculated directly using the B matrix loss formula,

$$P_L = \sum_i \sum_j p_i B_{ij} p_j + \sum_i B_{i0} p_i + B_{00} \tag{25}$$

where B = loss coefficients.

6. SIMULATION RESULTS

In this chapter, the proposed method is applied to a three unit test system, the example system contains a set of three generators in a simple six bus power system[6]. The system data are as Table 1.

The results about average run time obtained from the proposed method (PM) are compared with the results of the classical method (CM) optimization technique which is based on the sequential-quadratic-programming method. Results that for the six larger test systems are given in Table 2 for a single load level in each case. Training sample number is 18.

Table 1 Example unit data

parameter	unit 1	unit 2	unit 3
maximum	600MW	400MW	200MW
minimum	100MW	100MW	50MW
cost coefficients			
a_i	0.00156	0.00194	0.00482
b_i	7.92	7.85	7.97
c_i	561	310	78

Table 2 Simulation results

load	average run time(s)	
(GW)	CM	PM
5	34.9	1.8
10	295.8	3.1
20	1837.7	9.7
31.5	6037.5	34.9
40	43688.1	67.5
60	312189.9	290.1

Table 2 shows the average run time of proposed method are shorter than classical method, and the classical method sometomes is not iterative.

7. CONCLUSIONS

In this chapter an application of fast synthetic genetic algorithm combined with BP to short-term economic dispatch of hydrothermal power systems problem is proposed. simulation tests are performed. The results demonstrate that the proposed method is effective.

REFERENCE

[1] Zhang Ke and Liu Guizhong., "Selecting crossover site with unequal probability in genetic algorithms," *Information and Control.*, vol. 26, no. 1, February 1997, pp. 53 - 60

[2] Dinabandhu B, Nikhil R. P. and Sankar K. P., "Directed mutation in genetic algorithms," *Information Sciences.*, vol. 79, 1994, pp. 251- 270

[3] Jin Xidong and Li Zhi., "Genetic - catastrophic algorithms and its application in nonlinear control system, " *System Simulation Journal.*, vol. 9, no. 2, June 1997, pp. 111 - 115

[4] Wang Yaonan, "A neural network adaptive control based on rapid learning method and application," *Advances In Modelling & Analysis*, AMSE Press, 1994, 46(3), pp. 27-34

[5] D. J. Montana and L. Davis, "Training feed forward neural network using genetic algorithm," in Proc. of the *Eleventh International Joint Conference on Artificial Intelligence*, 1989, pp. 762-767

[6] Nikola Rajakovic, and Slobodan Ruzic, "Sensitivity analysis of an optimal short term hydro- thermal schedule," IEEE Tranctions on Power Systems, Vol. 3, No. 3, August 1993, pp. 1235-1332

16

PARALLEL PLATFORM DESIGN AND KINEMATICS ANALYSIS USING EVOLUTION STRATEGY VIA REGENERATION

A. E. KANARACHOS and K.S. ROUSSIS
Mechanical Design and Control Systems Division
Department of Mechanical Engineering
National Technical University of Athens
Zografou Campus, P.O. Box 64078, 15710 Athens, Greece

1. Introduction

A lot of studies have been made in order to improve the behavior of open-loop chain robots, by designing light and stiff robotic arms, developing light with high power actuators and by investigating new control techniques of flexible arms {1}. On the other hand a new approach of the existing problem for high precision and accurate movement in many applications lead to the design of parallel mechanisms a totally new idea with a lot of advantages. Many different types of parallel mechanisms have been developed classified in two general categories, planar and spatial mechanisms. The first refer to all parallel mechanisms whose links move in a plane and the second to those whose links move in the three dimensional space. Two more sub-categories of spatial mechanisms are know as 3DOF and 6DOF parallel platforms depending on the number of degrees of freedom which determine the position and orientation of the mechanism.

Inverse and direct kinematics are the two problems concerning parallel manipulators. Analytical solutions for the workspace definition and trajectory design have been proposed but without taking into account physical constraints of the mechanism {2}. Due to this disadvantage, optimization techniques become necessary in order to incorporate the system physical demands (spherical and revolute joints tolerances, minimum and maximum leg lengths, etc.). Quasi Newton, a general stochastic optimization method, was successfully used for the determination of all solutions of the forward displacement analysis of the generalized Steward platform {3}. A new approach to solve the inverse kinematics problem of redundant robot manipulators was proposed by A. C. Nearchou {4} and modified GA algorithm show the effectiveness of the methodology. An example of a 6DOF parallel mechanism ,which was used to move an equestrian simulator platform equipped with a saddle, were developed and inverse geometrical and kinematical model along with its control system were studied {5}. The architecture of a 5DOF parallel platform was presented and inverse kinematic analysis was demonstrated both with the determination of the singularity loci {6}. Finally solutions for the direct and inverse kinematic problems, of a

S.G. Tzafestas (ed.), Advances in Intelligent Systems, 171–178.

3DOF parallel platform, were given and kinematic simulation and computer aided design were discussed {7}.

In this chapter the inverse kinematics model is studied in order to determine the leg lengths of parallel platforms actuators, so as to succeed desired positioning and orientation of the moving platform. All system kinematics and design parameters are taken into consideration in the objective function which will be minimized. The choice of Evolution Strategy satisfies the requirements for a simple easy to program and be implemented algorithm witch also provides high speed converge and can override possible local minima. Although, even the two membered Evolution Strategy (e.g. one parent and one offspring) fails to find the global minimum. This is demonstrated for three different types of parallel platforms (3DOF and 6DOF mechanisms) whose the results revealed, are getting «stack» at local minima.

Therefore a novel technique called «Evolution Strategy with Regeneration» has been developed so as to overcome the above problem, showing great improvement of the optimization performance. Stepping over local minima during optimization procedure leads to the desired position and orientation of the moving platform while all kinematic and structural constraints are not violated. Results reveal that the two membered Evolution Strategy with Regeneration is an efficient method and works very well in a variety of parallel platform designs. In designing time Evolution can be used for the evaluation of a totally new design simply by simulating the specific design and analyze it's behavior through test paths due to the very close relationship between the real model and the simulated one.

2. Optimization procedure

Rechenberg's two membered Evolution Strategy can be successfully used to deal with unconstrained optimization problems. It is the minimal concept for an imitation of organic evolution and the two principles of mutation and selection, which (Darwin 1859) recognized to be most important, are taken as rules for variation of the parameters and for recursion of the iteration sequence respectively. Fitness of individuals is measured by means of an objective function to be optimized. The process begins by creating one offspring by means of mutation from one parent. The better of the parent and offspring is selected deterministically to survive to the next generation a mechanism witch characterizes this two membered or (1+1) Evolution Strategy. Unequal constraints are considered ($g_j : \Re^n \rightarrow \Re$ ($j \in \{1,...v\}$)) of the search space, an objective function $f : M \subseteq \Re^n \rightarrow \Re$ and individual vectors $x(t) \in \Re^n$ where t denotes the generation counter. The following algorithm defines two membered evolution method:

step 1: t=0

step 2: Initialization of $P(0)=\{x(0)\}$ such that $\forall j: g_j(x_p(0)) \geq 0$)

step 3: Mutation

$$P(t) : x'(t) = x(t) + \sigma \cdot z(t) \quad (1)$$

with probability density

$$P(z_i(t)) = \frac{1}{\sqrt{2\pi}} e^{(-\frac{1}{2}(z_i(t))^2)} \qquad (2)$$

step 4: Evaluation of $P(t)$: $f(x(t)), f(x'(t))$

step 5: Selection

$P(t+1)$ from $P(t)$

If ($f(x'(t)) \leq f(x(t))$ and $\forall j: g_j(x'(t)) \geq 0$)

then $x(t+1) = x'(t)$

else $x(t+1) = x(t)$

step 6: $t=t+1$

step 7: If termination criterion fulfilled then END

else repeat step 3.

The same standard deviation (σ in eq. 1) is often applied to each of the component vectors x(t) during mutation, while the variation of the step size control of the algorithm is performed according to Rechenberg's 1/5 success rule:
" The ratio of successful mutations to all mutations should be 1/5. If it is greater, increase; if it is less, decrease the standard deviation σ "

In many problems this rule proves to be extremely effective in maintaining approximately the highest convergence rate. According to Schwefel 1981 {8}, the frequency of successful mutations should be measured over intervals of 10n trials and the increase and decrease factors should be (1/0.85) and 0.85, respectively. The adjustment should take place every n mutations. Also the search is terminated if the value of the objective function have been improved by less than a given value ε in the course of 20*n* trials.

However, the "asymptotic" convergence of the Evolution Strategy towards the optimum, is a very difficult to measure property in practical applications: The not appropriate choice of a parent vector and of the initial magnitude of the standard deviation, very often leads to convergence towards local minima. Thus the credibility of the method suffers strongly, as the optimization may fail to reach the optimum. At this point regeneration process is been activated by locally increasing the value of standard deviation σ (two or three times more than its initial value) and by executing Evolution Strategy algorithm from start assuming the last found vector *x*(t) as the initial one. In this way regeneration can be assumed to be an extension of the Rechenberg's 1/5 rule by increasing the search space of the problem (more than 1/5 rule) in order to ensure that the given minimum is a global minimum by giving the algorithm the potential of a more accurate search-space scanning. As a consequence of this procedure it is possible for Evolution to step over local minima even if 1/5 rule fails to override the problem. Another critical issue concerning optimization in general is the choice of the initial vector, the parent vector. In order to have fast converge and to avoid trapping at local minima an intelligent guess of the parent vector must be done in order to lead the procedure to a global minimum. This depends on the physical problem characteristics and the physical system behavior and of course the understanding of these two critical parameters. In this chapter and for the determination of the parent vector the minimum change in leg length is considered and the initial values of displacements are chosen to be nearly zero. Also the standard deviation σ and mutation number m are considered to have the values of σ=0.3 and m=5 respectively

3. Objective function determination

In general the problem of minimizing function $f(\mathbf{w})$ (w=weights vector) subject to the constraints $g_j(\mathbf{w})$ and $h_i(\mathbf{w})$:

$$\begin{aligned} & f(\mathbf{w}) = \text{minimum} \\ \text{subject to :} \quad & g_j(\mathbf{w}) \le 0 \quad j=1,\ldots,n_g \\ & h_i(\mathbf{w}) = 0 \quad i=1,\ldots,n_e \end{aligned} \tag{3}$$

can be formulated as an unconstrained optimization problem, where the complex function $F(\mathbf{w})$ needs to be minimized:

$$F(w) = f(w) + \sum_{j=1}^{n_g} A_j \cdot G\left(g_j(w)_+\right) + \sum_{i=1}^{n_e} B_i \cdot H\left(\left|h_i(w)\right|\right) = \text{miminum} \tag{4}$$

where G is a function of the term $g_j(\mathbf{w})_+$, H a function of the term $h_i(\mathbf{w})$ and A, B weighting factors (A_j, $B_i \ge 0$).

In this chapter revolute and spherical joints tolerances, leg lengths minimum and maximum limits and finally tolerances of the desired moving platform position and orientation are considered to be penalty functions of the objective function. In this way for every successful step of the ES all kinematics constraints are considered and satisfied both with geometrical characteristics of the platform. It must be noted that minimum and maximum leg lengths are known while revolute and spherical joints constraints are calculating as follows:

$$\text{grad} = \tan(\varphi) = \frac{{}^2x_1}{{}^2y_1} \quad \text{for revolute joints}$$

$$\text{and } L_{SMIN} = \sqrt{(x_{C2} - {}^1x_1)^2 + (y_{C2} - {}^1y_1)^2 + (z_{C2} - {}^1z_1)^2} \quad \text{for spherical joints}$$

where x_{C2}, y_{C2}, z_{C2} are the coordinates of the center of local system C_2, ${}^2x_1$, ${}^2y_1$, ${}^2z_1$ are the coordinates of the joint B_1 in the global coordinates system C_1, Grad is the gradient of the leg L_1 in the plane x_1y_1 and L_{SMIN} is the minimum allowed length for the spherical joint B_1. Penalty functions are corresponding to these constraints within the predefined and desired tolerances. For example the distance between A_1 and C_2 points cannot violate the L_{SMIN} value with a tolerance of ±0.001 rads. Figures 1(a) and (b) refer to revolute and spherical joint respectively:

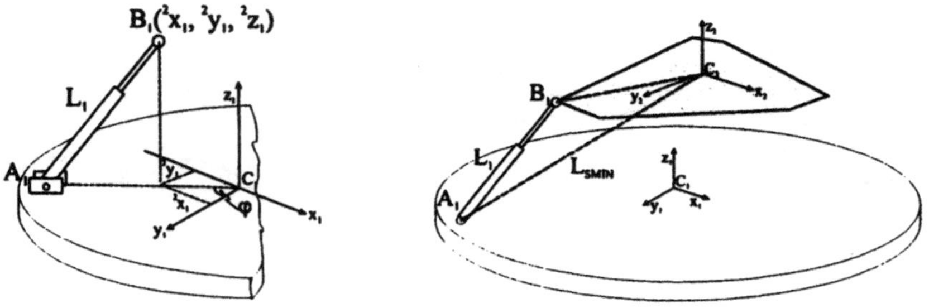

(a) (b)

Fig. 1. Revolute joints model (a) and Spherical joints model at point B_1

4. Simulation

4.1 Problem description

Three different types of parallel platforms are used in order to find the leg lengths of three predefined paths concerning all geometrical characteristics and mechanical tolerances. The used types are shown in Fig. 2. where the first is a 3DOF parallel platform with three revolute and three spherical joints, the second is a 3DOF parallel platform with a passive link, two revolute and six spherical joints and finally the third is a 6DOF parallel platform with six spherical joints.

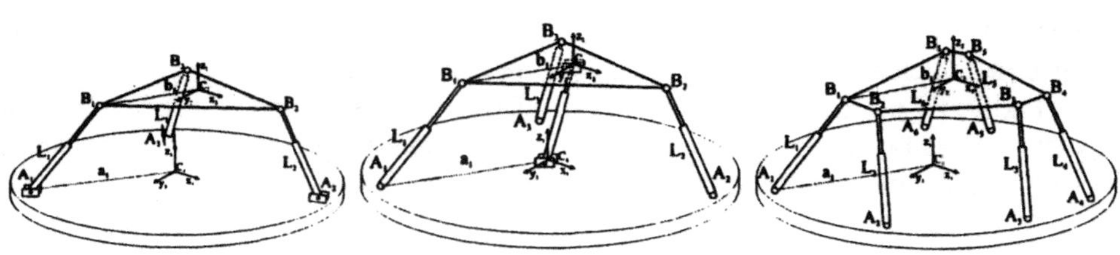

(a) (b) (c)

Fig. 2. 3DOF parallel platform (a), 3DOF parallel platform with passive link (b) and 6DOF parallel platform (c).

Geometrical Characteristics	
Base Diameter mm	R_{BASE}=800
Moving Platform Diameter mm	$R_{PLATFORM}$=400
Initial Position (C_1 Global Coordinates Origin) mm	X_{C1}=0, Y_{C1}=0, Z_{C1}=300
Displacement tolerance (mm)	0.01
Initial Orientation (C_2 Local Coordinates Origin) rad	φ=0, θ=0, ψ=300
Orientation tolerance rad	0.001
Revolute joints tolerance rad	0.001
Spherical joints tolerance rad	0.001

Table 1.

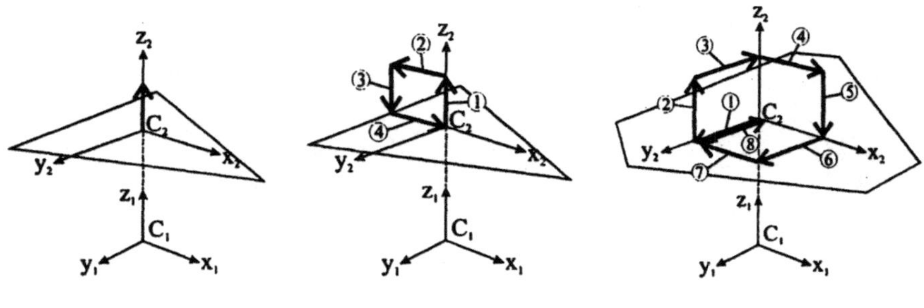

(a) (b) (c)

Fig. 3. Desired paths for the three platform types : 3DOF (a), 3DOF with passive joint (b) and 6DOF (c).

Geometrical characteristics, mechanical tolerances and desired paths are shown in the Table 1. and Fig. 3. respectively. In Fig. 3. (a) the first desired path is a single straight section having the length of 1 mm, in Fig. 3. (b) a simple square with edge length of 1 mm and in Fig. 3. (c) a three dimensional path with edge lengths equal to 1 mm. All values refer to local coordinates system C_1.

4.2 Results

Considering the above geometrical characteristics and desired paths results reveal that the method works very well. In Fig. 4. the leg lengths of the 3DOF platform and displacements in directions x, y and z shows that the desired path is followed, final platform positioning and orientation and leg lengths changes do not violate any of the specified tolerance ranges (see Table 1.). The same levels of accuracy are demonstrated in Fig 5 and 6. considering the cases of 3DOF with passive joint and 6DOF parallel platforms showing that despite the complexity of the desired paths Evolution Strategy with Regeneration can be applied with great success.

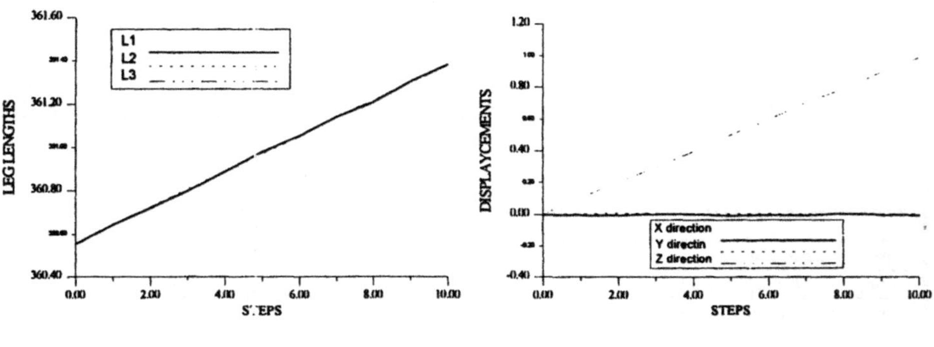

Fig. 4. Leg lengths and displacements of the moving platform center of 3DOF parallel platform.

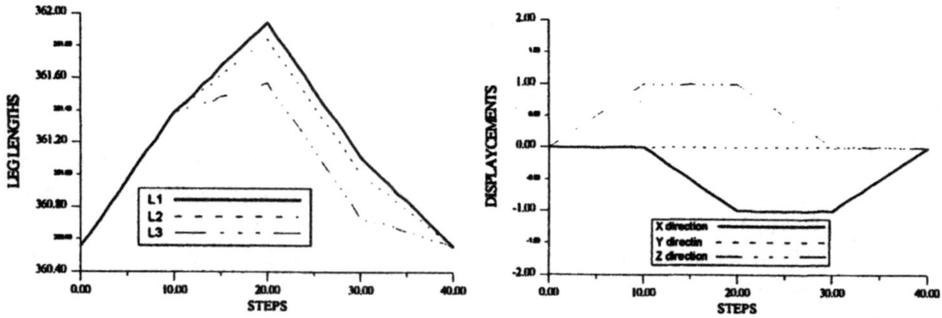

Fig. 5. Leg lengths and displacements of the moving platform center of 3DOF parallel platform with passive link.

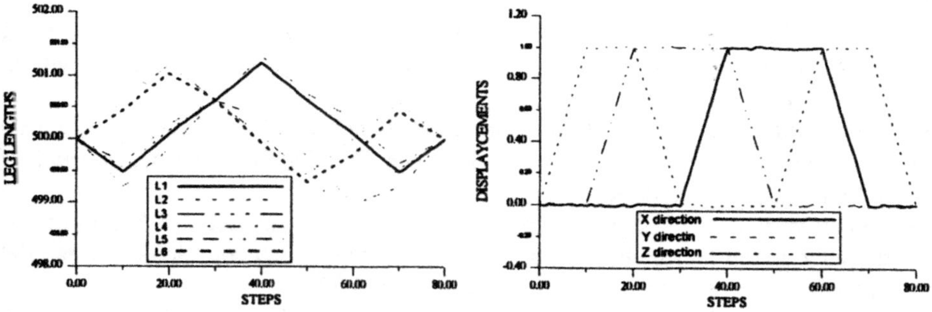

Fig. 6. Leg lengths and displacements of the moving platform center of 6DOF parallel platform.

5. Conclusions

As it was demonstrated inverse kinematics problem was solved and desired paths were followed satisfying leg lengths minimum and maximum bounds, revolute and spherical joints tolerances and the step by step position and orientation tolerances of the moving platform. In this way no significant difference exists between the simulated and the real model. Thus Evolution Strategy with regeneration is an efficient and very easily applicable procedure, giving high accuracy results. Also results reveal that any existent type of parallel manipulator can be evaluated through test paths showing the potentials of each type and it was demonstrated for three classical types of platforms. The greatest advantage of this method is that design changes can be performed without any particular difficulty, simply by changing the objective function terms and new, from scratch, designs can be simulated and evaluated.

REFERENCES

[1] Hiroaki Funabashi, 1994, In-parallel actuated mechanisms as a new robotic mechanism, *Advanced Robotics*, **8**, (6), 535-544.

[2] Shin-Min Song and Ming-Dong Zhang, 1995, A Study of Reactional Force Compensation Based on Three-Degree-Of Freedom Parallel Platforms, *Journal of Robotic Systems*, **12**, (12), 783-794

[3] Z. Sika, P. Kocandrle and V. Stejskal, 1998, An investigation of properties ot the Forward Displaycement Analysis of the Generalized Stewart platform by means of general optimization methods, Mech. Mach. Theory, **33**, (3), 245-253.

[4] Andreas Nearchou, 1998, Solving the Inverse Kinematics problem of Redundant Robots operating in complex environments via a Modified Genetic Algorithm, Mech. Mach. Theory, **33**, (3), 273-292.

[5] Y Amirat, C. Francois, G. Fried, J. Pontnau and M. Dafaoui, 1995, Design and Control of a new Six DOF Parallel Robot: application to equestrian gait simulation, *Mechatronics*, **6**, (2), 227-237.

[6] Jiegao Wang and Clement M Gosselin, 1997, Kinematic Analysis and Singularity Representation of Spatial Five-Degree-of Freedom Parallel Mechanisms, *Journal of Robotic Systems*, **14**, (12), 851-869

[7] Clement M. Gesselin, Louis Perreault and Charles Vaillancourt, 1995, Simulation and Computer-Aided Kinematic Design of Three-Degree-of-Freedom Spherical Parallel Manipulators, *Journal of Robotic Systems*, **12**, (12), 783-794

[8] H. P. Schwefel, *Numerical Optimization of Computer Models* (Wiley, Shishester 1981)

PART IV

HYBRID INTELLIGENT TECHNIQUES IN SYSTEM ANALYSIS AND CONTROL

17

A HYBRID NEURAL-CLASSICAL STRUCTURE FOR THE MODELING OF A BIOPROCESS

A. HANOMOLO*, PH. BOGAERTS*, J. GRAEFE*,
M. CHERLET**, J. WÉRENNE** and R. HANUS*
** Service d'Automatique et d'Analyse des Systemes*
*** Service de Biotechnologie des Cellules Animales*
Université Libre de Bruxelles
50, Av. F.D. Roosevelt, CP 165, 1050 Bruxelles, Belgique

INTRODUCTION

Biological processes exhibit a wide range of dynamic behavior and offer many challenges to modeling as a result of the presence of living organisms (cells). Two essential problems are encountered in this kind of processes. One consists in the fact that the kinetics of these processes are poorly understood making it difficult to develop an analytical model. The other is related to the lack of cheap and reliable sensors capable of providing direct real-time measurements; typically only some of the variables are determined by slow infrequent off-line laboratory analysis [1]. Due to their ability to capture nonlinear characteristics neural networks (NN) have been used also for the bioprocess monitoring. In the same time they served as software sensors for the estimation of biological variables difficult to measure on-line.

Neural networks have typically been used as black-box models that is, no prior knowledge about the process was assumed; the goal was to develop a process model based only on observations of its input-output behavior. The ability of neural networks to learn approximations to arbitrary functions is their strength but is also a weakness: there is a risk of overfitting and interpretation of such models is difficult [2]. A different approach has focused on combining *NN*s with prior knowledge (first principles, process constraints) about the process. This is called a hybrid configuration and it incorporates on one side a classical model describing the known part of the process and on the other, a neural network modeling what is unknown about the process.

Both black–box and hybrid approaches have been employed for the modeling of biological processes. The black box approaches in bioengineering represent process variable estimators and usually involve multiperceptron *NN*s as static models

S.G. Tzafestas (ed.), Advances in Intelligent Systems, 181–192.

between secondary variables (on-line measurements) and primary biological variables [3, 4]. A hybrid structure describing a bioprocess puts together neural networks and mass balances. Some hybrid configurations have been implemented and applied to fed-batch fermentation processes and animal cell cultures for state prediction [2, 5].

This chapter proposes a hybrid structure - applied to a batch process consisting in an animal cell culture - whose aim is to reconstruct the trajectories of the key biological variables from only the initial conditions. The hybrid feature of the structure is due to the use of classical mass balances in which only the pseudo-stoichiometric and kinetic part are described with a *NN*. A hybrid training algorithm is proposed: an unsupervized technique for the centers choice in order to extract some knowledge from the experimental data and a nonlinear optimisation technique for the weights and biases by using only the real and asynchronous data. In order to have a faster convergence a preprocessing of the data is carried out by using interpolated data for the weights and biases initialisation.

DESCRIPTION OF THE PROCESS

The application consists in an animal cell line (CHO-K1) growing on microcarriers in spinner flasks. Seven experiments were carried out and for each four signals are measured asynchronously: biomass, glucose, lactate and glutamine concentrations. These experiments differ only by their initial concentrations in the main substrates (glucose and glutamine) as can be seen in fig.1.

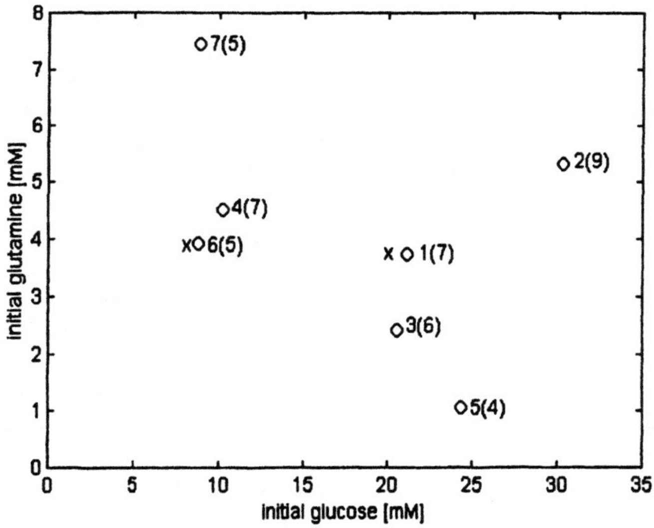

Fig.1 Initial conditions for the available experiments (in brackets, the number of available data points; crosses indicate test data)

Biomass was determined with a hemacytometer, while the other variables were measured by enzymatic methods. Only the measurements coresponding to instants for which glucose and glutamine are still present in the medium are taken into consideration. 43 samples are available which were divided in a training set (31) and a test set (12) marked with crosses in fig. 1.

MODELING OF A BATCH PROCESS

As bioprocesses involve living organisms, their dynamics are often poorly understood, strongly nonlinear and non-stationary. A general dynamic state space model – valid for batch, fed-batch and continuous operations - has been proposed in [6]:

$$\frac{d\vec{\xi}}{dt} = K\vec{\varphi}(\vec{\xi},t) - D\vec{\xi} - \vec{Q}(\vec{\xi}\,) + \vec{F} \tag{1}$$

The right term of the relation can be separated in two terms:

1) $K\vec{\varphi}(\vec{\xi},t)$ - the pseudo-stoichiometry and the nonlinear kinetics of biochemical reactions where:

$\vec{\xi}$ - vector of component concentrations;

$\vec{\varphi}(\vec{\xi},t)$ - reaction rates;

K – yield coefficient matrix

2) $-D\vec{\xi} - \vec{Q}(\vec{\xi}\,) + \vec{F}$ - the transport dynamics of the components through the bioreactor where:

D - dilution rate;

$\vec{Q}$ – vector of mass outflow rates in gazeous phase;

$\vec{F}$ – vector of feed mass flow .

The reaction rates are often very complex functions of the operating conditions and of the state of the process. The analytical modeling of these functions is usually cumbersome and is the subject of continuing investigation.These badly known, nonlinear and static relations among the component concentrations will be described here with a neural network. More precisely only the term $K\vec{\varphi}(\vec{\xi},t)$ from relation (1) will be replaced by a *NN* with $\vec{\xi}$ as as inputs as can be seen in fig. 2.

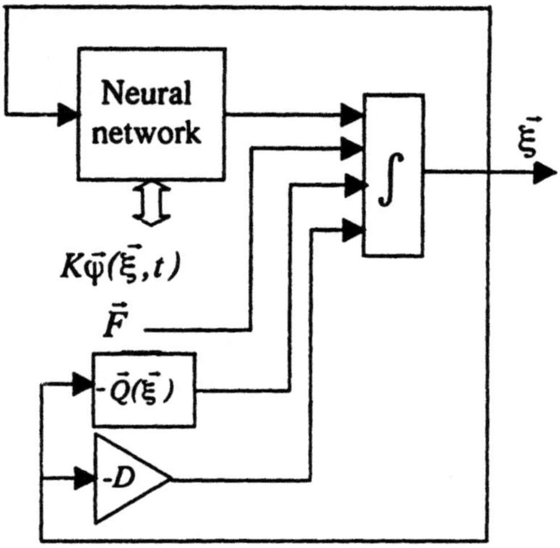

Fig. 2 General structure of the simulator

THE SIMULATOR STRUCTURE

In our case batch culture data were available, and the model (1) becomes:

$$\frac{d\vec{\xi}}{dt} = K\vec{\varphi}(\vec{\xi},t) \tag{2}$$

Following the relation (2) the hybrid structure reduces to an architecture which incorporates a neural network (describing the pseudo-stoichiometry and the kinetics) and an integrator (for the dynamics representation of the process) [7]. Discretizing the model (2), the latter becomes in "terms" of neural networks:

$$\vec{\xi}(k+1) = NN(\vec{\xi}(k)) + \vec{\xi}(k) \tag{3}$$

with:

$\vec{\xi}$ - the vector of the concentrations to estimate

NN – the neural network

Only the initial conditions of the four biological variables are to be provided to the simulator and the structure (fig.3) will have to reconstruct the whole trajectories of these variables.

Fig.3 Simulator structure

The procedure for identifying the *NN* consists in the following steps:

1. *choice of the neural structure* (type of activation function, number of input, output and hidden nodes). In our case radial basis functions are used as activation functions and the numbers of input and output nodes are the same and equal to the number of variables (4). The number of hidden neurons is varied in order to get the best cross-validation error;
2. *learning:* applied to the training data. First of all, the centers and the widths are computed. With these values fixed, the weights and biases are adapted by means of a supervised algorithm;
3. *validation*: the ability to reproduce training data is tested;
4. *cross validation*: the ability to reproduce test data is checked.

NEURAL NETWORK STRUCTURE

A radial basis *NN* (fig.4) has been employed in the hybrid structure from fig.3 firstly because it has proven to be a very good nonlinear function approximator and secondly because of its structure (the presence of the linear output layer) which allows a special case of linear training. The hidden layer is characterised by radial basis functions (in our case, gaussian functions).

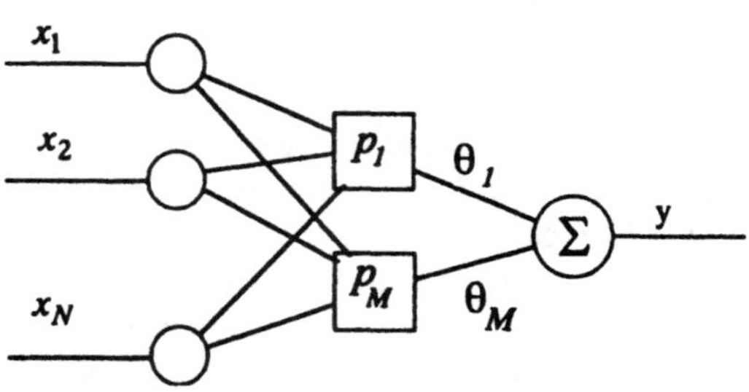

Fig. 4 The structure of a radial basis *NN* (case of N inputs and one output)

- p_i = gaussian functions

$$p_k = exp\left(-\sum_{i=1}^{N}\frac{\left\|x_i - c_{ki}\right\|^2}{\sigma_k^2}\right) \qquad k = 1..M$$

- M = number of centres
- N = number of inputs
- $c_{k,i}$ = the i-th component of the k-th centre
- σ_k = the width of the k-th node

$$y = \theta_0 + \sum_{k=1}^{M}\theta_k p_k$$

- θ_i = i-th weight
- θ_0 = bias.

Training algorithms

There are different training strategies for this kind of structure depending on how the centers of the radial basis functions are specified. In order to benefit from the linearity of the output layer a hybrid training algorithm was used:

1. unsupervised for the centers;

The centers of the gaussians should be placed in the data space so that to represent possible groups and the widths should be chosen in such a way that gaussians intersect among themselves and cover the whole training data space. The width of a hidden neuron was chosen as the mean value of the distances between the center of the respective neuron and the other gaussian centers.

For the choice of the centers, as nothing is known about the structure of the data a clustering procedure has to be applied. The objective is to classify the data according to similarities among them and to organize these latter into groups. There are more possibilities for such techniques and the following were tried in order to obtain the best cross validation error: k-means, fuzzy c-means, self-organizing maps.

The k-means is an iterative optimization [8] procedure which consists in putting together elements in such a way that a sum-of-squared-error criterion is minimized. A criterion with an euclidian distance as well as a Mahalanobis distance were tested. As starting point the number of centers can be chosen randomly or by means of a hierarchical clustering procedure which merges clusters according to a similarity measure (in this case it was the mean distance between two clusters).

The fuzzy clustering assumes a fuzzy partitioning of data: the transitions between classes are rather gradual than abrupt. The procedure develops in a way similar to the k-means with the difference that the objective function is weighted with the membership functions [9].

A nonlinear clustering was also employed, precisely the Kohonen maps which are self organising structures that can concentrate the information contained in the inputs, into local clusters [10]. Inputs that are similar from an euclidian distance point of view will be projected onto close (in the sense of a defined neighborhood) neurons. The neurons that are adjacent will have similar synaptic weights. The network is trained by means of an unsupervised algorithm: only the weights of the winning neurons are adapted (*winning neurons*: the neuron closest to the current input + those in the neighborhood). The weights of the winning neurons will represent the centers of the classes.

2. supervised for the weights and biases

Once the centers and the widths are computed they are fixed and a supervised algorithm is applied to determine the weights and the biases of the output layer. As the aim is to reconstruct the whole trajectories of the measured concentrations, starting from the initial conditions, the cost function is defined for all the measurement moments of all training series:

$$J = \sum_{i=1}^{s} \sum_{k_i \in K_i} [(\vec{\xi}_m(k_i) - \vec{\xi}(k_i))^T \cdot (\vec{\xi}_m(k_i) - \vec{\xi}(k_i))] \qquad (4)$$

s – number of training sets

$\vec{\xi}_m(k_i)$ - the measurement vector at moment k_i;

$\vec{\xi}(k_i)$ - the output vector of the simulator at moment k_i;

k_i - the measurement moments for the i-th training series.

K_i - the set of k_i for the i-th training series.

Due to the linearity of the output layer a linear training algorithm could have been performed if the inputs and the outputs of the *NN* have been known at any moment. In our case only the outputs of the integrator are known and they are the asynchronous samples of the measured key biological variables. This leads to a cost function nonquadratic in the parameters which implies a nonlinear algorithm for the parameter optimisation accomplished with the Levenberg-Marquartd algorithm. In order to improve the convergence rate of the optimisation procedure, the hidden layer is initialised with interpolated data obtained by means of a cubic spline interpolation which provides approximated synchronous samples.

RESULTS

Data well covering the input space were chosen for the learning phase (d2, d3, d4, d5, d7) which makes 31 samples and the rest (12 samples) were used for the cross validation. In order to avoid overfitting, for each type of unsupervised algorithm several structures with different numbers of hidden nodes have been tried and the architecture with the smallest cross validation error is selected as the best. The cross validation error was computed as the sum of the mean errors for the test series:

$$J = \sum_{i=1}^{2} \frac{1}{N_i} \sum_{k_i \in K_i} [(\vec{\xi}_m(k_i) - \vec{\xi}(k_i))^T \cdot (\vec{\xi}_m(k_i) - \vec{\xi}(k_i))] \qquad (5)$$

where N_i is the numbe: of measurements available in the test data set *i*.

K-means (both with random and hierarchical clustering initialization) provided a best structure with 6 hidden neurons (see table 1). In the second case a faster convergence was noticed. If a random initialisation is employed the results are highly dependent on th: initial values: the cross validation error varies substantially. In fig. 5 and 6 are represented the simulation results in validation and cross validation with a structure with k-means (hierarchical initialisation).

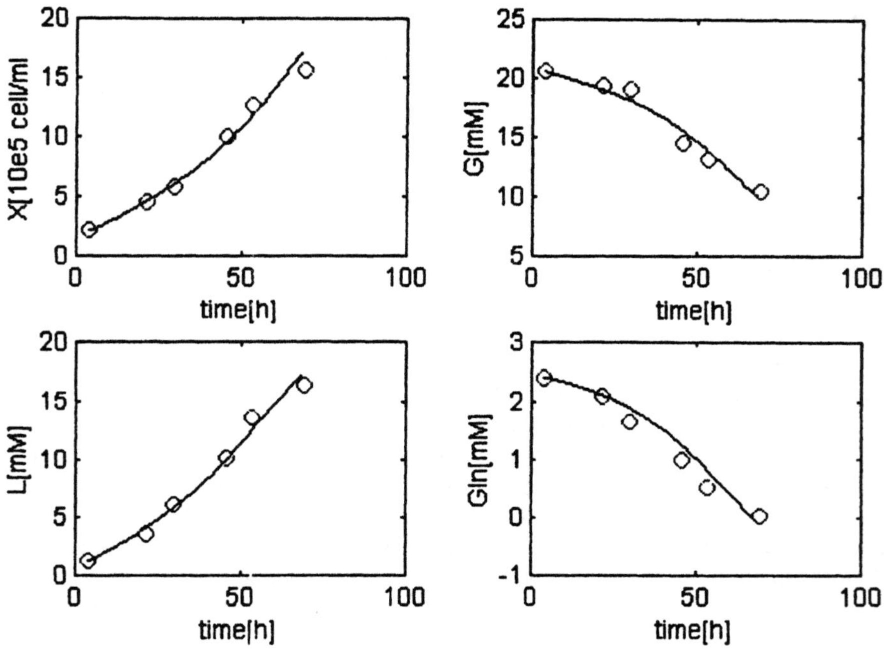

Fig.5 Validation results (d3)
X – biomass concentration,G – glucose concentration
L- lactate concentration, Gln – glutamine concentration

The fuzzy clustering provided a good validation error and a bad cross validation error. A clustering technique with the Kohonen map gives results close to the clustering technique with *k*-means. In any case, even if some differences can be noticed with respect to the validation and cross validation errors, in what concerns the best structure, the same result is always obtained: 6 hidden neurons.

No of neurons	5	6	7	8
Validation error	96	42.07	43	37
Cross validation err.	7.2	6.9	18.79	26.45

Table 1 Results with k-means (euclidian distance and hierarchical initialisation)

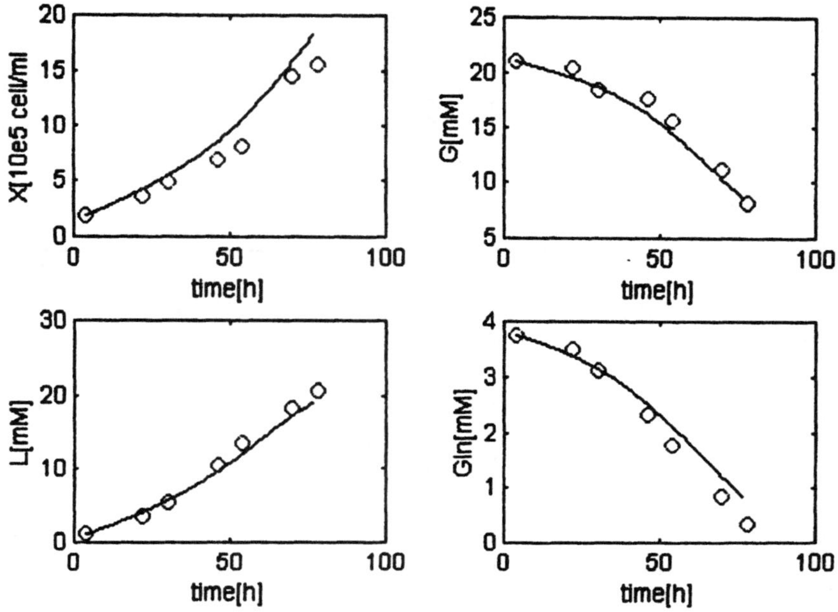

Fig.6 Cross validation results (d1)

As measurements of biological data are associated with a lot of errors, an error based analysis is further carried out. It is supposed that all signals are associated with a white noise measurement error normally distributed (with zero mean). The standard deviation is estimated for each variable, at each instant. The cost function will be modified according to these errors and in the expression (4) each term of the sum will be weighted by the variance ($\sigma_\alpha^2(k_i)$):

$$J = \sum_{i=1}^{s} \sum_{k_i \in K_i} \sum_{\alpha \in M} \left(\frac{\xi_\alpha(k_i) - \xi_{m\alpha}(k_i)}{\sigma_\alpha(k_i)} \right)^2 \qquad (6)$$

where $\sigma_\alpha(k_i)$ is the standard deviation associated to the concentration α belonging to the set of concentrations M, at moment k_i from series i. Starting with the 6 hidden nodes configuration obtained previously (with a k-means initialised with a hierarchical procedure) a new training is performed. Some simulation results are presented in fig.7 (validation) and in fig.8 (cross validation). The circles represent the measurement points, the "+" the measurement points plus 3 times the standard deviation and the "*", the measurement points minus 3 times the standard deviation. Each interval "x +" corresponds to a 99% confidence interval for the measured value "o". It was noticed that in validation as well as in cross validation, for almost all experimental points, the simulated values are located inside the confidence intervals.

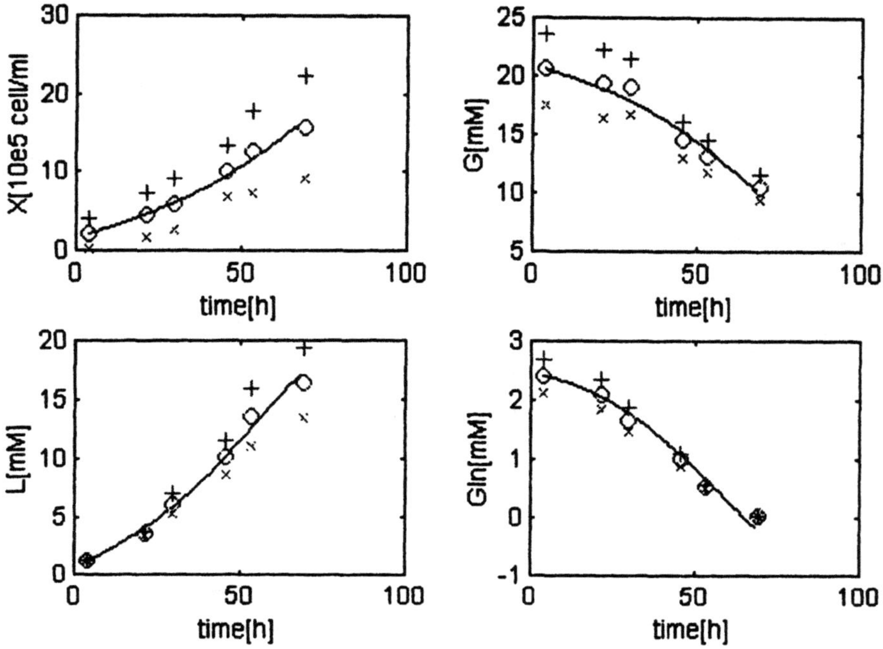

Fig.7 Validation results (d3)

CONCLUSIONS

A hybrid (classical-neural) structure was proposed for the implementation of a simulator for a batch animal cell culture. A neural network whose role was to model the stoichiometry and the kinetics of the process was combined with mass balances. A radial basis function was chosen and a hybrid training algorithm was employed for the computation of the network parameters. So far, with the available data and with a good choice of the training set some interesting results have been obtained even in cross validation. For different clustering procedures (for the choice of the gaussian centers) the optimal neural structure has 6 hidden neurons and the smallest errors are obtained by a *k*-means clustering with hierarchical initialisation which also provides a faster convergence of the supervised algorithm. Only the real asynchronous data were used for the training phase; for a better starting point to the weights and biases identification an initialisation was performed by using interpolated data.

Due to the important measurement errors encountered in bioprocesses, an error based analysis was also carried out starting with the "best" configuration obtained from the analysis without measurement errors. Good results in validation and cross validation were also obtained: most of the simulated values were situated inside the confidence intervals.

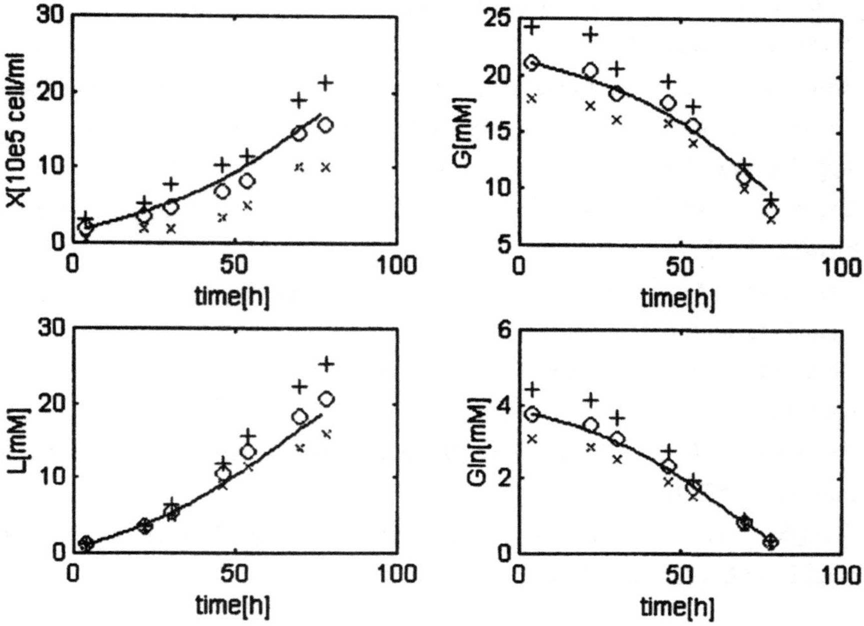

Fig. 8 Cross validation results (d1)

Based on these promising results, the hybrid structure will be tested and if necessary adapted for fed-batch cultures. Moreover, it would be interesting to extend the simulation model so as to take into account the effect of environmental and experimental conditions (e.g. temperature, vitamin concentration) which were constant in the present study.

Bibliography

[1] Zhang B.S. e.o. *Improved Quality & Productivity in Secondary Metabolite Fermentation through Estimation, Control & Scheduling*, IFAC, San Francisco, USA, 1996

[2] Psichogios D.C., Ungar L. H. A Hybrid Neural Network-First Principles Approach to Process Modeling, AIChE Journal, vol. 38, 1992

[3] Di Massimo C. e.o., *Towards improved Penicilin Fermentation via Artificial Neural Networks*, Comp. Chem. Eng., vol. 16, no. 4, pp. 283-291, 1992

[4] Willis M.J. e.o. Ar i'icial Neural Networks in Process Estimation and Control, *Automatica*, vol.28, 1992

[5] Dors M., Simutis R., Lubert A., *Advanced Supervision of mammalian Cell Cultures Using Hybrid Process Models*, Preprints of the 6th Int. Conf. on Comp. Appl. In Biotechnology , Germany, 1995

[6] Bastin G., Dochain, *On-line Estimation and Adaptive Control of Bioreactors*, Elsevier, 1990

[7] Graefe Jorg, *Modélisation et simulation de cultures de cellules animales par réseaux de neurones*, thesis, 1997

[8] Duda R.O, Hart P.E., *Pattern Classification and Scene Analysis*, John Wiley & Sons, 1973

[9] Babuska R., *Fuzzy Clustering in Modelling and Identification*, PhD thesis, Univeristy of Delft, 1997

[10] Haykin S., *Artificial Neural Networks. A Comprehensive Foundation*, MacMillan Collge Publishing Company, 1994

18

ESTIMATION OF STATE VARIABLES IN POWER SYSTEM COMBINING THEORETICAL KNOWLEDGE WITH RBF NEURAL NETWORK

T. ATANASOVA*, V. LUBENOVA*, J. ZAPRIANOV*
and H.A. NOUR ELDIN**
**Institute of Control and Systems Research*
Bulgarian Academy of Sciences
Acad. G. Bonchev Str., bl. 2, 1113, Sofia, Bulgaria
***Group of Automatic Control and Technical Cybernetics*
University of Wuppertal, Wuppertal, Germany

1. Introduction

The neural network (NN) approach has been demonstrated by numerous researches as useful tool for automatic control of nonlinear systems. The application of neural network architectures to nonlinear identification and control has been investigated widely [1, 2]. In [3] it is suggested to design an intelligent controller on a base of NN that learns and identifies the system dynamics. In [4] it has been noted that NN are suitable for control of large scale nonlinear systems, such as power systems, due to the specific NN features as (i) parallel information processing; (ii) ability of modeling nonlinear systems; (iii) training by data samples. In power system applications NN approach is exploited mainly in design of power stabilizer [4]. In [5] a nonlinear state estimator using a feed forward neural network and stochastic approximation is proposed. In considered cases multilayer backpropagation nets are basically used. This kind of neural architecture expressed stability problems and time consuming training. Radial basis function networks (RBF) represent different neural structures that have recently attracted considerable attention due to some features such as local adjustment of the weights and mathematical tractability [6, 7, and 8]. However developing a pure neural model without using of the existing theoretical knowledge is obviously not very effective [9], since a great deal of information about the plant already compiled in form of the theoretical

S.G. Tzafestas (ed.), Advances in Intelligent Systems, 193–200.

model is discarded while building a neural model. So it is desirable effectively employ the available prior knowledge and empirical data.

The rest of the chapter is organized as follows. First a mathematical model of the power system is briefly described. Then the RBF network based δ-estimator is designed on a base of the model. The estimator is tested through simulations.

2. Description of the mathematical power system model

A simplified dynamic model of power system, namely a single generator connected through two parallel transmission lines to a very large network approximated by an infinite bus. The model is shown in fig. 1.

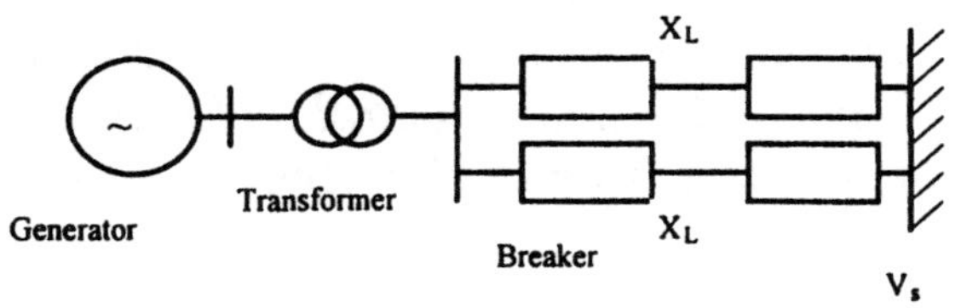

Figure 1. Single machine-infinite bus model.

A plant model considered can be written as follows [10,11]:
Mechanical equations:

$$d\frac{\delta(t)}{dt} = \varpi(t)$$
$$d\frac{\omega(t)}{dt} = -\frac{D}{H}\varpi(t) + \frac{\sigma_0}{H}(P_m - P_e(t)), \tag{1}$$

Generator electrical dynamics:

$$d\frac{E_q'(t)}{dt} = (E_f(t) - E_q(t))\frac{1}{T_{d0}} \tag{2}$$

Electrical equations:

$$E_q(t) = \frac{x_{ds}}{x'_{ds}}E'_q(t) - \frac{x_d - x'_d}{x'_{ds}}V_s\cos\delta(t)$$
$$E_f(t) = k_c u_f(t)$$
$$P_e(t) = \frac{V_s E_q(t)}{x_{ds}}\sin\delta(t)$$
$$I_q(t) = \frac{V_s}{x_{ds}}\sin\delta(t) = \frac{P_e}{x_{ad} I_f}$$
$$Q(t) = \frac{V_s}{x_{ds}}E_q(t)\cos\delta(t) - \frac{V_s^2}{x_{ds}} \tag{3}$$
$$E_q(t) = x_{ad} I_f(t),$$

$$V_t(t) = \frac{1}{x_{ds}}\{x_s^2 E_q^2(t) + V_s^2 x_s^2 + 2x_s x_d x_{ds} P_e(t) ctg\,\delta(t)\}^{1/2} \tag{4}$$

where $\delta(t)$ is the power angle of the generator, $\omega(t)$ the relative speed of the generator.

For details of the other symbols used in the above equations, reader can refer to the Ref. [11]. The parameters of the synchronous generator and the line parameters are given in the Appendix.

The measurable physical variables are $P_e(t)$, $Q(t)$, $I_f(t)$ and $\omega(t)$. The power angle $\delta(t)$ should be available for designing an appropriate feedback law for the system. In [12] it is shown that the introducing of the power angle $\delta(t)$ in a feedback controller can improve transient stability of the system greatly. Since this state variable is difficult to measure it is need to construct the power angle $\delta(t)$ from the measurements available.

In the next section we test the RBF network abilities to model and reconstruct the power angle $\delta(t)$.

3. RBF network based δ estimator

3.1 Radial Basis Function

Radial basis functions share many of the advantages of conventional feed forward (backpropagation) neural networks. Both types of networks with enough neurons and training data can approximate an arbitrary nonlinear function to a certain level of accuracy. RBF networks offer some other advantages in comparison with backpropagation networks [13]. When the basis function are fixed, the outputs of networks are linear in the coefficients (the network weights). Then the results of the theory of linear systems can be applied to the weights adaptation and RBF networks integration in control design.

One of the commonly used radial basis function networks is the Gaussian radial basis function networks (GRBF), which is expressed by

$$f^{NN}(x) = \Sigma \alpha_k R_k(x),\ k = 1, 2, \dots m, \tag{5}$$
$$R_k(x) = exp\,[-v_k^2 \,|x-c_k|^2], \tag{6}$$

where $R_k(x)$ -is the Gaussian activation function for the kth node, c_k - center, v_k -width, α_k - amplitude.

The results of [7] indicate that the RBF NN are capable of on-line approximation of nonlinear function, the nonlinearities that are investigated are mild (cubic and sine function).

3.1 RBF based δ-estimator

An effective way to involve neural networks in a control design is to use them for the identification of the system's dynamics [8]. The partition of the NN to subnets, each one identifying a part of the dynamics, seems very reasonable and has been discussed by several researches.

In the chapter it is proposed to evaluate $\delta(t)$ using neural network approach combined with existing knowledge about the functional dependencies in the power system.

First, a dynamic simulation model of the system has been developed using mathematical equations that describe the behavior of the power system. Then on a base of this model a neural network was trained as an observer to the system. The inputs of the neural network x are δ - values obtained from a dynamic simulation model. The outputs y are ω - values. There is function $\omega = f(\delta)$, and the RBF network is constructed so as to produce a fitting surface to the function f. The function f has some degree of smoothness to allow a fitting surface to be producing.

The phase plot $\omega = f(\delta)$ may be interpreted as the generator of the observed data. The locations on the phase plot produce the $(x, y) = (\delta, \omega)$ input/output pairs.

The Gaussian RBF network model of f can be represented by

$$\omega = \sum_{i=1}^{N} \theta_i z_i$$

$$z_i = \exp\left[-\frac{1}{2}\left(\frac{(\delta_1 - c_{i1})^2}{\sigma_{i1}^2} + \frac{(\delta_2 - c_{i2})^2}{\sigma_{i2}^2} + \ldots + \frac{(\delta_n - c_{in})^2}{\sigma_{in}^2}\right)\right] \tag{7}$$

where ω represents network output, δ represents the network input, c_i, σ_i, θ_i are network parameters: center, width and weights.

The network was trained using a gradient descent procedure. The positions and the spreads of centers were not trained, because the simulations indicated satisfactory performance without such a computation-costly training. They were selected before training. Thus the knowledge embodied in the center and width parameters gives to the fixed neural architecture the ability to learn the dependence $\omega = f(\delta)$.

To obtain δ value from ω - measurement it is need to invert the NW representation of the function f. Since f is nonlinear, uniquely determining of δ is, in general, not possible. In conventional multilayer networks gradient descent is often used to iteratively solve for the input. RBF NW suggests a closed form solution; no iteration is required.

4. The digital simulation results

The power system used for the simulation is described in section 2. The model has been implemented in an emulator by means of SIMULINK on MATLAB.

In the real power system there are always disturbances which can be used for training the neural network. In this chapter the following disturbances have been used in the simulation:

- 20% reduction in mechanical power input;
- 50% increasing of the transmission line parameter values;
- measurement noise – white noise + $0.1\sin(10t)$;
- three phase short circuit is applied at the generator terminal for 0.15 sec and the fault is switched off afterwards.

On Fig. 1 the simulation MATLAB model of power system is shown. Fig. 2 represents the function approximation $\omega = f(\delta)$ in transient period after feeding of the generator with 0.9 p.u. mechanical power.

Fig. 2 represents the phase plot $\omega = f(\delta)$. However, $\delta = \int \omega\,(t)$ is not suitable for RBF NW approximation. Approximation error of ω with 20% reduction of mechanical power in t=5sec. is shown on fig. 3.

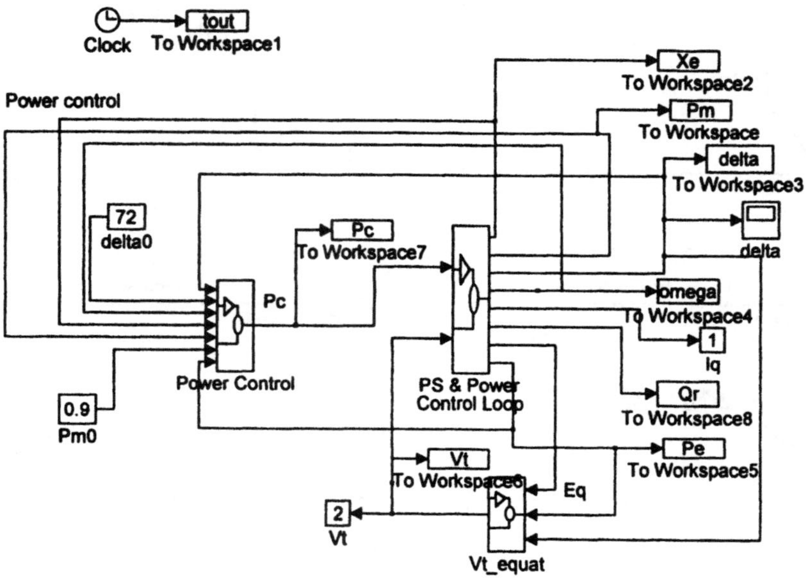

Fig. 1.

4.1 Parameter variation and measurement noise

The parameter values for the PS model are nominal values. It is possible that the values could be incorrectly determined or that the values could change with time.

The transmission line parameter values were increased by 50% during simulations and system response explored. In addition to this disturbance the measurement noise was included (*white noise* + $0.1\sin(10t)$).

Figures 4(a, b) represent the phase plot with different levels of noise.

For noisy data it is need more neurons in RBFN structure to model the system dynamic and state estimation. Figure 5 shows that good approximation results can be obtained during different time intervals and periods of operation. The sum-squared error during training epochs is presented on figure 6.

Figure 7 represents estimation error when parametric disturbances and measurement noise were added to ω - values.

Under constrained disturbances the generalizing abilities of neural network allow to reconstruct $\delta\,(t)$ from on-line measurements of $\omega\,(t)$.

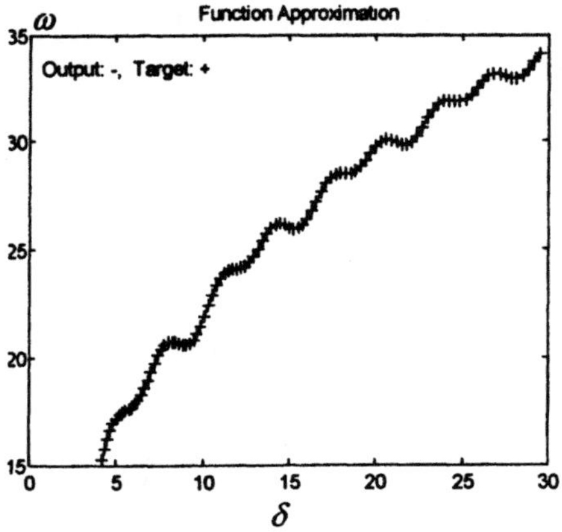

Figure 2.

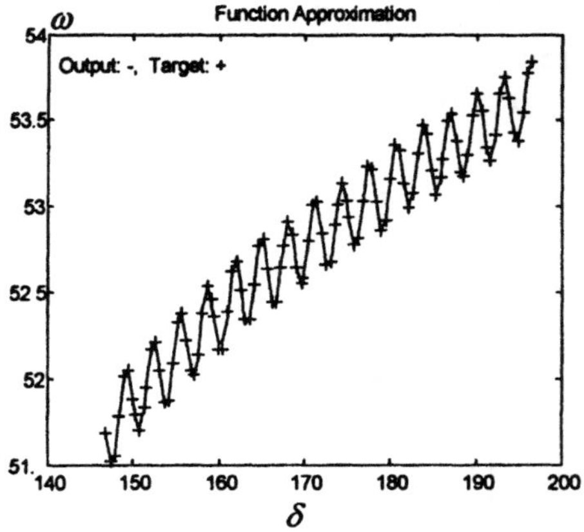

Figure 3.

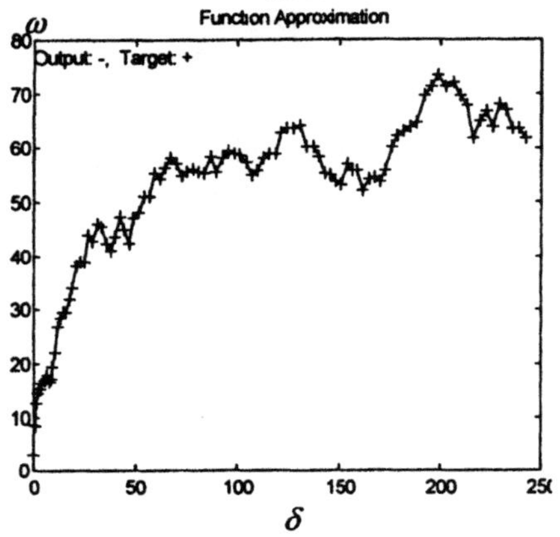

Figure 4 (a)

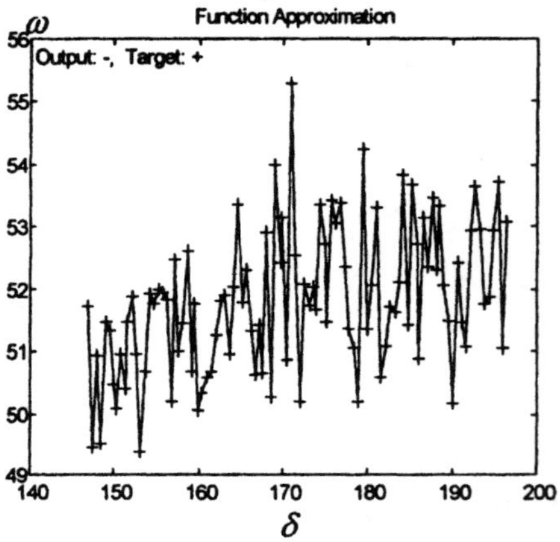

Figure 4 (b)

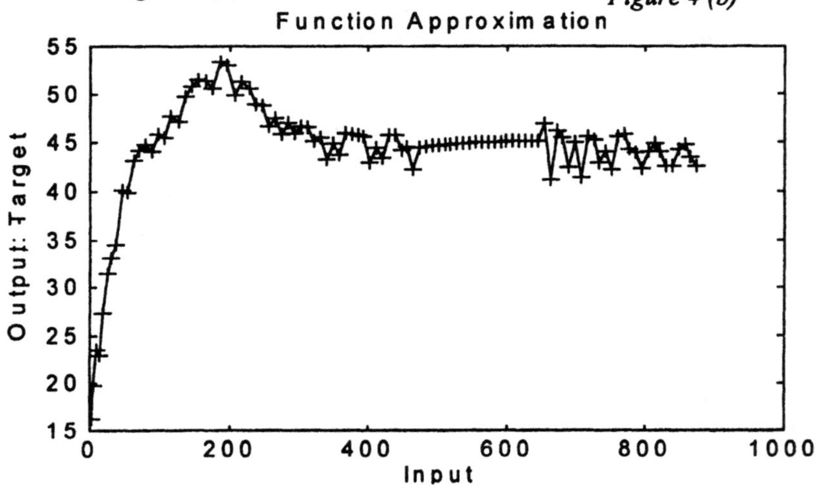

Fig. 5.

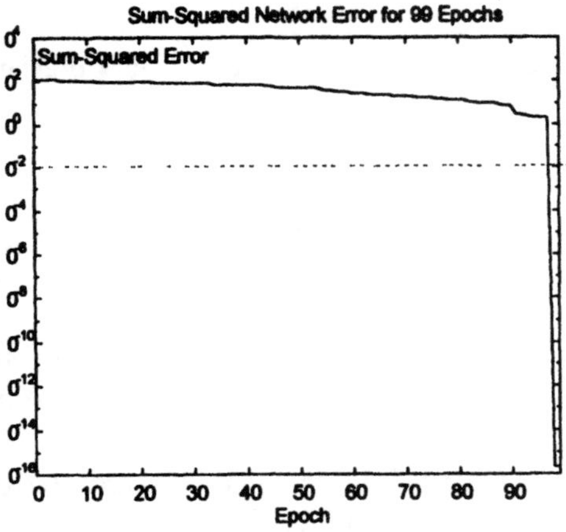

Figure 6.

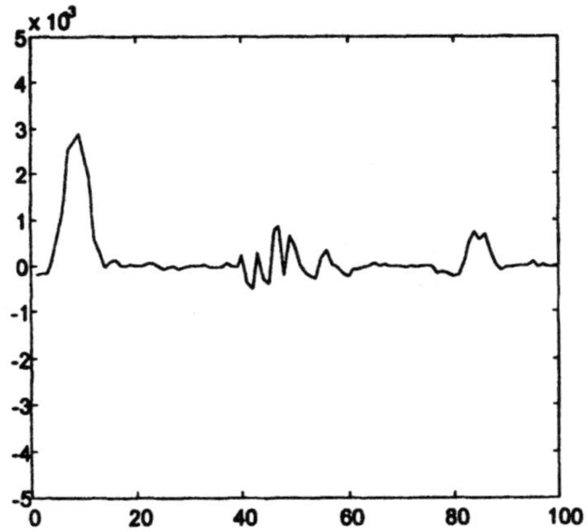

Figure 7.

5. Conclusion

Radial basis function neural networks provide an attractive method for non-parametric models for use in nonlinear control. They are faster to train than conventional feed forward networks with sigmoidal activation nets. The RBF model structure is better suited for adaptive control, since if the basis functions are fixed, the model is linear in the coefficients.

The RBF network abilities to model and estimate $\delta(t)$ have been tested by digital simulations.

The simulation results indicate that a neural estimator composed of radial basis activation functions is capable to provide very small estimation error. The influence of parameter variations and measurement noise on system response was explored. For noisy data it is need more neurons in RBF network structure to model the system dynamic and to estimate the state variable.

The RBF NW based estimator can be considered as only subnet of more complex hierarchical neurocontrol architecture for power system, and it is a subject of future investigations.

Appendix

Example system parameters used in simulation studies are as follows:
ω_0 =314.159; D=5.0; H=8.0s; T_{d0}=6.9s; k_c=1; x_d=1.863; x'_d=0.257; x_T=0.127; x_L=0.4853.
$x_{ds}=x_T+0.5x_L+x_d$; $x'_{ds}=x_T+0.5x_L+x'_d$; $x_s=x_T+0.5x_L$;.

The physical limit of the excitation voltage is max $|k_c u_f(t)|$=1.8 p.u.

The operating point is δ_0=72°, P_{m0}=0.9 p.u., V_{t0}=1.0 p.u.

Acknowledgment

The financial support of the National Foundation of Scientific Researches is appreciated.

References

1. Christodoulou, M. A. "Neural Networks in Control". Proceedings of the IFAC Workshop on Mutual Impact of Computing Power and Control Theory "MICC'92", Prague, Czechoslovakia, pp. 169-174 , 1992.
2. Levin, A.U., K. Narendra, "Control of Nonlinear Systems Using Neural Networks - Part II. Observability, Identification and Control", *IEEE Transactions on Neural Networks*, 7(1):30-42, 1996.
3. Shoureshi, R. "A Structure for Intelligent Control". In: Intelligent Control Systems, Plenary Volume of Preprints, 13 World Congress of IFAC, San Francisco, CA, USA, June 30-July 5, pp. 101-102, 1996.
4. Cheng, S., R. Zhou, L. Guan. "An On-line Self-learning Power System Stabilizer Using a Neural Network Method", *IEEE Trans. on Power Systems*, vol. 12, no. 2, May, pp. 926-930, 1996.
5. Alessandri A., and T. Parisini. "Nonlinear Modeling and State Estimation in a Real Power Plant Using Neural Networks and Stochastic Approximation", Proceedings of the American Control Conference, Seattle, Washington, pp. 4265-4269, 1995.
6. Sanner, R.M., J.-J.E. Slotine. "Gaussian networks for direct adaptive control". *IEEE Trans. Neural Networks*, 3:837-863, 1992.
7. Shah, M.A., P.H. Meckl, "On-line control of a nonlinear system using radial basis function neural networks". Proceedings of the American Control Conference, Seattle, Washington, pp. 4265-4269, 1995.
8. Tzafestas, S.G., P. A. Procopiou, C. S. Tzafestas. "Telemanipulator Neurocontrol Using Multiple RBF Networks". Proceedings of the 12th IEEE International Symposium on Intelligent Control, 16-18 July, Istanbul, Turkey, pp. 257-262, 1997.
9. Johansen, T. A.., "Identification of non-linear systems using empirical data and prior knowledge - an optimization approach", *Automatica,* 32(3):337-356, 1996.
10. Bergen, A. R., "*Power System Analysis",* Prentice-Hall, Englewood Cliffs, NJ, 1986.
11. Wang, Y., D. Hill, R. Middleton, L. Gao, "Transient Stabilization of Power Systems with an Adaptive Control Law", *Automatica,* 30(9):1409-1413, 1994.
12. Wang, Y., D. Hill, R. Middleton, L. Gao, "Transient Stability enhancement and voltage regulation of power systems", *IEEE Transactions on Power systems*, 8(2): 620-627, 1993.
13. Ungar, L.H., R. D. De Veaux. "EMRBF: A statistical basis for using radial basis functions for process control". Proceedings of the American Control Conference, Seattle, Washington, pp. 1872-1876, 1995.

19

ROUTE SELECTION IN A NEURO-FUZZY VEHICLE NAVIGATION SYSTEM

G. K. H. PANG
Dept. of Electrical and Electronic Engineering
The University of Hong Kong
Pokfulam Road, Hong Kong

1. Introduction

A route guidance system is a routing system that provides instructions to drivers based upon "optimum" route solutions. A driver can make the destination known to the system. A dynamic route guidance (DRG) system would route drivers using the current traffic conditions such as congestion and roadworks. The system can provide actual routing advice to the driver in light of the real-time traffic conditions. It will be based on real-time information regarding conditions and incidents of the traffic network. It is conceived as the integration of the routing and the traffic control functions. With the recent advances in the micro-electronics technology and informatics, the development of DRG systems has been made possible.

One objective of such a dynamic route guidance system is to balance the level of service on all major network links so as to increase the efficiency, speed, safety and quality of travel (e.g. to minimize travel time). Such a system would be particularly useful when accidents or roadworks occurred in the traffic network. Also, the system is highly beneficial to the motorist when driving in unfamiliar areas. A DRG system would act as the driver's assistant and try to reduce his tension.

The issue of driver behaviour in terms of route choice and response to guidance is complex. One focus in this chapter is on the modeling of driver behaviour in route choice on the DRG system. Here, we assume that the DRG system has a voluntary choice scheme and the driver has the option of not following the advice from the DRG system. The driver's preference is modeled as a fuzzy expert system, and his reaction to the advice and information provided by a DRG system is stored. The

S.G. Tzafestas (ed.), Advances in Intelligent Systems, 201–210.

previous choices of the driver, in particular the deviation from the recommendation, are then used for the training of the system so that it is made adaptive to the driver.

2. Decision Support in Route Selection

The success of a DRG system will depend on how the human being, as a driver, will be taken into account from the very early stage of the work. Hence, topics like driver characteristics, driver decision-making, driver response to guidance advice and driver capacity are important, and must be treated. The most modern information technology will allow the introduction of real-time data exchange and real-time control as new functions in road traffic. The information system must also support the driver effectively in decisions (i.e. information in the form of advice for proper action). The general information should be screened for decision-making whilst driving, so that the system does not overload the driver with information.

In this chapter, the focus will be on route selection, i.e. the capability to support the driver effectively in deciding on an optimum route to his preference. Figure 1 describes such a navigation system. The core of such a system is an adaptive route selection algorithm based on a hybrid fuzzy-neural approach. Each route candidate has a set of attributes associated with it. The attributes are correlated and the final decision (choice) by the driver is perceived as a nonlinear function of the attributes.

3. Routing Algorithm

So far, the idea of "optimum" has been taken in a very limited sense. Most route guidance systems nowdays compute the "best" route for the driver based on either the shortest time [1,2] or the shortest distance [3]. Some systems would provide information on congestion of the road as well [4]. Hence, the route selection function makes use of the distance data (static) and information on average travel speed (dynamic). In the Vehicle Information and Communication System (VICS) project announced by the National Police Agency in 1991, optimum route selection was a focus but the focus was based only on travel distance and travel time reduction based on empirical and real time information. Also, it was not clear how the two criteria can be resolved or compromised to give the optimum route. For example, in the driver's route selection logic in [5], it is not clear how to tradeoff the relative importance of minimum distance route and minimum time route.

From a survey carried out in 1989 [6], respondents were asked to choose between guidance systems that chose routes on the basis of shortest time, shortest distance or some combination of the two. In London, 42% of the respondents would prefer a system that can perform a tradeoff between travel time and travel distance. 56% said they would choose route based on the shortest travel time. A similar result is obtained in Paris. In Munich, 71% would prefer a tradeoff between travel time and travel distance, and only 27% would choose route based just on the shortest travel time.

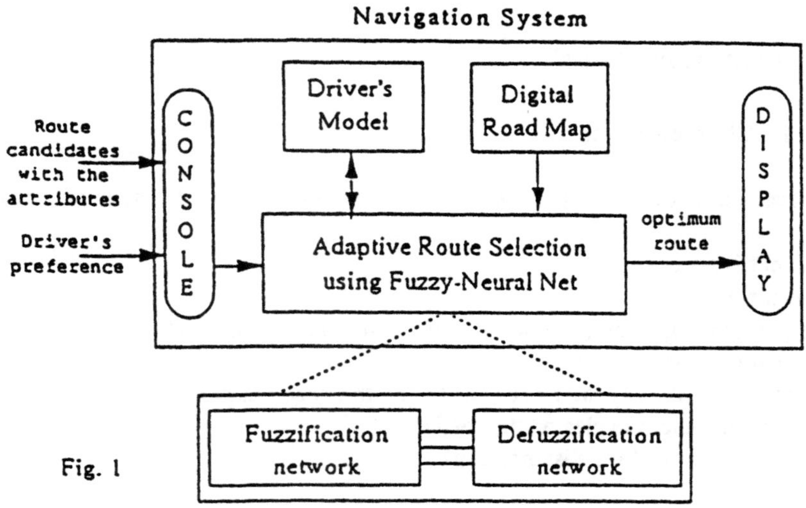
Navigation System
Route candidates with the attributes
Driver's preference
CONSOLE
Driver's Model
Digital Road Map
Adaptive Route Selection using Fuzzy-Neural Net
optimum route
DISPLAY
Fuzzification network
Defuzzification network

Fig. 1

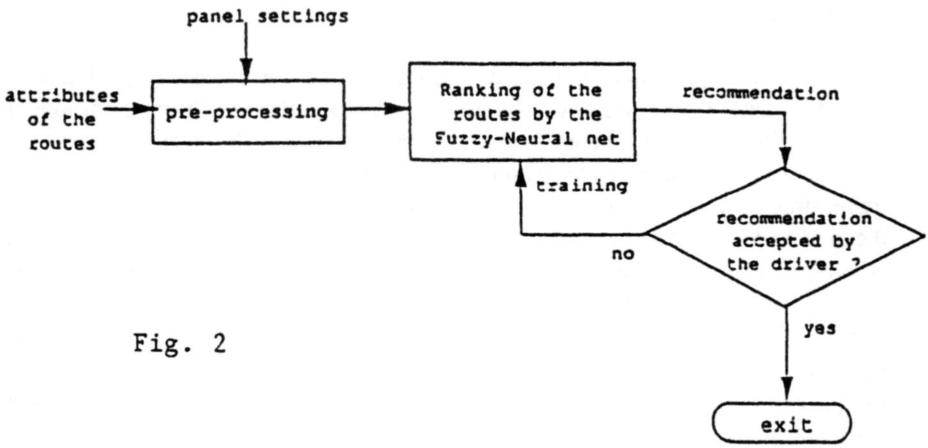
panel settings
attributes of the routes
pre-processing
Ranking of the routes by the Fuzzy-Neural net
recommendation
training
recommendation accepted by the driver ?
no
yes
exit

Fig. 2

Therefore, it can be observed that a routing algorithm that can accommodate the various route selection criteria and their tradeoff would be highly desirable.

4. System Description

4.1 Route Characteristics

It is perceived that a driver may select a route based on many different factors which include :

- travel distance
- travel time
- degree of congestion (number of cars on the road, queue length)
- toll (of expressway or highway)
- degree of difficulty of travel (width of the road, number of lanes, number of pedestrians and bicycles on the road etc.)
- scenery (esp. for long distance trip)

4.2 Route Attributes

It is perceived that a candidate route has many different attributes. These attributes coincide with the factors which are used by the driver in route selection. Below is a set of some of the most important attributes of a candidate route. Note that each attribute has a range from zero (0) to one (1).

- travel distance : 1 denotes the route with the shortest travel distance, relative to the set of candidate routes. 0 can be used to denote routes which are x km longer than the shortest route, where x is a system parameter. The attribute value for other routes can be decided based on a linear scale.
- travel time : 1 denotes shortest travel time, relative to the set of candidate routes. 0 can be used to denote routes which are y minutes longer than the quickest route, where y is a system parameter. Again, the attribute value for other routes can be decided based on a linear scale.
- degree of congestion : 0 denotes no congestion at all. 1 denotes the worst situation.
- toll (of expressway or highway) : denotes no toll and no highway at all. 1 denotes the worst situation.
- difficulty of travel (narrowness and winding of the road, number of traffic lights, road work, number of pedestrians and bicycles on the road etc.) : 0 denotes the ideal road situation, very easy to drive. 1 denotes the worst situation.
- scenery (esp. on long distance trip) : 1 denotes the best scenery. The higher, the better.

4.3 Types of attribute

It can be noticed that for the attributes travel distance, travel time and scenery, a driver would like those attribute scores to be as close to 1 (i.e. as large) as possible for an ideal route. These three attributes are therefore called ``positive attributes". As for attribute scores for congestion, toll and degree of difficulty, a driver would prefer them to be as close to 0 ( i.e. as small ) as possible and they are called ``negative attributes".

4.4 Decision Support (Figure 2)

It is perceived that at a particular instance of time, a number of different candidate routes which have different set of attributes should be considered by the driver. The driver has to make a decision based on the relative importance of the different factors for route selection. Each decision is based on a combination of different factors. There could be some heuristics in route selection but some preferences could be difficult to express in words. The objective here is to design an optimum route search function in a typical in-car navigation system so that it will have the following characteristics :

- It is a decision-making assistant to the driver in route selection. In other words, it embodies a route selection algorithm.
- It can model the behaviour of the driver by storing his preference and previous decisions/choices.
- It can adapt and learn from the recent decisions of the driver. It can also be perceived that some attributes of a candidate route are dynamic while some can be considered as static. The dynamic ones are travel time, degree of congestion and degree of difficulty of travel. The static ones are travel distance, toll and scenery.

5. Fuzzy-Neural Approach

5.1 Artificial Neural Net (ANN)

Neural networks [7] can be developed to model the driver behaviour. It is chosen for this study for their ability to learn from examples, to generalize, to predict and to cope with incomplete input data. A neural network is a parallel distributed information processing system. It consists of a large number of highly interconnected but very simple processing elements known as neurons. Each neuron has a number of inputs and one output which branches out to inputs of other neurons. The output of a neuron is a nonlinear function of the sum of all inputs through the weighted links. Hence, the knowledge of a network is distributed throughout the weighted links.

For our application, the inputs will be the various attributes of a route and the output will be an acceptance measure of the route. The ANN can be trained off-line. The real-time execution of the ANN will be extremely fast. It also has the ability to adapt to different users of the car. Any new user can train the network to learn his preference.

5.2 Fuzzy systems

It is well-known that the decision-making of the driver is fuzzy. The aim is to develop a fuzzy expert system [8] which can be used to rank the route candidates. At the beginning of the system, heuristic rules on how the driver evaluates the route attributes can be specified.

Advantage of Fuzzy systems
Rule-based fuzzy systems are based on fuzzy theory, with expert knowledge represented explicitly using a set of fuzzy if-then rules. They offer a high degree of transparency into the system being modeled.

Advantage of Neural systems
Neural networks have very strong learning capability. Given some numerical data on how a system should behave, there are many neural network algorithms which try to learn this system behaviour. Knowledge is stored implicitly in the weights of a neural network. However, the neural network offers no insight to internal dynamics and relationships.

A Hybrid Approach
A hybrid fuzzy-neural approach can combine the advantages of both approaches. This will further enhance the intelligence of the DRG system, especially in the modeling of the driver behaviour.

The ideas are as follows :

1. A rule-based fuzzy system is developed which represents a preliminary model of the driver.
2. The rule-based fuzzy system is then implemented using a neural network. A method of constructing a neural network which is equivalent to the fuzzy system is developed. It is constructed so that the procedures and membership functions of the fuzzy system can be retrieved from the implementation of the neural network.
3. A special learning algorithm is then used to learn and adapt itself to the recent choices of the driver. The weights of the network will be adjusted. The derivation of the learning algorithm is based on a gradient descent algorithm.
4. After the training procedure, the modified membership functions of the fuzzy systems can be retrieved. This fuzzy system with modified membership functions represent the latest model of the driver. The fuzzy system has now been tailored to the particular driver. The model of the driver can also be represented by a set of weights of the equivalent neural network.

5.3 Construction of Fuzzy-Neural Network

In this section, the architecture of a kind of fuzzy-neural network is described. The network is essentially a parallel implementation of a fuzzy system using a particularly structured neural network. The structure involves the construction of a fuzzification sub-network and a defuzzification sub-network. The two sub-networks will be integrated in such a way that the structure and decision-making process of the original fuzzy system can be fully retrieved from its network implementation. The corresponding neural network should have similar performance as the original fuzzy system.

Figure 3 shows the structure of the system. Each sub-network performs a different function explained as follows :

- fuzzification sub-network : to represent the membership functions of the linguistic terms of the input attributes;
- defuzzification sub-network : to generate a defuzzified value which represents the acceptance level of a candidate route

6. Modeling of Driver Behaviour

6.1 Fuzzy ranking rules

Each driver would have his own perspective of a desirable route. The system is designed so that a driver can specify his preference using fuzzy rules with some pre-defined linguistic terms. Below is an example of a set of fuzzy rules that may be used by a typical driver for route ranking :

If **travel distance** is **averagely** short, then route is **good.**

If **travel time** is **averagely** short, then route is **very good.**

If **congestion** is **averagely** heavy, then route is **very bad.**

If **toll** is **averagely** high, then route is **not good.**

If degree of difficulty is **averagely** high, then route is **bad.**

If **scenery** is **averagely** good, then route is **fair.**

The linguistic terms are in brackets. In the rule antecedent part, the driver can choose from the term set {very, averagely, slightly}.

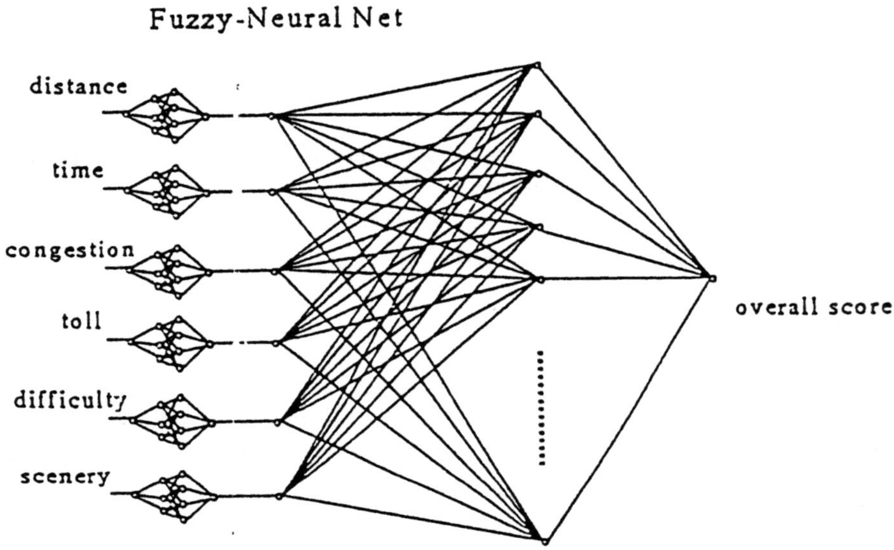
Fuzzy-Neural Net
distance
time
congestion
toll
difficulty
scenery
overall score

Fig. 3

6.2 Linguistic terms and rule firing

The numeric value of a route attribute is the crisp input to the associated fuzzy rule. The fuzzification sub-network will compute the degree of match between the crisp input and the fuzzy set describing the meaning of the rule-antecedent. The fuzzy set describing the meaning of the rule-antecedent is scaled to the same degree to which the rule-antecedent has been matched by the crisp input. Finally, the scaled fuzzy sets of each rule are aggregated to obtain the defuzzified value, which is a performance measure of the route candidate.

In graphical representation, the firing of each fuzzy rule will contribute an area to the defuzzification procedure. The algorithm adopted in this chapter is the center-of-area (COA) method, which is the best well-known defuzzification method.

6.3 Training

A set of learning rules similar to the backpropagation learning algorithm can be derived from the gradient descent method. The back-propagation learning algorithm developed for standard multilayer feedforward neural networks can be extended for the training of the fuzzy-neural network.

7. Summary of the Intelligent Route Selection Procedures

1. The system is setup by a driver inputing his fuzzy rules on route selection using a set of pre-defined linguistic terms. This is the initial setup of the system and a fuzzy-neural network is formed. The weights of the fuzzy-neural network are determined from the shape of the linguistic terms.
2. When planning a trip or during a trip, the driver can modify the relative importance of the various route attributes using some settings on a panel. This is a convenient way for specifying driver's preference, which could be useful for planning a special purpose trip.
3. The driver inputs his origin and destination to the system, and a set of route candidates is obtained.
4. For each route candidate, the attribute scores are inputs to the fuzzy-neural network, and the output is an overall score of that route candidate. With the computation of this overall score, a ranking of the set of route candidates is performed.
5. The driver can accept the recommendation from the system. Alternatively, he can choose an alternate route. Any derivation from the recommendation will be stored, and this information is used for forming the training pairs of the fuzzy-neural network. Hence, the system can be made adaptive to the decision-making of the driver.

8. Conclusions

A useful routing system should have the capability to support the driver effectively in deciding on an optimum route to his preference. In this paper, an optimum route search function in a typical in-car navigation system is developed. The routing algorithm is orientated on the driver's preference. A fuzzy-neural approach is used to represent the correlation of the attributes with the driver's route selection. A recommendation or route ranking can be provided to the driver. Based on a training of the fuzzy-neural net on the driver's choice, the route selection function can be made adaptive to the decision-making of the driver. This methodology paves the way for more intelligent navigation systems.

References

[1] N.B. Hounsell, M. McDonald and R.A. Lambert, ``The integration of SCOOT and dynamic route guidance", Proc. Sixth Int. Conf. on Road Traffic Monitoring and Control, IEE, pp. 168-172, April 1992.
[2] G. Hoffmann and J. Janko, ``Travel times as a basic part of the LISB guidance strategy", Proc. Third Int. Conf. on Road Traffic Control, IEE, pp. 6-10, May 1990.
[3] T. Saito, J. Shima, H. Kanemitsu, Y. Tanaka, ``Automobile navigation system using beacon information", Proc. 1st International Conf. on Vehicle Navigation & Information Systems (VNIS), Toronto, Canada, pp. 139-144, 1989.
[4] K. Sakai, H. Fukushima, H. Fujii, ``Characteristics of drivers' route selection and survey system", Meeting on Road Traffic and Automobile, Institute of Electrical Engineers of Japan, RTA-92-23, 1992.
[5] M. Shibata, T. Matsumura, Y. Tanaka and N. Imacho, ``The economic effect of in-car navigation system", The 3rd International Conference on Vehicle Navigation & Information Systems (VNIS), Norway, Sept 2-4, 1992, pp. 632-637.
[6] P.W. Bonsall and T. Parry, ``Drivers' requirements for route guidance", Proc. Third Int. Conf. on Road Traffic Control, IEE, pp. 1-5, May 1990.
[7] P. K. Simpson, *Artificial Neural Systems*, Pergamon Press, 1990.
[8] B. Kosko, *Neural networks and Fuzzy Systems : a dynamical systems approach*, Prentice-Hall, 1992.

20

HIERARCHICAL POSITION CONTROL USING FUZZY LOGIC: OPPORTUNITIES AND LIMITATIONS

A. ZILOUCHIAN, T. JORDANIDES**
and D.W. HOWARD*
**Department of Electrical Engineering*
Florida, Atlantic University
Boca Raton, FL 33431, U.S.A.
*** Department of Electrical Engineering*
California State University
Long Beach, CA 90840, U.S.A.

1. Introduction

During the last four decades, most of controls problems have been formulated by the objective knowledge of the given systems (e.g. mathematical model). Many of these "model-based" approaches have found their way into practice and provided satisfactory solutions to spectrum of complex problems[1]. However, there are many knowledge based systems which cannot be merely described by the traditional mathematical representations [2]-[4]. Such relevant subjective knowledge of a system is often ignored by a designer at the front end, but often utilized in the last phase in order to evaluate the design methodology. In general, so called "knowledge based" approaches provide a framework for both type of these knowledge. These methods are much closer to human thinking than traditional classical approaches[3]-[6].

In a broad perspective, knowledge based approaches underlie what is called *"soft computing"*. In traditional hard computing, the prime objectives of the computations are precision and certainty. However, in soft computing, the precision and certainty carry a cost. Therefore, it is realistic to consider the integration of computation, approximate reasoning, and decision making as various partners in a consortium in order to provide a framework for the trade-off between precision and uncertainty. This integration of methodologies provide a foundation for the conceptual design and deployment of intelligent systems. The principal partners in such consortium are fuzzy logic, neural network computing, generic algorithms and probabilistic reasoning. Furthermore, these methodologies in most part are complimentary rather than competitive [5], [7].

In contrast to analytical methods, soft computing methodologies mimic consciousness and cognition in several important respects as outlined in the literature [3]-[7]. Increasingly, these approaches are also utilized in combination which are commonly

S.G. Tzafestas (ed.), Advances in Intelligent Systems, 211–222.

referred to as "hybrid systems" [7]-[9]. Hybrid intelligent systems are likely to play a critical role for many years to come.
The fuzzy logic has been the area of heated debate and much controversy during the last three decades. The first paper in fuzzy set theory, which is now considered to be the seminal paper of the subject was written by Zadeh[3], who is also considered as the founding father of the field. In that work, Zadeh was implicitly advancing the concept of human approximate reasoning to make effective decisions on the basis of available imprecise linguistic information. The first implementation of Zadeh's idea was accomplished by Mamdani [10], which demonstrated the viability of Fuzzy Logic Controller(FLC) for a small model steam engine. After this pioneer work, many consumer products and industrial applications using fuzzy technology have been developed and are currently available in Japan, U.S. and Europe. A survey of recent industrial applications of FLC may be found in [11]-[13].
A FLC utilizes the fuzzy logic in order to convert the linguistic information based on the expert knowledge into an automatic control strategy. In order to use the fuzzy logic for control purposes, a front-end "fuzzifier" and a rear-end "defuzzifier" are added to the usual input-output data set. A simple fuzzy logic controller contains four components: rules, fuzzifier, inference engine, and defuzzifier. Once the IF and THEN rules have been established, the controller can be implemented as a nonlinear mapping from the input to the output [4]- [6],[10].

In this chapter, the hierarchical position control of a robot manipulator using FLC is addressed. The control of a manipulator is hampered by complex kinematics and non-linear motion. The proposed solution in this chapter will solve these problems by using: (i) the simple forward kinematics equations to train a Fuzzy Associative Memory (FAM) to map the inverse kinematics solution, and (ii) test data from the DC motor to train the fuzzy controller. The detailed controller designed, as well as inverse kinematics simulations results can be found in [14]. Adaptive Neuro-Fuzzy Inference System (ANFIS)[9] is utilized for inverse kinematics solution as well as controller design of Microbot. ANFIS system generated by the fuzzy toolbox available in Matlab allows for the generation of a standard Sugeno style fuzzy inference system or a fuzzy inference system based on sub-clustering of the data [15].

The chapter is organized as follows. In section 2, solution of inverse kinematics using ANFIS is presented. In section 3, design of a conventional PID controller as well as a hierarchical controller structure for a Microbot[16] are discussed. Section 4 is pertaining to controller design of the robot with ANFIS. Finally, section 5 include the conclusions and remarks related to the proposed method.

2. Solution Of Inverse Kinematics Using ANFIS

In this method, a set of training samples are collected from measurements. This is done by moving the robot to different desired end effector positions (P_x, P_y, P_z), and measuring the corresponding angles. By utilization of such data an ANFIS is trained with a sufficient number of data points. The block diagram of the proposed method is shown is Figure 1:

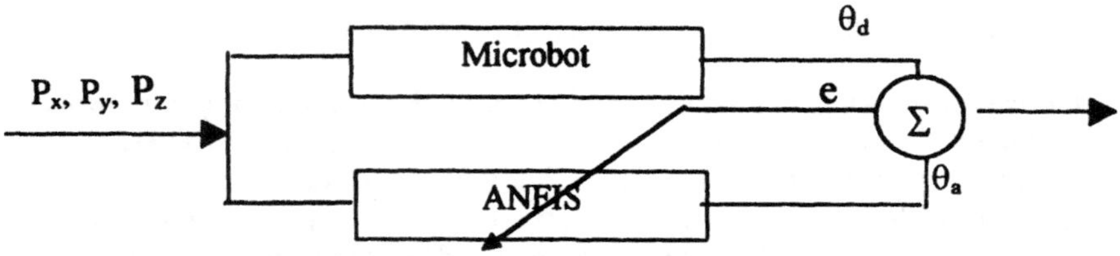

Figure 1. ANFIS Inverse Kinematics Block Diagram

In the training phase, the membership functions and the weights will be adjusted such that the required minimum error is satisfied. In sequel the trained ANFIS can be utilized in order to provide fast and acceptable solutions of the inverse kinematics of the Microbot for various applications, such as on-line control of the robot.
The method used to generate the training data is similar to using the Microbot manipulator to create it's own inverse kinematic solution. Instead of solving the inverse kinematic equation, various angles are applied to the robot and the resultant cartesian coordinates (P_x, P_y & P_z) would be determined. This output is directly related to the angles input, and is one solution to inverse kinematic equation. This method allows the creation of the inverse mapping based on actual information from the manipulator. For instance, in a Microbot the angle θ_1 is dependent on the X and Y only. This is confirmed by the physical configuration of the robot. In the simulation experiments for angle θ_1 of Microbot, the test data was created from all possible ranges of X and Y. similar procedures were followed for θ_2 and θ_3.
The following method was used to obtain the fuzzy mapping for joints angles. The first step was to allow the ANFIS algorithm to generate the membership functions and number of rules. Next, the ANFIS was trained with the test data and the resultant accuracy was determined. The initial fuzzy inference system resulted in a relatively high error in L2 norm sense which was not considered acceptable. In sequel, the membership functions were modified in order to obtained an accurate mapping. For detailed simulation results please see [14].

3. Hierarchical Position Controller Design

Robotic manipulator control is predominately motion control using classical servomechanism control theory [16], [17]. Due to the non-linearity of the manipulator motion a wide variety of control schemes have been devised. including computed torque, resolved motion, PID decoupled, model reference adaptive control to name a few [20]. Some of these schemes can be very complicated and require intensive computer resources. For instance, the computed torque technique uses the Lagrange-Euler equations of motion of the manipulator to determine the required torque to servo each joint in real time to track the desired trajectory [17]. On the other hand, model reference adaptive control requires an accurate reference model of the manipulator to compare against an adaptive mechanism to store the adjustable feedback gains. For comparison purposes, the simple PID decoupled control scheme will be used in this chapter.

3.1 Design of a Conventional Controller

In this section, a conventional controller for the Microbot is briefly described. The PID parameters for each individual joint were determined by considering each joint as uncoupled from the other, and calculating the approximate corresponding values. The values were then fine tuned for the optimum values. In a conventional controller, the designer would normally generate the required trajectory of the individual links based on the given end effector positions. Based on the physical limitations of the Microbot, the joints have a defined acceptable range. The base, θ_1, can rotate from -170° to 170°. The elbow joint θ_2 and the shoulder joint θ_3 can also be adjusted from -170° to 170°. However, due to the existence of the floor, the links and end effector have the additional constraint of not going below an elevation of zero. This is accounted for in the conventional controller by selecting the inverse kinematic solution of θ_2 to be the positive angle and using Matlab saturation modules.

To provide a comparison for the fuzzy controller, the Microbot manipulator will be controlled with a conventional proportional-derivative type controller. Based on the allowable ranges of robot joints, various position changes have been selected to test the fuzzy controller. These desired changes were simulated in Matlab and provide a wide range of change including multiple joint movements at one time. The plot of the simulation results for PID and fuzzy controllers are shown in section 4. The errors obtained from the conventional controller are also shown on Tables 1 and 2.

3.2 Fuzzy Controller Design

The use of fuzzy logic to aid in the control of robot manipulators has been studied by several researchers within the last five years[18]-[22]. In [18], [19], the control of robot manipulators using FL were investigated. In [20], a method to generate fuzzy mapping of the inverse kinematics equations for a robot arm is described. Furthermore, Kumbla and Jamshidi[21] evaluated the hierarchical control of a robotic manipulator using fuzzy logic which also included a fuzzy solution to the inverse kinematics equations. In [22], the implementation of fuzzy logic controller for the Remote Manipulator System(RMS) of the Space Shuttle was evaluated.

One of the advantages of Fuzzy Logic Control is that the controller can be designed in the same way as a person would think of doing the control. In the case of a robotic manipulator point to point or trajectory position control, one could think of an operator telling the end effector where to go by specifying the required joint angles. In sequel, each joint could individually move to the desired angle. Therefore, the proposed approach is a hierarchical control scheme. A fuzzy mapping of the inverse kinematics solution will supervise the individual joints [14]. A fuzzy controller will direct each individual joint to the desired position as shown in Figure 2.

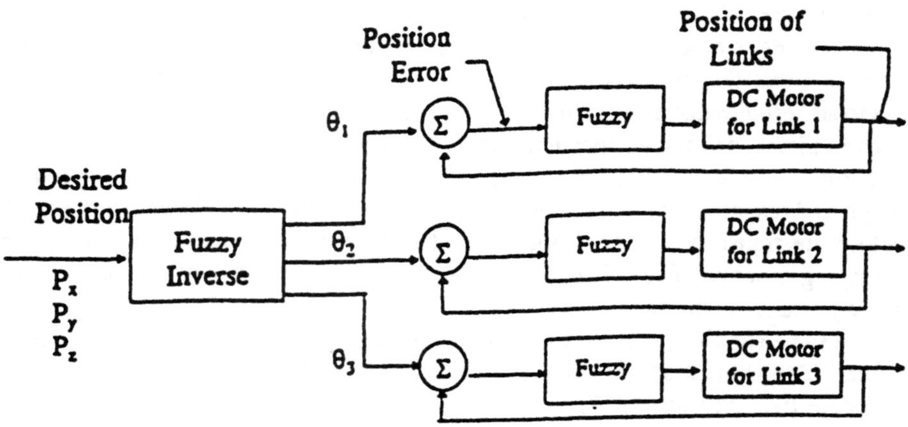

Figure 2 Hierarchical fuzzy controller

4. ANFIS Controller Design for Microbot

In this section, the ANFIS controller is proposed in conjunction with the switching curve for each joint of the robot in order to provide the desired end effector position of the Microbot. Due to the interaction and coupling between the links of the manipulator, applying a constant torque to each joint will simplify the control of the other joints. In addition, applying the maximum torque should produce the fastest response. To produce a fast response, the premise is to apply a maximum torque in the required direction, then at the last possible instance apply a braking torque to stop the motion. One method of determining when to apply the maximum torque is called the "switching curve".

The concept of sliding surface ("switching curve" or "variable structure") for nonlinear systems is well known and was originated by several Russian mathematicians (e.g. Aizerman and Gantmacher, Filippov). However, the application of the switching surface create chattering [23]. In order to smooth out such effect, the ANFIS algorithm has been utilized. For each joint of Microbot, the switching curve is a relation between the position error and the rate of change of position. The switching curve can be generated analytically if an accurate model of the motor and joint is known. However, in most of cases such model is unknown. The practical method is to determine the switching curve by using the actual manipulator. In this method, the maximum torque is applied to the manipulator and the resultant position error versus the rate of change of position is plotted. This plot is then inverted to obtain the switching curve as shown in Figure 3. For a DC motor application, for all conditions on one side of the curve, a maximum positive voltage is applied. The other side of the switching curve receives a maximum negative voltage. The switching curve is therefore a very steep transition surface that could induce instability or chattering in the controller. To reduce the possibility and effects of this chattering, a transition region is created. The transition region is essentially twice the distance that the manipulator can travel during one sampling period. The size of the transition region can be determined from the data that created the switching curve. The next step in the algorithm is to generate test data and produce the fuzzy equivalent of the switching curve for each joint. The resultant surfaces are shown in Figure 2. The switching curve is the basis of the proposed fuzzy controller, but to enable the FLC to handle a dynamic system with improved response, gains are applied to the inputs of the FLC. These gains essentially allow the transition region (not curve) to rotate and account for a fast or slowly changing process. The value of the gains will be determined using the functioning system and tuned on-line.

The same algorithm was used to create the fuzzy inference systems for the remaining joints. The simulation results for point to point control of Microbot (cases 1-3) compare the conventional PID controller to the fuzzy logic controller of the robot utilizing the switching curve surfaces.

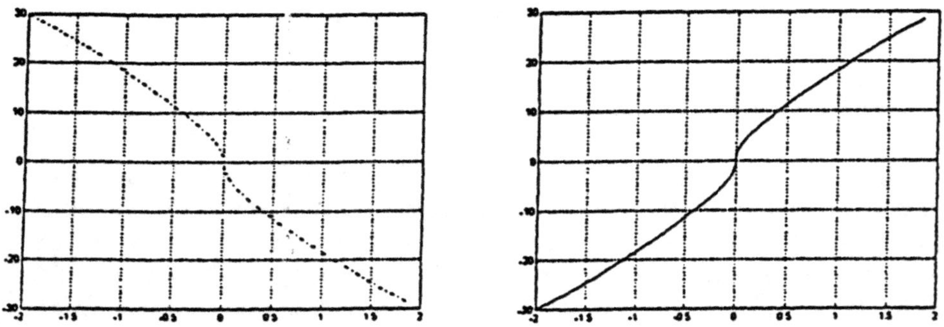

Figure 3-a. Switching curve for base joint and motor

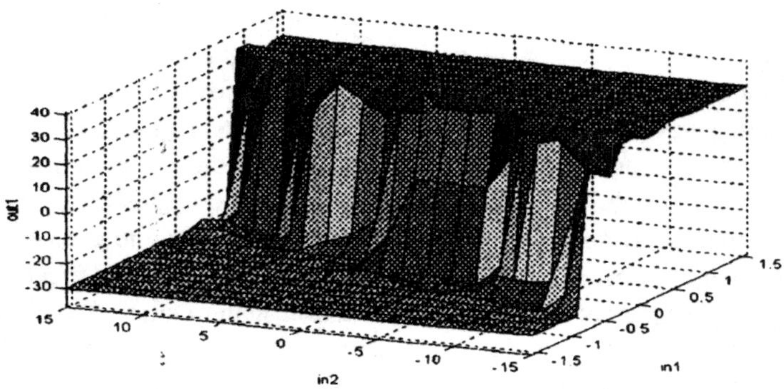

Figure 3-b. 3-D fuzzy switching surface for base joint and motor

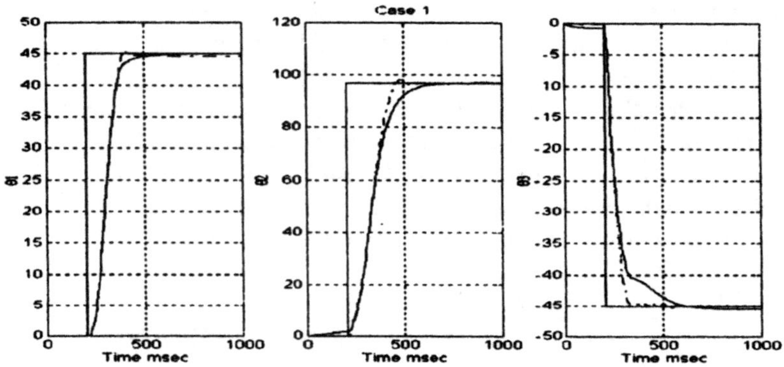

Figure 4. Simulation results- Case 1

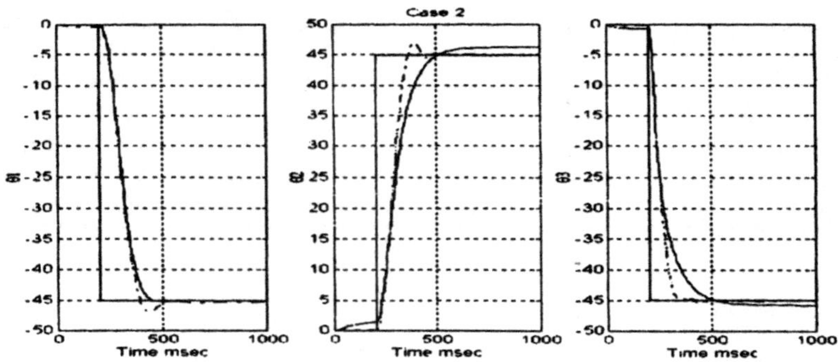

Figure 5. Simulation results- Case 2

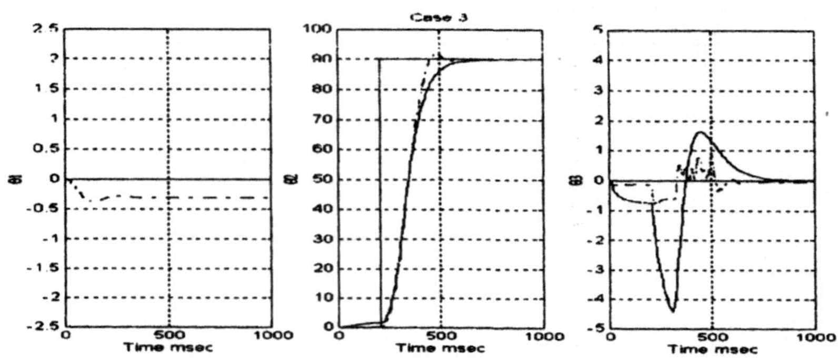

Figure 6. Simulation results- Case 3

The results of comparing the P-D controller with the fuzzy logic controller (shown in Figures 4 through 6) show that the FLC performs well . However, the simulation also displays that coupling exists between the elbow and the shoulder joints of the Microbot manipulator. One method of accounting for this coupling is to use a feedforward controller for each of the joints as described in [13].

To evaluate the FLC in terms of robustness and sensitivity, additional case studies were performed. The original case 1 was performed again with a 100% increase in mass of link 3, which was designated as case 4. Additionally, case 1 was performed again with a change in the time constant (25%) of the elbow joint motor, this was called case 5. Tables 1 and 2 also tabulates the results of these two cases

Table 1 Point to point control of the Microbot

	Desired final position			Actual final position with PID controller			Actual final position ith ANFIS controller		
Case	X	Y	Z	X	Y	Z	X	Y	Z
1	0.0	0.15	0.812	.1504	.0154	.816	.1517	.1521	.8095
2	0.433	-.433	0.412	.4296	-.4296	.4199	-4334	.4315	.4076
3	0	0	0.9	0.	0	.899	.0003	0	.9
4	0.15	0.15	0.812	.1504	.1504	.8102	.1534	.1518	.8091
5	0.15	0.15	0.812	.1499	.1499	.8111	.1525	.1509	.8101
6	0.0	0.5	0.5	0.0	.4972	.2056	.0016	.4943	.1933

Table 2 Position error (L2 norm)

	Desired final position			PID error	ANFIS error
Case	X	Y	Z		
1	0.15	0.15	0.812	.0048	.0044
2	0.433	-.433	0.412	.0125	.0063
3	0	0	0.9	0.009	.0003
4	0.15	0.15	0.812	.0023	.0058
5	0.15	0.15	0.812	.0011	.0039
6	0.0	0. 5	0.5	.0116	.0178

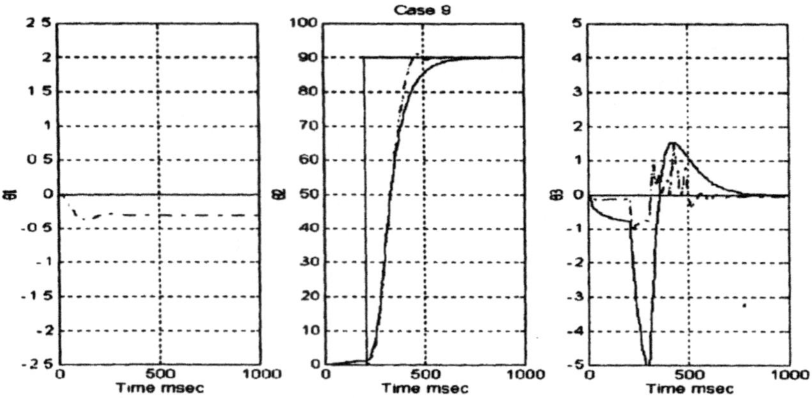

Figure 7. Simulation results - Case 4

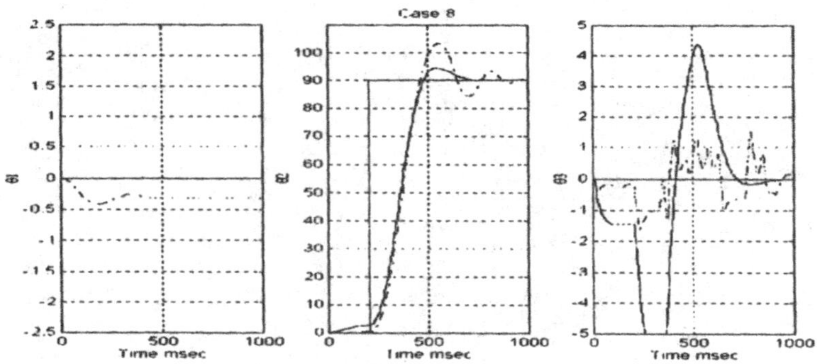

Figure 8. Simulation results - Case 5

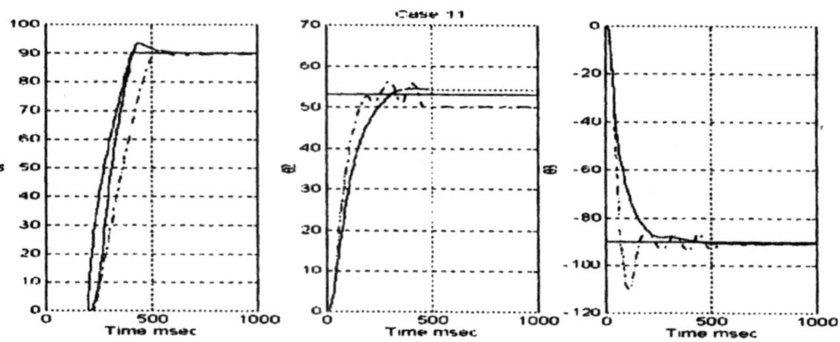

Figure 9. Case 6: Simulation results (angle plots)

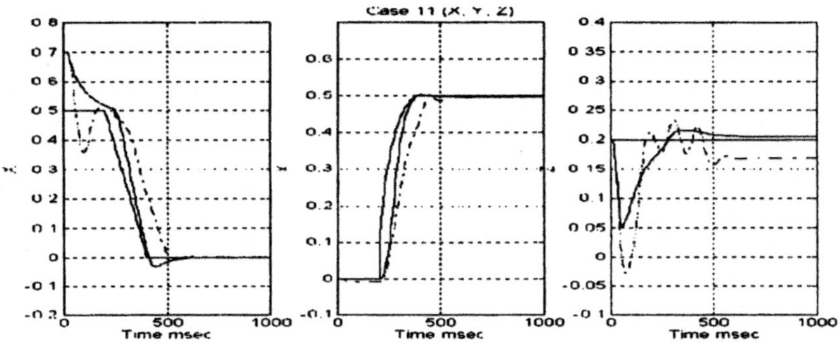

Figure10. Case 6: Simulation results (X, Y, Z plots)

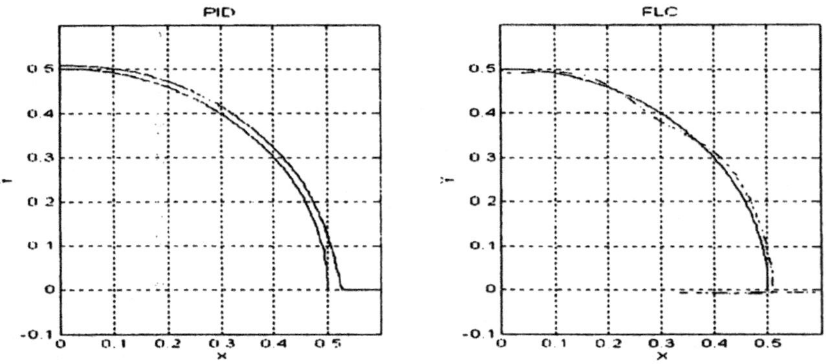

Figure 11. Case 6: P-D vs. FLC with fuzzy inverse (trajectory plot)

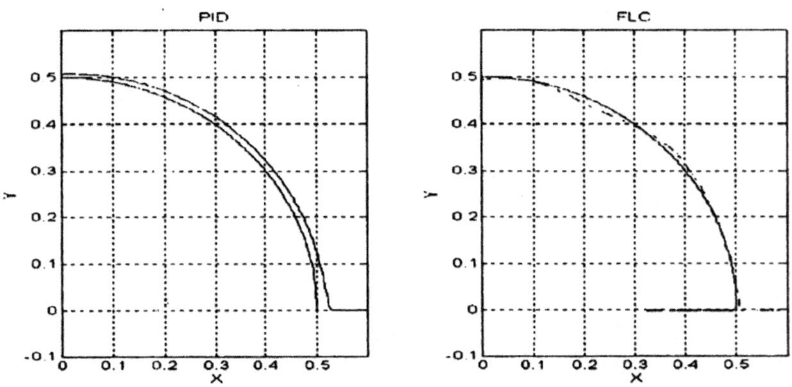

Figure 12. Case 6: P-D vs. FLC with calculated inverse (trajectory plot)

The cases 1 through 5 demonstrate the controller performance for the end effector point-to-point control of the Microbot. In order to determine the overall performance, case 6 was created using a planned trajectory. The desired trajectory in this case is the movement of the end effector from the home position (0.7, 0, 0.2) to an intermediate position (.5,0, .2), then to the final position (0, 0.5, 0.2) via a circular trajectory.

Case 6 simulated a simple trajectory following scenario, figure 11 displays the PID controller against the FLC with the fuzzy inverse kinematics. While Figure 11 a simulated the same trajectory except FLC utilized the same calculated fuzzy inverse kinematics as the P-D and figure 11 displays the results. Figure 11 shows that while the end point error is slightly higher during the transit, the FLC is actually superior to the P-D controller. Analysis of both figures (11 and 12) indicates the significant difference is the fuzzy inverse kinematics portion which could be solved analytically in order to increase the accuracy of the mapping.

5. Conclusions and Remarks

In this chapter, ANFIS was used to create a simple algorithm for the control of a Microbot with minimum theoretical modeling. The method was based on using the actual manipulator to create the supervisory portion (fuzzy inverse kinematics) and the controller (individual joint, switching curve FLC). Specifically, the implementation of an ANFIS to produce a working controller for a multiple degree-of-freedom robotic manipulator with essentially no modeling of the manipulator has been investigated. Finally, the intention of the chapter was to provide a simple algorithm to control a robotic manipulator with minimal or no modeling of the system. Hierarchical control was chosen since this method is very similar to the human thought process. ANFIS was chosen due to the adaptive nature and use of training data to create a fuzzy inference system. Overall the algorithm is: (i) build the desired manipulator, (ii) use the manipulator to generate training data for the mapping between the Cartesian space and the joint angle space, (iii) use the manipulator to create the switching curves for the control of the individual joints and (iv) fine tune the controller to achieve the desired performance. The simulation results related to kinematics is not shown here and can be found in [14]. However, the simulation results related to ANFIS and PID controller were shown.

References

[1] Antsaklis, P.J., and Passino, K.M. (editors): *An Introduction to Intelligent and Autonomous Control*, Kluwer Academic Publishers, Norwell, MA (1993).

[2] Zadeh, L.A.: A critical View of Our Research in Automatic Control : *IRE Trans. on Automatic Controls*, AC-7, pp. 74 (1962).

[3] Zadeh, L.A.: Making the Computers Think Like People, IEEE Spectrum (1994).

[4] Gupta, M. and Sinha, N.(editors): *Intelligent Control Systems: Theory and Applications*. IEEE Press, Picataway, N.J. (1996).

[5] Yager, R. and Zadeh,A. (editors): *An Introduction to Fuzzy Logic Applications in Intelligent Systems*, Kluwer Academic Publishers, Boston (1992).

[6] Kosko, B. : *Fuzzy Engineering*, Prentice Hall, Upper Saddle River, N.J (1997).

[7] Kosko, B. : *Neural Networks and Fuzzy Systems*, Prentice-Hall, Englewood Cliffs, N.J. (1991).

[8] Jang, J., Sun, C. and Mizutani, E. : *Neuro Fuzzy and Soft Computing*, Prentice Hall, Upper Saddle River, N.J. (1997).

[9] Jang, J. R. : Self-Learning Fuzzy Controllers based on Temporal Back Propagation, IEEE Transactions on Neural Networks, Vol. 3, No. 5, (1992).

[10] Mamdani, E. H. : Application of Fuzzy Algorithms for Control of Simple Dynamic Plant Proceeding of IEE, Vol. 121, No. 12 (1974).

[11] Surgeno, M. (editor): *Industrial Applications of Fuzzy Control*, North-Holland, Amsterdam (1985).

[12] Ralescu, A. (editor) : *Applied Research in Fuzzy Technology*, KluwerAcademic Publishers, Boston (1994).

[13] A. Zilouchian, F. Hamono and T. Jordanides: "Recent Trend and Industrial Applications of Intelligent Control System Using Artificial Neural Networks and Fuzzy Logic", in: Spyros Tzafestas (editor):*Method and Application of Intelligent Control*, Kluwer Academic Publishers, Boston (1997).

[14] Howard, D. and Zilouchian, A.: Application of Fuzzy Logic for the Solution of Inverse Kinematics and Hierarchical Controls of Robotic Manipulators, International Journal of Robotic and Intelligent (1998).

[15] *Fuzzy Logic Toolbox User's Guide*, MathWorks, Inc, MA (1997).

[16] Wolovich, W. A., *Robotics:* Basic Analysis and Design, Holt, Rinehart and Winston, New York (1987).

[17] F.L. Lewis, C.T. Abdallah and D.M. Dawson, *Control of Robot Manipulator* Macmillan, NY (1993).

[18] Lim, C. M. and Hiyama, T. : Application of Fuzzy Logic Control to a Manipulator, IEEE Trans. on Robotics and Automation, Vol. 7, No. 5 (1991).

[19] Nianzui, Z., Ruhui, Z., Maoji, F.: Fuzzy Control Used in Robotic Arm Position Control, IEEE International Conference on Fuzzy Systems (1994).

[20] Kumbla, K. K. and Jamshidi, M.: Control of Robotic Manipulator Using Fuzzy Logic, Proceeding of IEEE International Conf. on Fuzzy Logic (1994).

[21] Lea, R. N., Hoblit, J., Yashvant, J. : Fuzzy Logic Based Robotic Arm Control. In: Mark, R.(ed.) *Fuzzy Logic Technology and Application*, (1994).

[22] Martinez, J., Bowles, J., Mills, P.: A Fuzzy Logic Positioning System for an Articulated Robot Arm, IEEE Int.l Conference on Fuzzy Systems, (1996).

[23] Slotine,J.J. and W. Li: *Applied Nonlinear Control* , Prentice-Hall, Englewood Cliffs, New Jersey (1991).

21

IMPEDANCE CONTROL FOR MULTI-DOF MANIPULATOR BY LEARNING ENVIRONMENT MODELS

F. NAGATA*, K. WATANABE**, K. SATO**
and K. IZUMI***
**Interior Design Research Inst., Fukuoka Ind. Tech. Center*
Agemaki – 405-3, Ohkawa-shi, Fukuoka 831-0031, Japan
*** Department of Advanced Systems Control Engineering*
Saga University, 1-Honjomachi, Saga 840-8502, Japan
**** Dept.Mech. Engrg.,Faculty of Science and Engineering*
Saga University, 1-Honjomachi, Saga 840-8502, Japan

1. Introduction

Problems to be resolved in applying an impedance control[1-4] to an industrial robot are how to simply realize mechanical impedance on the end-effector of the robot using its servo system and to design the desired compliance in a simple manner. To solve these problems, a position-based impedance control for an industrial robot with an open architecture controller was proposed. Moreover, we proposed a tuning method which generated the desired time-varying compliance to be in contact with an object at critical damping, according to the changes of the inertia and Jacobian matrices of the manipulator. It was shown that the proposed method was very effective, if the physical parameters, such as stiffness and viscosity, of the environments were known. On the contrary, in a practical application, the physical parameters of the environment can not be known correctly. Therefore, we proposed a position-based impedance control using fuzzy environment models[5-6]. When used this method, the sufficient performance for the force control can be realized even if the environments are not known precisely. The fuzzy environment models estimate the each directional stiffness of object by using force sensor measurements and yield the desired damping for a stable response. Using the desired damping generated by the fuzzy environment models, the position-based impedance control can achieve the contact force suppressing overshoots and oscillations. It should be noted, however, that the ability of the fuzzy environment models is determined by the construction of the fuzzy rules, and they must be suitably tuned corresponding to the each task and environment. In general, the suitable design of the fuzzy rules depends on the heuristic knowledge. In this chapter, we present an automatic tuning method of the fuzzy rules using genetic algorithms(GA)[7-10]. Simulation results show that the proposed methods are very effective even if the environment has not been learned and has different characteristics in each direction.

S.G. Tzafestas (ed.), Advances in Intelligent Systems, 223–232.

2. Position-Based Impedance Control

2.1 Proposition of Position-Based Impedance Control

It is well-known that the desired impedance equation for a robot manipulator can be represented by

$$M_x(\theta)\ddot{x}+B_d(\dot{x}-\dot{x}_d)+K_d(x-x_d) = SF+(E-S)K_f(F-F_d) \tag{1}$$

where x, $\dot{x}$ and $\ddot{x}$ are the position, velocity and acceleration vectors, respectively. $M_x(\theta)$ is the inertia matrix, F is the force-moment vector acting on the end-effector defined by $F^T=[f^T\ n^T]$, where f and n are the force and moment vectors, respectively. K_f is the force feedback gain matrix. x_d, $\dot{x}_d$ and $F_d^T=[f_d^T\ n_d^T]$ are the desired position, velocity and force vectors; B_d and K_d are the coefficient matrices of desired damping and stiffness, respectively. S and E are the switch matrix diag $(S_1, \ldots, S_6)$ and identity matrix. It is assumed that K_f, B_d and K_d are positive-definite diagonal matrices. It should be noted that eq. (1) has no switch matrix S applied to the desired stiffness in Hybrid Compliance/Force Control(HCC)[1-2]. We consider here a position control strategy in the direction for force control, if the object dynamics is known. Note that if $S = E$, then eq. (1) becomes a compliance control system in all directions; on the other hand if $S = 0$, it becomes a force control system in all directions. The manipulator dynamic model in Cartesian space is expressed by

$$M_x(\theta)\ddot{x} + H_x(\theta,\dot{\theta}) + G_x(\theta) = J^{-T}(\theta)\,\tau + F \tag{2}$$

where θ is the vector of generalized joint coordinates describing the pose of the manipulator, $\dot{\theta}$ is the vector of joint velocity, $H_x(\theta,\dot{\theta})$ is the Coriolis and centrifugal forces in Cartesian space, $G_x(\theta)$ is the gravity term, $J(\theta)$ is the Jacobian matrix, and τ is the joint driving torque. We assume that the displacement of eq. (1) is very small and the inertia term can be ignored. Defining $X := x - x_d$, then eq. (1) can be rewritten as

$$\dot{X} = AX + Bu \tag{3}$$

where

$$A := -B_d^{-1} K_d,\ B := B_d^{-1}$$
$$u := \{SF + (E-S)K_f(F-F_d)\}$$

In general, eq. (3) can be solved as

$$X = e^{At} X(0) + \int_0^t e^{A(t-\tau)} B\, u(\tau)d\tau \tag{4}$$

In the following, we will consider the form in the discrete time k using a sampling time Δt. It is assumed that u, A and B are constant at $\Delta t(k-1) \le t < \Delta t\, k$. Defining $X(k)=X(t)|_{t=\Delta t k}$ $(k=0,1,\cdots)$, then eq. (4) is transformed into the discrete time domain description:

$$X(k) = e^{A\Delta t k}\left(X(0) + \int_0^{\Delta t k} e^{-A\tau} B\, u(\tau)d\tau \right)$$

$$= e^{A\Delta t} X(k-1) + e^{A\Delta t k} \int_{\Delta t(k-1)}^{\Delta t k} e^{-A\tau}\, d\tau\, B\, u(k)$$

$$= e^{A\Delta t} X(k-1) + \left(e^{A\Delta t} - E\right) A^{-1} B\, u(k) \tag{5}$$

or alternatively, $X(k) = x(k) - x_d(k)$ leads to

$$x(k) = x_d(k) + e^{-(B_d^{-1} K_d)\Delta t} \{x(k-1) - x_d(k-1)\}$$
$$-\{e^{-(B_d^{-1} K_d)\Delta t} - E\}\, K_d^{-1} \{SF + (E - S)\, K_f\, (F - F_d)\} \tag{6}$$

2.2 A Position-Based Impedance Control Considering the Dynamics of an Environment

Figure 1 shows that the end-effector is made to contact with an environment: it is tried to control the contact force from the vertical direction of the slope to converge to the reference ${}^S f_d$, in which the left over superscription S means the value in sensor coordinates system. The sensor coordinates system is given by rotating the base coordinates system with Z-Y-Z Euler angles ($\theta, \varphi, 0$). In this case, the rotation matrix ${}^B R_s$ in sensor coordinates system based on base coordinates system is expressed by

$${}^B R_s = \begin{pmatrix} \cos(\theta) & \sin(\theta) & 0 \\ -\sin(\theta) & \cos(\theta) & 0 \\ 0 & 0 & 1 \end{pmatrix} \times \begin{pmatrix} \cos(\varphi) & 0 & -\sin(\varphi) \\ 0 & 1 & 0 \\ \sin(\varphi) & 0 & \cos(\varphi) \end{pmatrix} \tag{7}$$

Then, the desired force vector f_d in base coordinates system is obtained by

$$f_d = {}^B R_S\, {}^S f_d \tag{8}$$

For example, ${}^s f_d^T = [\ 0\ 0\ {}^s f_{dz}\ ]$ is transformed as follows:

$$f_d^T = {}^s f_{dz} [-\cos(\theta)\sin(\varphi)\ \ \sin(\theta)\sin(\varphi)\ \ \cos(\varphi)] \tag{9}$$

It is assumed that the contact force F acting on the end-effector is represented by

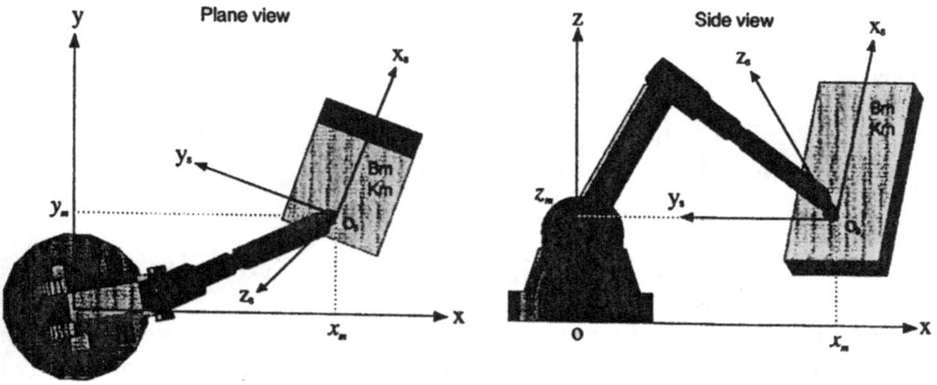

Fig. 1 Plane and side views of a manipulator in force control

$$F = -B_m \dot{x} - K_m (x - x_m) \quad (10)$$

where B_m and K_m are the damping and stiffness coefficients of the object to be positive-definite diagonal matrices, and $x_m^T=[x_m\ y_m\ z_m]$ is the position of the object. The desired damping $\tilde{B}_{di}$ considering the critical damping condition[11] of eq. (1) is given by

$$\tilde{B}_{di} = 2\sqrt{(J^{-T}MJ^{-1})_i\left[K_{di} + \left\{S_i K_{mi} + (E-S)_i K_{fi} K_{mi}\right\}\right]} - \left\{S_i B_{mi} + (E-S)_i K_{fi} B_{mi}\right\} \quad (11)$$

Here, i ($i = 1,\ldots,6$) denotes the i-th diagonal element of each matrix and M is the inertia term in the joint space. In our proposed position-based impedance control, $\tilde{B}_{di}$ is used as a reference of the desired damping so that sufficient performance for the force control can be realized without overshoots and oscillations. It should be noted that in this method, the physical parameters of the environment must be known to calculate $\tilde{B}_{di}$.

3. Fuzzy Environment Models Learned with a Genetic Algorithm

It is known that when the environment is hard, the stiffness of environment has an important effect on the response of the force control, while the viscosity has a minor effect on the stability[12]. By this reason, in order to realize a stable force control without overshoots and oscillations, it is enough to estimate the stiffness of environment. Therefore, we propose the fuzzy environment models (hereafter, it is described as 'FEMs') that can estimate the stiffness of environment in each direction.

3.1 Fuzzy Environment Models (FEMs)

The FEMs estimate the stiffness of environment using the simple reasoning method with single input/output. In this section, it is described on how to construct the FEMs giving an example of force control issue shown in **Fig. 1**. The position vector of the end-effector and the force vector acted on the end-effector are $f^T=[f_x(t)\ f_y(t)\ f_z(t)]$ and $x^T=[x(t)\ y(t)\ z(t)\ \phi(t)\ \theta(t)\ \psi(t)]$ respectively. It is assumed that each reasoning in the x-, y- or z-direction is conducted independently, in which each input for the reasoning is defined by

$$x^*(t) \triangleq f_x(t)/(x_m - x(t)) \quad (12)$$

$$y^*(t) \triangleq f_y(t)/(y_m - y(t)) \quad (13)$$

$$z^*(t) \triangleq f_z(t)/(z_m - z(t)) \quad (14)$$

Following the fuzzy model approach; the environment models are described by

Rule 1 IF $x^*(t)$ is $\tilde{A}_1$ THEN $B_m^1\dot{x} + K_m^1(x - x_m) = -F$

Rule 2 IF $x^*(t)$ is $\tilde{A}_2$ THEN $B_m^2\dot{x} + K_m^2(x - x_m) = -F$

⋮

Rule L IF $x^*(t)$ is $\tilde{A}_L$ THEN $B_m^L\dot{x} + K_m^L(x - x_m) = -F$

if the information on only x-direction is used, where $\tilde{A}_i$ ($i = 1,\ldots, L$) is the antecedent fuzzy set and L is the label number. If the x-directional element of stiffness coefficient matrix K_m, K_{m1}, is estimated in parallel with above fuzzy environment models, the reasoning rules can be obtained by

Rule 1	IF	$x^*(t)$ is $\tilde{A}_1$	THEN	$K_{m1} = K_{m11}$
Rule 2	IF	$x^*(t)$ is $\tilde{A}_2$	THEN	$K_{m1} = K_{m12}$
		⋮		
Rule L	IF	$x^*(t)$ is $\tilde{A}_L$	THEN	$K_{m1} = K_{m1L}$

where K_{m1i} ($i = 1,\ldots, L$) denotes the consequent constant. Using these rules, the estimated value $\hat{K}_{m1}$ in the x-direction is computed by the centroidal method expressed as

$$\hat{K}_{m1} = \sum_{i=1}^{L} p_i\left(x^*(t)\right) K_{m1i} \tag{15}$$

where $p_i\left(x^*(t)\right)$ denotes the normalized confidence given by

$$p_i\left(x^*(t)\right) = \frac{\mu_{\tilde{A}_i}\left(x^*(t)\right)}{\sum_{j=1}^{L} \mu_{\tilde{A}_j}\left(x^*(t)\right)} \tag{16}$$

with the antecedent fitness $\mu_{\tilde{A}_j}\left(x^*(t)\right)$ described by the following Gaussian membership function

$$\mu_{\tilde{A}_j}\left(x^*(t)\right) = \exp\left\{\log(0.5)\left(x^*(t) - \alpha_j\right)^2 \beta_j^{\,2}\right\} \tag{17}$$

in which α_j is the center of membership function and β_j is the reciprocal value of standard deviation. In the same manner, the estimated values $\hat{K}_{m2}$ and $\hat{K}_{m3}$ in the y- and z-direction can be computed by giving eq. (13) and eq. (14) to the input of fuzzy reasoning, respectively. **Figure 2** shows the block diagram of position-based impedance control with the distinct FEMs in each direction. The FEMs estimate the stiffness of environment using the force sensor measurements and provide the desired damping $\tilde{B}_d$ to maintain the system in the critical damping. The position-based impedance control generates the reference $x(k)$ for the servo system with $\tilde{B}_d$.

As can be seen from the above arguments, the key for the design of the FEMs is to suitably determine the design parameters α_j, β_j and γ_n, in which γ_j denotes the consequent constant. However, the appropriate design method for such parameters has not been established. This means that if we want to use the FEMs then we must adjust the parameters by trails and errors through many simulations or experiments. Therefore, in the remainder of this section, we propose a tuning method for the FEMs using a GA concept. It is expected that the design parameters are automatically tuned, if the GA is applied successfully.

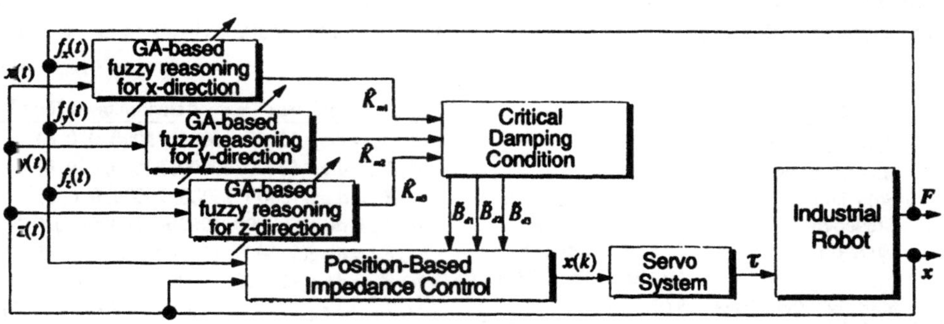

Fig. 2 Position-based impedance control with fuzzy environment models

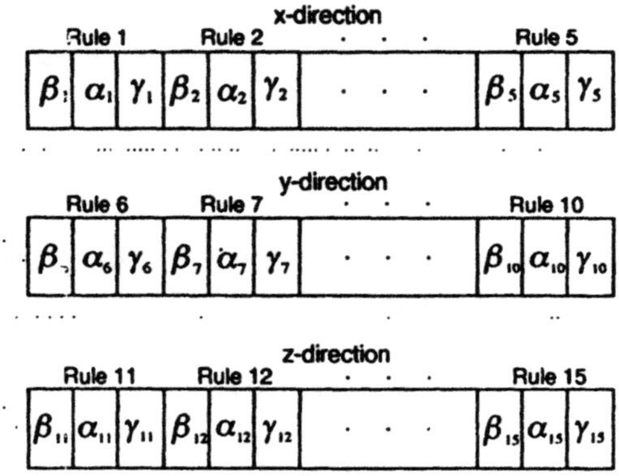

Fig. 3 Fuzzy rules encoded into an individual

3.2 The Learning of FEMs with a Genetic Algorithm

The phenotypes of fuzzy rule parameters are α_i, β_i and γ_i, and their genotypes are obtained by transforming each phenotype to 16 bits binary code. **Figure 3** shows that an individual is composed of 15 rules($L = 5$) and it has 720 bits. Each individual is evaluated by applying the proposed method shown in **Fig. 2** to the force control issue given in **Fig. 1**. However, we substitute $-\pi/6$ and $\pi/6$ into θ and φ with right-hand rule respectively and give [0 0 1] into ${}^s f_d^T$. An evaluation E_v of each individual is the sum of error between the force f acting on the end-effector and the reference f_d at every sampling time Δt, which is given by

$$E_v = \min\left\{ \sum_{k=1}^{T/\Delta t} \left(\left| f(k\Delta t) - f_d \right| \right) \right\} \tag{18}$$

where T is the total simulation time. This is a minimization problem and therefore if the individual has the minimum fitness then it can survive as an elite. When an overshoot, oscillation or a noncontact state occurs, a large dummy fitness(penalty) is given to the individual to be excluded, because such an individual is undesirable in the force control. The parameters used in the GA operations are tabulated in **Table 1**. Additionally, we have substituted a relatively larger value (i.e. from 1/240 to 1/10) into the mutation rate so that the population can keep the diversity of search space through generation replacements and the appearance of premature convergence can be suppressed. To learn the FEMs, the eight kinds of environments shown in **Table 2** were selected. Because of the aforementioned reason, the viscosity of the environments was fixed as $\boldsymbol{B}_e = \mathrm{diag}(15, 15, 15, 0, 0, 0)$ [Ns/m]. In addition, it was assumed that the stiffness and viscosity of the environments have the same characteristics in x-, y- and z-

Table 1 Parameters for GA operations

Population size	60
Number of elite	6
Selection	Tournament selection
Crossover	Uniform crossover
Mutation	Random mutation
	(1/240 < rate < 1/10)
Maximum generation	470

Table 2 Stiffness of learned environments [N/m]

(1) $\boldsymbol{K}_e = \mathrm{diag}(25000, 25000, 25000, 0, 0, 0)$
(2) $\boldsymbol{K}_e = \mathrm{diag}(50000, 50000, 50000, 0, 0, 0)$
(3) $\boldsymbol{K}_e = \mathrm{diag}(75000, 75000, 75000, 0, 0, 0)$
(4) $\boldsymbol{K}_e = \mathrm{diag}(100000, 100000, 100000, 0, 0, 0)$
(5) $\boldsymbol{K}_e = \mathrm{diag}(125000, 125000, 125000, 0, 0, 0)$
(6) $\boldsymbol{K}_e = \mathrm{diag}(150000, 150000, 150000, 0, 0, 0)$
(7) $\boldsymbol{K}_e = \mathrm{diag}(175000, 175000, 175000, 0, 0, 0)$
(8) $\boldsymbol{K}_e = \mathrm{diag}(200000, 200000, 200000, 0, 0, 0)$

directions, respectively. In the simulations, we set up the manipulator as the PUMA560[13] and its dynamics was solved by Runge-Kutta method. In this case, a resolved acceleration control was used in the servo system shown in **Fig. 2**. The control law is represented by

$$\tau = M(\theta)\, J^{-1}(\theta) \left\{ \ddot{x}_d + K_v \left(\dot{x}_d - \dot{x} \right) + K_p \left(x_d - x \right) \right\} + H_x \left(\theta, \dot{\theta} \right) + G_x (\theta) \tag{19}$$

where, K_v and K_p are the gains of velocity and position, respectively.

3.4 The Learnings of FEMs

Figure 4 shows the relation between the stiffness K_{e1} (or K_{e2}, K_{e3}) of environment and the estimated stiffness $\hat{K}_{e1}$ (or $\hat{K}_{e2}$, $\hat{K}_{e3}$) in steady state which has been obtained from the results using eight elite individuals. **Figure 5** shows the relation between the stiffness of environment and the desired damping given from (11) with $\hat{K}_e$. It is observed that there is a proportional relation between the real stiffness and the desired damping coefficient. At present stage, eight FEMs exist as many as the learned environments. Of course, it is not guaranteed that the desired performance can be achieved except for the learned environments. The most important function required for the FEMs is that the desired damping should be generated for the stiffness within assumed range. Therefore, in the following section it is considered on how to use the learned FEMs.

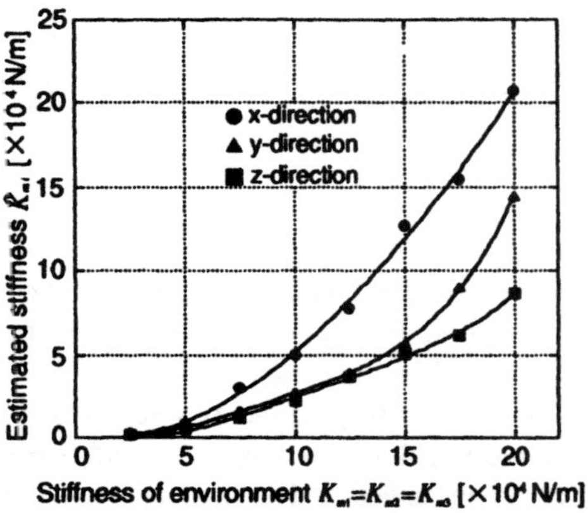

Fig. 4 Estimated stiffness for each direction

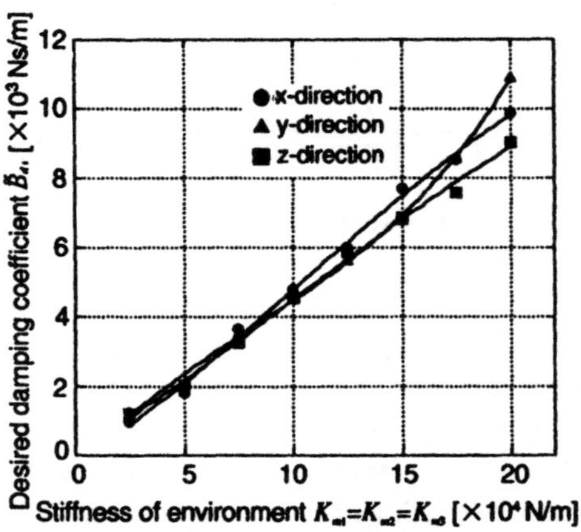

Fig. 5 The relation between stiffness of environment and desired damping coefficient

4. The Generalization of FEMs

Each FEMs which has been learned in section 3 has five fuzzy rules. If those are simply combined, then the fuzzy rules become 40 and the antecedent membership functions are placed on the top of one another on the support set. Using the combined FEMs, it might be happened that a suitable force control can not be realized even under the learned environment. The reason is that the fuzzy rules influence to each other. Therefore, we propose a design method which can make the FEMs work well for any environments if only the stiffness is within the range(e.g. from 25000 to 200000) shown in **Fig. 4**. The design method is that the antecedent membership functions as many as the learned FEMs are placed within the range with the same interval and the consequent constants are set by the estimated values shown in **Fig. 4**. In the following, the design method is concretely described.

First of all, the antecedent membership functions are composed of eight Gaussian membership functions ($L = 8$). K_{e1}(or K_{e2}, K_{e3}) shown in **Table 2** (1)-(8) are used as the center of each membership function so that they are allocated on the set [25000, 200000] with an equal interval. Furthermore, we give 12500 to the standard deviation in order that membership functions disposed side by side can intersect at the fitness 0.5. When the consequent part is designed, consequent models as many as the antecedent labels are prepared. The estimated stiffness coefficients, which are obtained by giving the center of the antecedent membership functions to the horizontal value shown in **Fig. 4**, are used for the consequent constants. **Fig. 6** and **Table 3** show the designed antecedent membership functions and consequent constants, respectively.

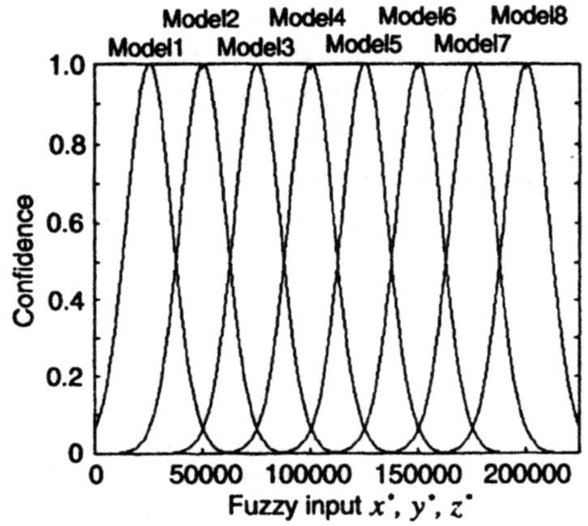

Fig. 6 Designed antecedent membership functions

Table 3 Consequent constants for each direction

	x-direction	y-direction	z-direction
Model1	2622	2392	1823
Model2	7885	5718	4920
Model3	29936	15837	12080
Model4	50134	27072	22349
Model5	84504	74442	33219
Model6	163440	101300	56890
Model7	178020	185030	71230
Model8	209780	279670	85700

Next, the property of robustness is confirmed for the position-based impedance control with the designed FEMs. Here, the force control simulation shown in **Fig. 1** is examined under the condition of K_e=diag(190000, 110000, 30000, 0, 0, 0). The force control result is shown in **Fig. 7**. The changes of the estimated stiffness and desired damping coefficient are shown in **Fig. 8** and **Fig. 9**, respectively. It is found that the each directional force acting on the end-effector desirably follows the changes of the reference without overshoots and oscillations. From these results, it is recognized that the proposed method is very effective even if the environment has not been learned and has different characteristics in each direction.

5. Conclusions

The conventional impedance control strategies is unpractical to generate the desired damping for the stable force control. To cope with this problem, we have proposed a position-based impedance control using fuzzy environment models learned with a genetic algorithm. Using the fuzzy environment models, it has been possible to estimate the each directional stiffness of environment and to yield the desired damping removing the overshoots and oscillations. Simulation results have shown that the proposed method works appropriately to the unknown and unlearned environment, which has different characteristics in each direction, by generalizing the fuzzy environment models learned under the several fixed environments.

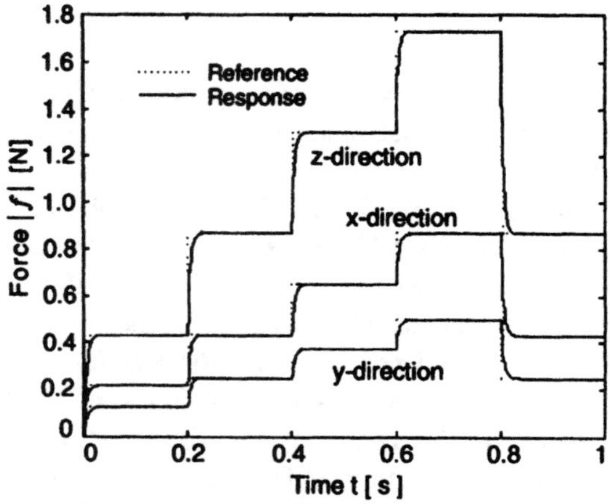

Fig. 7 Force control results for an unlearned environment

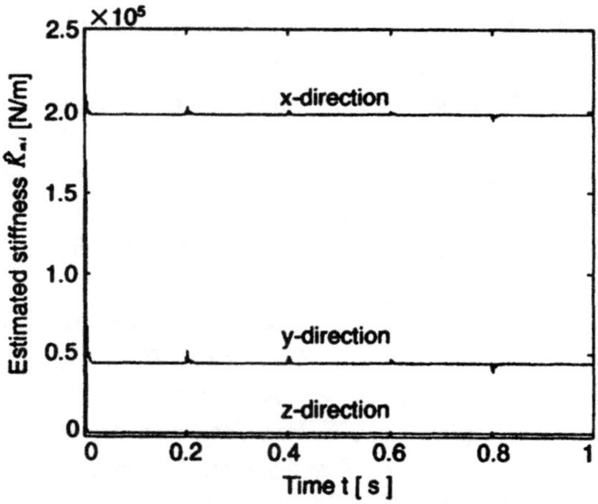

Fig. 8 Estimeted stiffness generated by FEMs

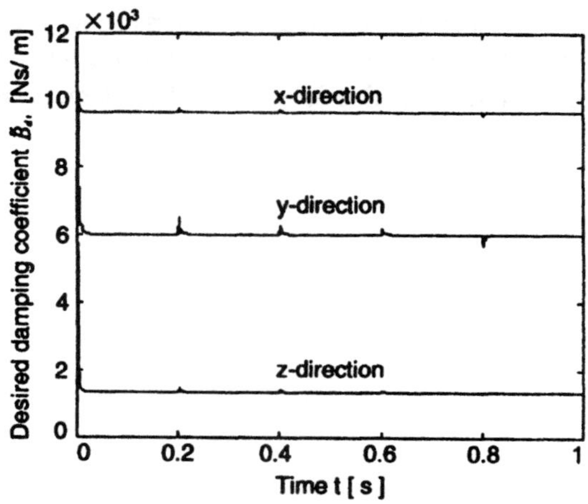

Fig. 9 Desired damping with estimated stiffness

6. References

1) Katsuragawa T., Ioi K., Kubota N. and Noro O., Application of Hybrid Compliance/Force Control for Industrial Robot, *Journal of the Robotics Society of Japan*, **12**- 6, 893- 898 (1994) (in Japanese)
2) Katsuragawa T., Yasumoto F., Ujiie H., Itoko T., Ogimoto K., Hirai S., Wakita Y. and Machida K., Application of Hybrid Compliance/Force Control to Super Long Distance Teleoperation, *The International Journal of the Robotics Society of Japan*, **11**-2, 199-212 (1997)
3) Sasaki T. and Tachi S., Contact Stability Analysis on Some Impedance Control Methods, *Journal of the Robotics Society of Japan*, **12**-3, 489-496 (1994) (in Japanese)
4) Sasaki T. and Tachi S., A Stabilizing Method of Contact Tasks on Servo-based Impedance Control, *Journal of the Robotics Society of Japan*, **13**-5, 659-665 (1995) (in Japanese)
5) Watanabe K., Ohtsubo T., Force Control of One Degree-of-Freedom Manipulators Using Fuzzy Impedance Models, *Procs. of the 15th SICE Kyushu Branch Annual Conference*, 189-192 (1996) (in Japanese)
6) J. R. Layne and K. M. Passino: Fuzzy Model Reference Learning Control, *Journal of Intelligent and Fuzzy Systems*, **4**, 33-47 (1996)
7) Furuhashi T., Nakaoka K., Maeda H. and Uchikawa Y., A Proposal of Genetic Algorithm with a Local Improvement Mechanism and Finding of Fuzzy Rules, *Journal of Japan Society for Fuzzy Theory and Systems*, **7**-5, 978-987 (1995) (in Japanese)
8) Watanabe K., Izumi K., Akama S., Learning of Fuzzy Environment Model for an Impedance control using a GA, *Procs. of the 16th SICE Kyushu Branch Annual Conference*, 175-178 (1997) (in Japanese)
9) D. A. Linkens and H.O. Nyongesa: Genetic Algorithms for Fuzzy Control Part1: Offline System Development and Application, *IEE Proc. Control Theory Appl.*, **142**-3, 161-176 (1995)
10) D. A. Linkens and H.O. Nyongesa: Genetic Algorithms for Fuzzy Control Part2: Online System Development and Application, *IEE Proc. Control Theory Appl.*, **142**-3, 177-185 (1995)
11) Nagata F., Watanabe K., Sato K., Izumi K. and Suehiro T., Impedance Control for Articulated Robot of 6 Degree-of-Freedom in Consideration of Critically Damped Condition with an Object Dynamics, *Procs. of the 36th SICE Annual Conference*, 1119-1124 (1997)
12) H. C. An and J.M. Hollerbach, Dynamic Stability Issues in Force Control of Manipulators, *Procs. of IEEE International Conference on Robotics and Automation*, 890-896 (1987)
13) Corke P., A Robotics Toolbox for MATLAB, *IEEE Robotics & Automation Magazine*, **3**-1, 24-32, March (1996)

PART V

COMPUTATIONAL INTELLIGENCE IN ENGINEERING APPLICATIONS

22

ENGINEERING APPLICATIONS OF N-STEP-AHEAD NEUROCONTROL

A. HOUNTRAS, I. ANTONIADIS and A.KANARACHOS
Mechanical Design and Control Systems Division
Department of Mechanical Engineering
National Technical University of Athens
Zografou Campus, P.O. Box 64078, 15710 Athens, Greece

1. INTRODUCTION

The conventional methods for the control of dynamic systems, despite their general success, often face problems, such as the requirement for a-priori knowledge of the system's model and the difficulties in handling strong system non-linearities. Furthermore, industrial robot controllers mostly use reference models, that do not always exploit the potential of the system [1-2]. Neural Networks (NN), which can model, identify and control nearly any dynamic system, offer an alternative approach [3]. Typically, neurocontrol methods use the system output errors to train the neural controller, since the output errors of the controller are not directly accessible ('Credit Assignment Problem') [4]. A possible solution are Neural Emulators (NE), which provide Credit Assignment data through back-propagation calculations [5]. Usual neurocontrol methods develop One-Step-Ahead controllers by using for training information based only on the most recent time-step. However, since the short-term optimal control targets do not necessarily coincide with the

Fig. 1: Cartesian Hydraulic Manipulator

S.G. Tzafestas (ed.), Advances in Intelligent Systems, 235–246.

long-term optimal ones, N-step-ahead controllers would be more suitable [2]. Only few N-step-ahead controllers are known, such as [6] and the Truck-Backer-Upper example in [7], but they are implemented in simulated toy-cases only.

The main target of the present chapter is to establish an N-Step-Ahead neurocontrol method called 'Chain-Back-Propagation' (CBP) and to validate its performance in complex real-world applications [1]. The CBP method considers the dynamic system a Black-Box and models it with a NE. It solves the Credit Assignment problem in real-time and no "teacher" controller is needed. Various penalty functions , which relate to the *system orbit as a whole*, are used for training purposes, such as Final Error, Overshoot and Rise Time. Training is further assisted by local-in-time penalty terms and simple transformations, that improve the convergence. Only set-points are used, but the resulting controller is efficient even in continuous reference orbits.

The CBP method is demonstrated in two applications: i) a 2-axes articulated manipulator and ii) a *real-world* cartesian 3-axes heavy duty hydraulic manipulator (fig. 1) [1]. Both systems are highly non-linear and their axes of motion are strongly coupled. The CBP method is compared to hard-limited PD, extended PD and One-Step-Ahead neurocontrol and shows significant superiority.

2. THE CHAIN-BACK-PROPAGATION (CBP) METHOD

2.1. Enhancements of Error-Back-Propagation Neural Networks

The well established [8] Error-Back-Propagation (EBP) NN are used, as they are reliable, effective, and able to provide Credit Assignment data. As certain of the updating weight variances $\delta w_{i,j,k}$, calculated by the EBP procedure, are often too small in comparison to their mean value, the convergence is significantly decelerated. To compensate for this, the *Signed Square Root (or Signed-Sqrt) Transform* is introduced as

$$S_{SR}(x) = \sqrt[s]{x} = \mathrm{sign}(x) \cdot \sqrt{|x|} \; . \tag{1}$$

The signed-sqrt transform shortens the gap between the smallest and the largest calculated variances $\delta w_{i,j,k}$, which otherwise would be up to several orders of magnitude. Hence, the network weights are updated as

$$w_{i,j,k}(\mathrm{new}) = w_{i,j,k}(\mathrm{old}) + n_L \cdot S_{SR}(\delta w_{i,j,k}) \tag{2}$$

where n_L is the learning coefficient. The signed-sqrt transform will be exclusively used in weight-updating, since it ensures fast and stable convergence, as well as qualitatively better final results.

2.2. Neuro-Imitating Controllers (NImC) and Neural Emulators (NE)

In NImC applications a neural network is trained to emulate a specific control algorithm and replace it in the control loop. Essentially, the network *copies* the original controller. Training may take place in real time in parallel to the actual system in operation. The behavior of a NImC is improved by using the signed-sqrt transform (eqn 1) at its input and output nodes, i.e. by using terms of type

$$S_{SR}(e_{i,j}), \; S_{SR}(\dot{e}_{i,j}) \text{ and } S_{SR}(u_{0,j})_{ref} \tag{3}$$

where $e_{i,j}$, $\dot{e}_{i,j}$ are the error and error rate of DOF j at time $t_i = t - i \cdot \Delta t$, and $(u_{0,j})_{ref}$ is the output force of the 'teacher' controller for DOF j at time $t = t_0$, where $j = 1, \ldots, m$.

A Neural Emulator is a NN trained to imitate (model) the behavior of a specific dynamic system. In the general case of a system with coupled DOFs, one NE should be used for all DOFs, as in the present chapter. Since the recall phase of the NE as a discrete dynamic system in time includes feedback from the output nodes (previous states) to the input ones (next state), the output errors of the NE are accumulated and ,thus, amplified. To improve this situation, a larger time-step Δt_N is used for the NE, while the rest of the control loop continues to run on the usual Δt time-step. Additionally, the signed-sqrt transform is used at the input nodes corresponding to the system inputs. The relative inputs have the form

$$S_{SR}\left(u_{0,j}\right) \tag{4}$$

A special advantage of the CBP method is the fact that the NE does not necessarily have to emulate the dynamic system accurately, but only to provide a clear association of the effect of the control force input to its outputs.

2.3. Description of the Chain-Back-Propagation Method

NImC seems to be of little use, since it *assumes* the existence of a conventional controller as a "teacher". Also, its efficiency is *at most* equal to the efficiency of the "teacher". However, NImC is a secondary tool that the CBP method may use for controller network initialization instead of a "virgin" random network. The main tool of the CBP method is a NE, that the method uses to overcome the Credit Assignment problem. The (already trained) NE provides the CBP algorithm with Credit Assignment data via back-propagation calculations.

Supposing that:

i) An m-DOF dynamic system is to be guided from an initial state $x_N=x_{ref}$(previous) to a (target) final state $x_0= x_{ref}$(current), where $x_i=x(t_0-i\cdot\Delta t)$. This task must be accomplished in N time-steps.

ii) The orbit from the initial to the final state should satisfy a number of criteria that are expressed by a penalty function.

the CBP procedure may be briefly described as follows:

i) The weights of the neural controller are initialized either with random numbers or with the weights of an already trained NImC. The neural controller controls the actual dynamic system in *Real-Time*.

ii) N time-steps are executed (fig. 2). In each time-step the neural controller is supplied with information from the system and supplies the latter with its output, in order to proceed to the next time-step.

iii) The result of the *Global Penalty Function* is calculated, according to section 2.4.

iv) This result is *Back-Propagated* through the NE and the controller *alternatively*, starting by the former and ending to the latter, for N/m_N time-steps *backwards in time*. Simultaneously, the calculated variances of the controller network are accumulated in a temporary array. The calculated variances for the emulator network are discarded.

v) The controller network is *updated*. The emulator network remains *intact*.

vi) If necessary, the learning coefficients and the penalty function parameters are adjusted.

vii) After setting $x_0=x_N$, a new target point x_{ref} is selected and the CBP procedure continues from step (ii).

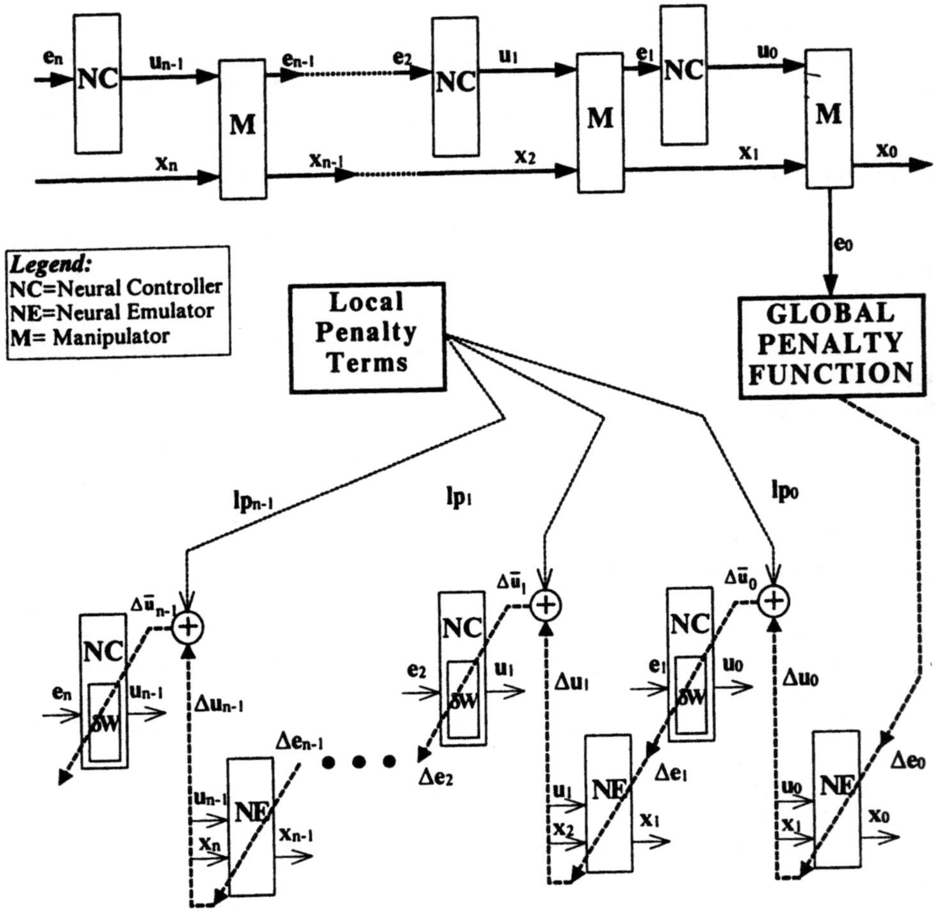

Fig. 2: Training Phase of Chain-Back-Propagation (CBP) Neurocontrol Method.

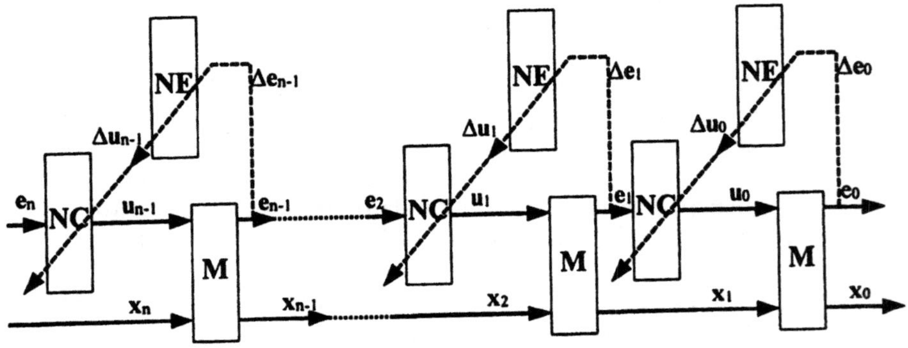

Fig. 3: Training Phase of Neural-Network-Pair (NNP) Neurocontrol Method.

According to step (iv) of the CBP method, as shown in fig. 2, the error-back-propagation learning procedure for a typical NN involves the successive utilization of the variances of the input layer of a network as variances of the output layer of a previous (in time) network. More specifically:

A) In the Neural Emulator:

The variances Δe_i of the error are used in order to compute the corresponding variances of the control output Δu_i.

B) In the Neural Controller:

the previously calculated values of Δu_i, added to a local penalty term lp_i, are used for the calculation of the variances of the neural controller's weights, as well as for the calculation of the variances of its inputs Δe_{i+1}.

The above calculations take place alternately, as the CBP procedure steps backwards in time from NE to controller and from controller to NE.

2.4. Penalty Functions for the CBP Method

2.4.1. Global Penalty Function

The utilized terms for the global penalty function (see fig. 2) can be divided in three basic categories:

- Functions depending on the error e_0 and the rate of error $\dot{e}_0$ at the final time-step of the N-step chain.
- *Qualitative* Criteria related to the shape of the whole N-step orbit of the system, such as Overshoot and Rise-Time. This particular category is emphasized, because implementation of such criteria is hardly found in relevant applications.
- Mathematical Criteria related to the entire N-step orbit of the dynamic system, such as Absolute Mean Error, etc.

The sum of the above functions weighted by appropriately selected factors constitutes the Global Penalty Function.

2.4.2. Conditioning of the Result of the Global Penalty Function

The result of the global penalty function can improve the whole procedure, if it is conditioned in one or more of the following ways:

- Set Upper and Lower bounds (Hard-Limits) to the result.
- Use only the sign of the result.
- Use the signed-sqrt transform of the result (eqn 1).

2.4.3. Local Penalties

During each time-step the Local Penalties lp_i are added to the variances Δu_i of the NE inputs (see fig. 2). In proportion to PD control parameters, the local penalties relate to the local error e_i and error rate $\dot{e}_i$. Thus, the variances of the neural controller weights and inputs are calculated, using the following variance for the network outputs (vector of network output errors)

$$\Delta z_{NC} = \Delta \bar{u}_i = \Delta u_i + lp_i = \Delta u_i + (k_P \cdot e_i + k_D \cdot \dot{e}_i) \tag{5}$$

where k_P, k_D are constant diagonal matrices.

2.4.4. Local Conditioning for Convergence Control of the CBP Procedure

During the evolution of the CBP procedure the back-propagated weight variances may be either amplified or diminished. Taking into account that this happens in a geometric progression of N/m_N steps, the procedure may diverge (overflow) in the former case or simply not converge in the latter, due to the negligible weight adjustments. Therefore, local conditioning for the chain-intermediate errors $\Delta\bar{u}_i$ and Δe_i is introduced. This conditioning, which ensures the stability and efficiency of the CBP learning procedure, involves:

- Setting of Upper and Lower bounds (Hard-Limits) to the intermediate (chain-local) back-propagated errors $\Delta\bar{u}_i$.
- Introduction of an Amplifying Multiplying Factor (>1) for very small intermediate (chain-local) back-propagated errors $\Delta\bar{u}_i$ or a Decay Multiplying Factor (<1) for very big intermediate back-propagated errors $\Delta\bar{u}_i$.

2.5. The Neural-Network-Pair (NNP) Control Method

A well-known neurocontrol method which produces *one-step-ahead* controllers is the NNP method [5]. As it may be considered a simplified version of the CBP method, it uses a neural controller in pair with a NE (see fig. 3). In brief, the system error is back-propagated through the already trained NE to provide the controller output error, in order to train the latter. The controller network is initialized with random weights. The results of the NNP neurocontrol method will be compared with the CBP method in the next paragraph.

3. APPLICATION OF THE CBP METHOD ON A 2-DOF ARTICULATED MANIPULATOR

3.1. 2-Axes Articulated Manipulator in the Vertical Plane

The CBP method is applied on a simulated 2-axes articulated manipulator operating in the vertical plane under the effects of gravity. Its dynamic equations of motion contain highly-nonlinear trigonometric terms. At first, the manipulator is controlled by a 'slow' hard-limited PD algorithm presenting slow and inaccurate control behavior. A second 'fast' hard-limited PD algorithm, which is empirically optimal for the manipulator, will be used for *comparison* purposes. Due to its very high gains ($>10^7$) the algorithm ends up to bang-bang control. Additionally, a NNP controller is trained to be compared with the CBP controller.

3.2. Training & Experimental Evaluation Procedures

Training takes place using set-points, that are uniformly random inside the envelope of the manipulator. The time-step Δt is 4msec.

The conventional and the neural controllers will be tested in:

- A *set-point orbit* defined by the 4 corners of a square ran in both directions. The manipulator has to reach each of the altogether 8 set-points in 1.2sec.
- Two *continuous reference orbits* defining a *circle* and a *square*. Both orbits have has to be covered in 4sec with constant end-point speed.

By the given times it can be realized, that special emphasis is given to the speed of motion for all of the above orbits.

3.3. Chain-Back-Propagation (CBP) Method

A NE using Δt_N=0.008sec is trained to imitate the system. When the CBP method is started with random weights, the resulting CBP controller has overshooting behavior and it is outperformed even by the 'slow' PD controller. Therefore, a NImC is trained to imitate the performance of the 'slow' PD controller

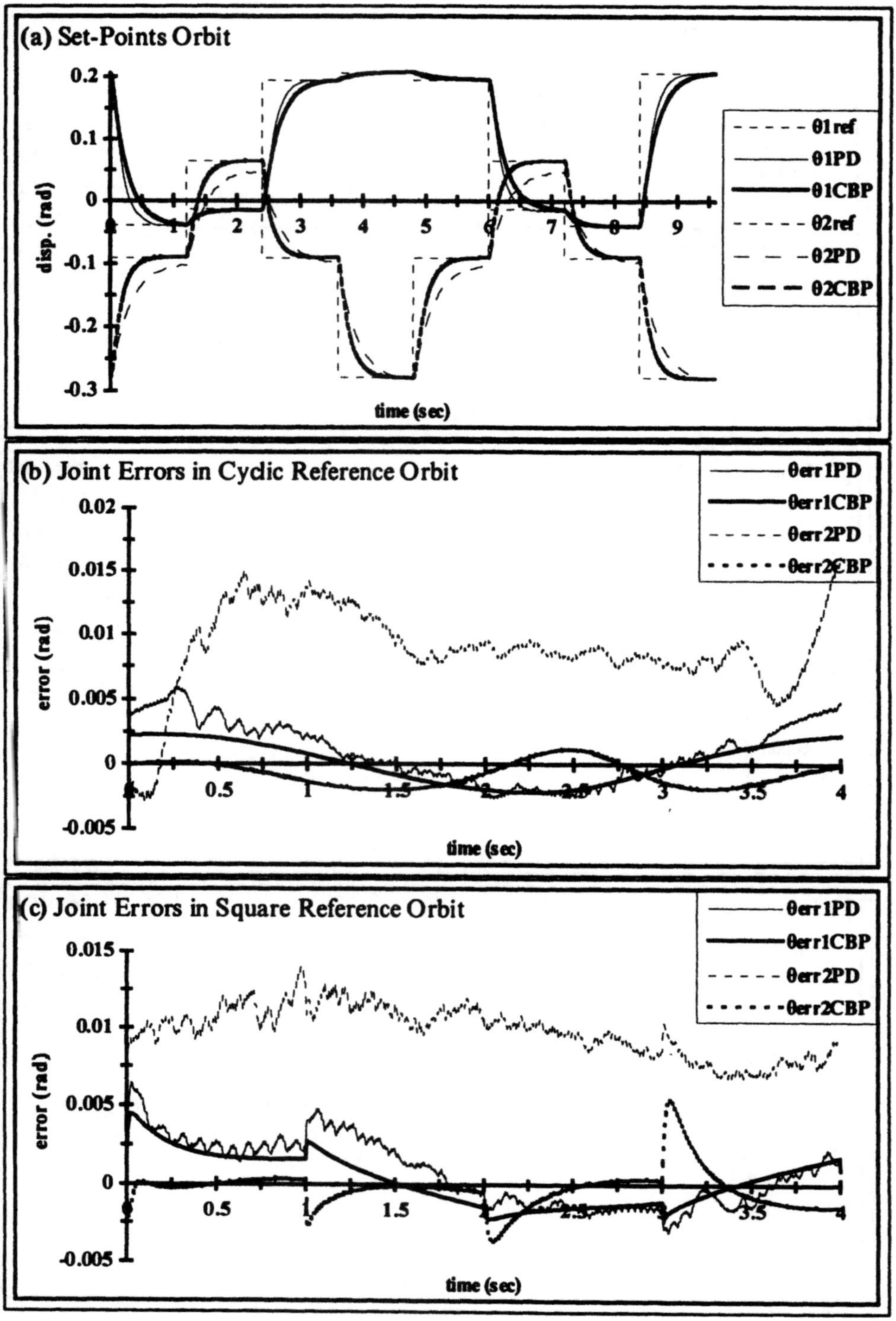

Fig. 4: CBP Control vs. Hard-Limited PD Control of Articulated Manipulator

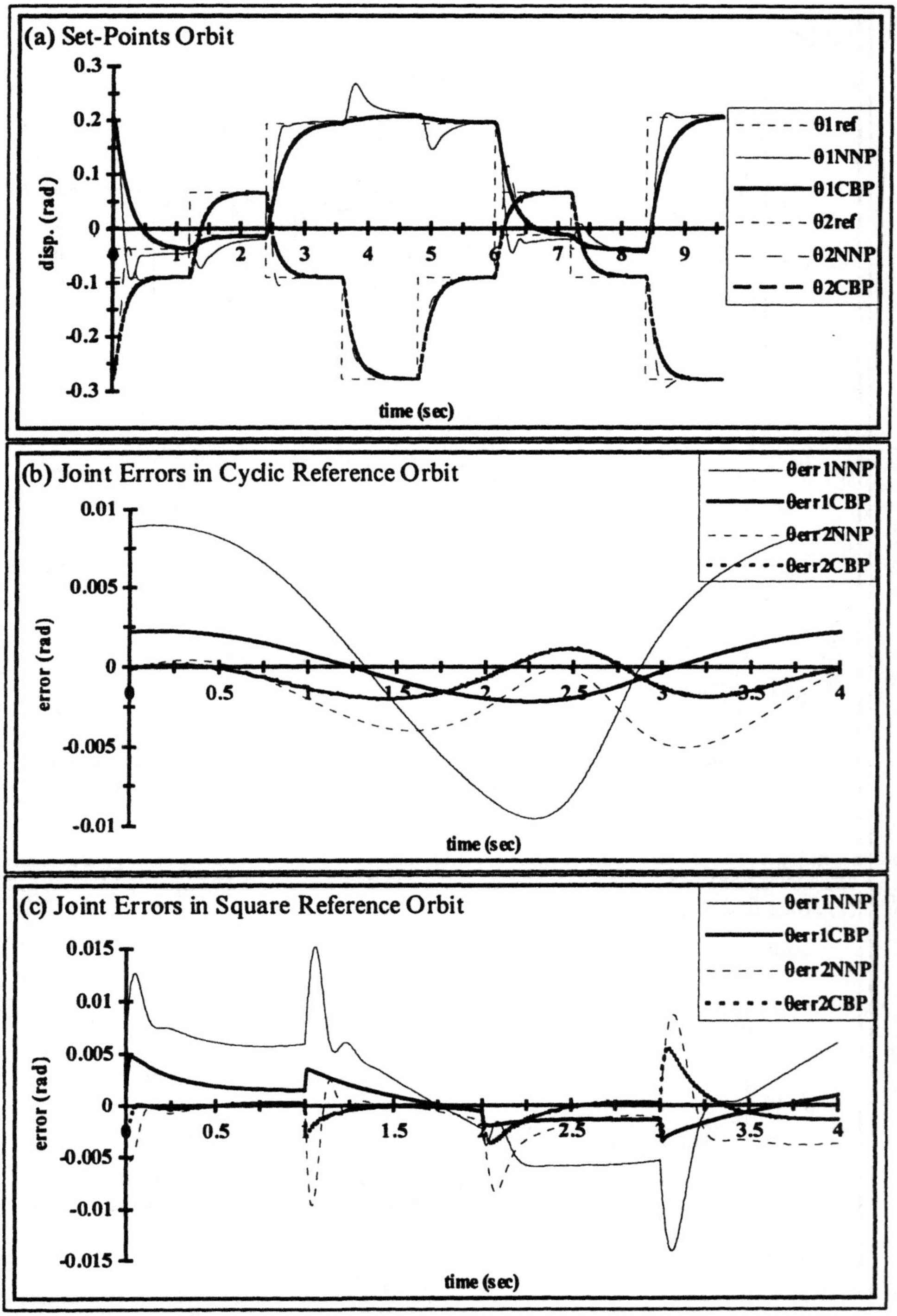

Fig. 5: CBP Control vs. NNP Control of Articulated Manipulator

and the CBP method is launched by using its weights. The majority of the available error penalty function terms are used and an amplification factor improves the convergence of the CBP algorithm. Local penalties are not used.

The resulting CBP controller is compared with the 'fast' hard-limited PD controller in fig. 4 and it presents 9 times better accuracy of position (final error). In addition, it is 25% faster and it achieves that with 5 times less effort (mean outputs). Similarly, in the cyclic and square continuous orbits the average error is 5 times smaller for the CBP method with about 6 times less effort.

The NNP method always led to a complete failure when training started with random network weights. Therefore, it was also initialized with the same NImC as the CBP method. The CBP controller is compared with the NNP controller in fig. 5. Despite the quite fast response in the set-points orbit, there are unacceptable overshoots and the final error is about 6 times greater than that of the CBP method. Concerning the continuous reference orbits, the average error was 3.3 times better for the CBP method with about the same output effort.

4. APPLICATION OF THE CBP METHOD IN A 3-AXES HEAVY DUTY HYDRAULIC MANIPULATOR

4.1. Description of the Manipulator

The hydraulic manipulator is of Cartesian type (see fig. 1) and presents a strongly non-linear dynamic behavior. Although identical in construction, all axes have highly different moving masses and maximum displacements. The speed, i.e., the oil flow in each axis is adjusted by a 4-way analog servo-valve, which is controlled by the DACs of the computer. Double Pressure Relief Valves (brakes) protect the system from high pressure peaks during intense deceleration or emergency shut-off. Activation of the brakes *rapidly increases* the system's non-linearities. Because the power source for all axes comes from the same hydraulic power-pack, the axes are coupled. Furthermore, excessive operation in high speeds overheats of the hydraulic circuits. This randomly alters the null point of the 4-way analog valves and, in consequence, the manipulator's behavior close to set-points. Therefore, the manipulator tends to present relatively large final errors.

In the experimental section there is a discrimination between low and high speed operation, because in the latter case the non-linearities of the system are greatly increased and the natural dumping of the system reduced. In low-speed mode a low-gain hard-limited PD controller is used. In high-speed mode the conventional PD is extended to a high-gain Double-PD algorithm. The Double-PD includes a change in the k_P, k_D constants, when the manipulator gets close enough to a set-point. To protect the hydraulics of the system hard-limiters are used for the input voltage and for the input voltage *changes* between consecutive time-steps.

4.2. Application of CBP Control

The time-step Δt is 0.01sec. The results are again poor, when the CBP method is initialized with random controller weights. Therefore, a NImC is trained to imitate the low-gain PD controller and its weights are subsequently used to launch the CBP controller.

In Low Speed Mode a NE is trained with Δt_N=0.08sec. Nearly all types of penalty functions and control variables for the convergence of the chain are used.

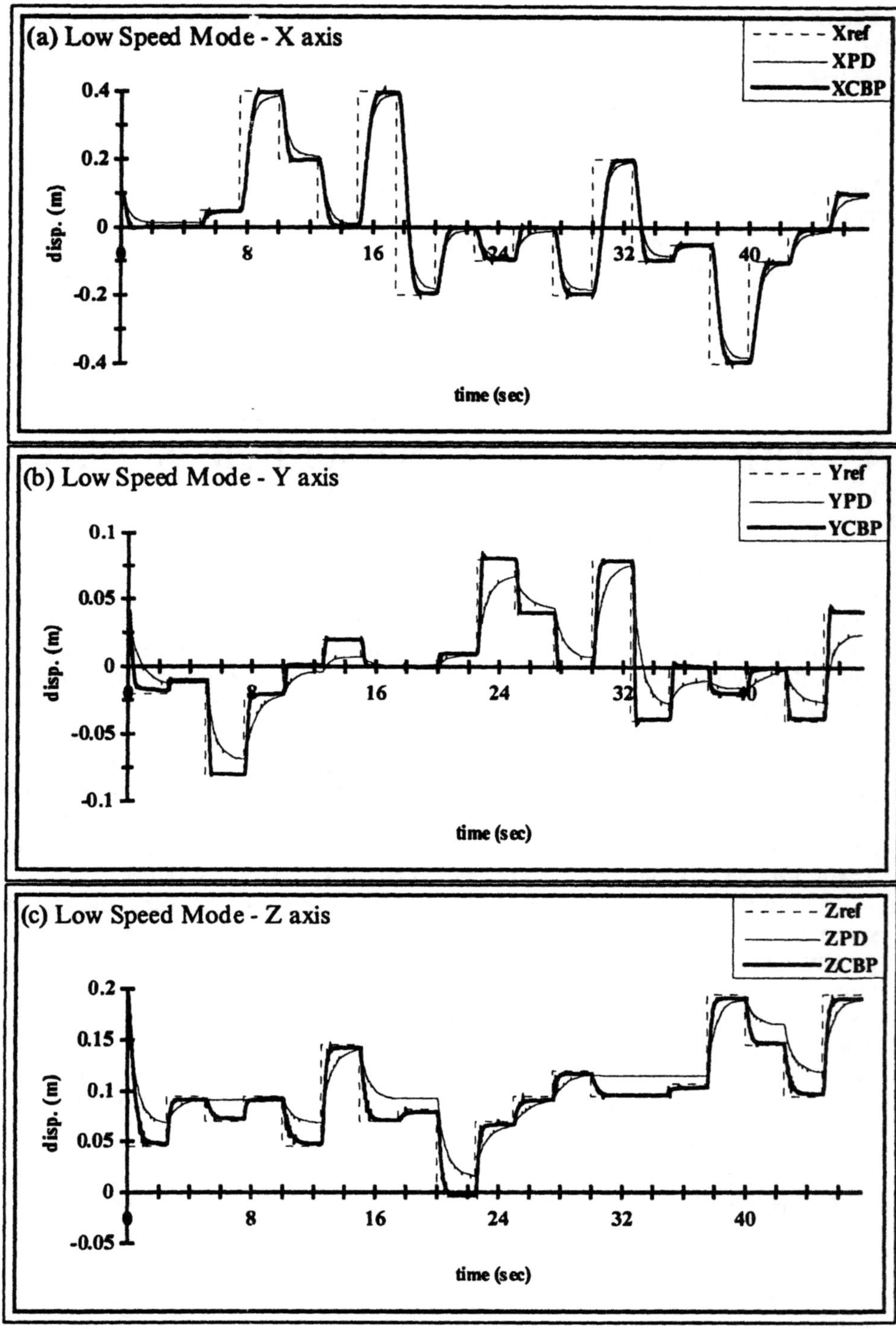

Fig. 6: CBP control vs. Hard-Limited PD Control of Hydraulic Manipulator

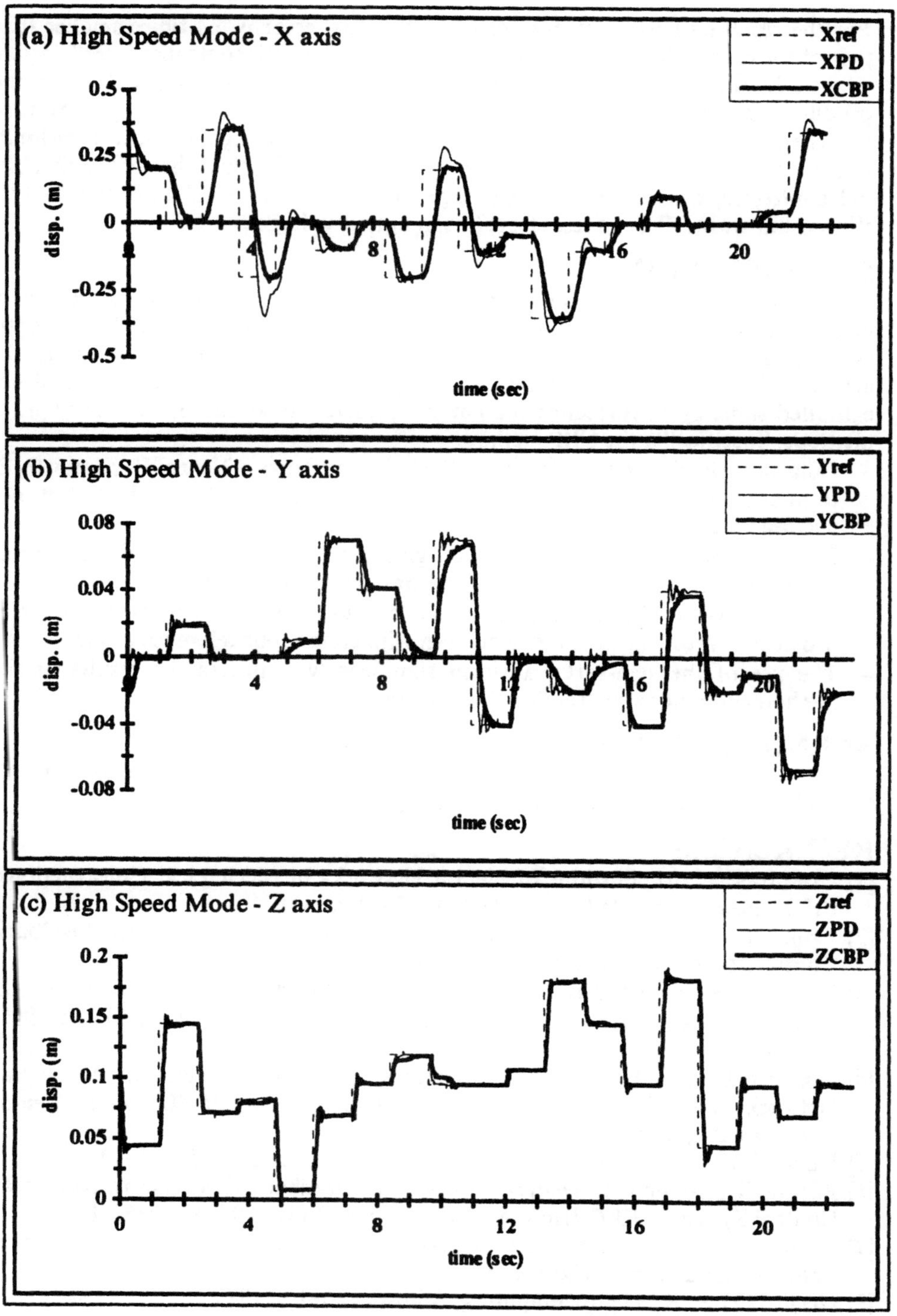

Fig. 7: CBP Control vs. Hard-Limited Double-PD Control of Hydraulic Manipulator

For the resulting CBP controller of fig. 6 the final error is 5.4 times smaller than that of the conventional PD controller. Moreover, the CBP controller is 50% faster.

In High Speed Mode a NE is trained with Δt_N=0.01sec. The CBP method produced the controller of fig. 7. Nearly all penalty error function types are used and the *convergence* of the CBP method is *stabilized* by a high and low hard-limits to the local penalty terms. The final error is 2 times smaller than that of the Double-PD controller, which also develops large overshoots. The CBP controller is 12% slower, but with 2. 7 times less effort (mean output voltage).

5. CONCLUSIONS

The Chain-Back-Propagation (CBP) is an N-Step-Ahead neurocontrol method, that has great potential in the control of non-linear multi-DOF robotic systems. It can be efficiently implemented in complex problems, such as the articulated manipulator or the real-world hydraulic robot of this chapter and has exhibited superior performance compared to conventional and other neurocontrol methods. Specifically, the CBP method:

- Considers the dynamic system as a Black-Box and uses a Neural Emulator (NE) to model it. The NE is necessary for the success of the method, but not necessarily accurate.
- It integrates numerous kind of orbit-global and orbit-local penalty functions and transformations, that improve and stabilize its convergence.
- Does not use a reference orbit in the control loop, only set-points. However, it provides controllers quite efficient, even with continuous reference orbits.
- The use of the sign-sqrt transform significantly improves the results of the method and makes the training time shorter.

ACKNOWLEDGEMENTS

Mr. A. Hountras has been supported partly for this work as a Ph.D. Candidate by the Greek National Scholarship Foundation.

REFERENCES

[1] Hountras A.: Applications of Neural Networks in Mechanical Constructions, PHD Thesis (Submitted to N.T.U.A. for Evaluation).

[2] Ichikawa Y., Sawa T.: Neural Network Application for Direct Feedback Controllers, *IEEE Trans. Neural Networks*, **3/2**, March 1992, 224-231.

[3] Narendra K. S., Parthasarathy K.: Identification and Control of Dynamical Systems Using Neural Networks, *IEEE Trans. Neural Networks*, March 1990, **1/1**, 4-27.

[4] Hoskins D. A., Hwang J. N., Vagners J.: Iterative Inversion of Neural Networks and its Application to Adaptive Control, *IEEE Trans. Neural Networks*, **3/2**, March 1992, 292-301.

[5] Kosko B.: Neural Networks for Signal Processing, Prentice-Hall, 1992.

[6] Parisini T., Zoppoli R.: Neural Networks for Feedback Feedforward Nonlinear Control Systems, *IEEE Trans. Neural Networks*, May 1994, **5/3**, 436-449.

[7] Nguyen D. H., Widrow B.: Neural Networks for Self-Learning Control Systems, *IEEE Control Systems Mag.*, April 1990, 18-23.

[8] Hecht-Nielsen R.: Neurocomputing, Addison-Wesley, 1989.

23

NEURAL CLASSIFICATION OF DANGEROUS SHIPS IN COLLISION SITUATIONS

J. LISOWSKI, A. RAK and A.GRZENIA
Gdynia Maritime Academy
83 Morska Str., 81-225 Gydnia, Poland

1. INTRODUCTION

An increasing traffic of marine crafts on the sea routes has been leading to considerably increasing requirements with respect to both qualifications of deck officers and application of technical arrangements supporting navigator in the process of making decision on a safe manoeuvre under a collision situation [1]. ARPA (Automatic Radar Plotting Aids) equipment is at present the basis for the development of most collision systems. They operate using the analysis of radar signals enabling to track the course and speed of selected objects and the determination of their maximum approach (DCPA - Distance of the Closest Point of Approach and TCPA - Time to the Closest Point of Approach) under the situation threatening a collision [2]. The analysis of radar signals is based on the assumption that the kinematics of the objects does not change. Navigator makes a decision on the type of manoeuvre to avoid collision considering the information originated from ARPA system and also basing on his experience and intuition. The latest trends in the development of anti-collision systems concentrate on the construction of a superior ARPA based system to determine safe manoeuvre which may be automatically performed after having been accepted by navigator.

S.G. Tzafestas (ed.), Advances in Intelligent Systems, 247–254.

The improvement of quality for the solutions of collision situation, offered by such a system may be attained by complementing a kinematic model of processing with the navigator's heuristic knowledge. This chapter presents the results of attempts to represent the navigator's experience in a form of an appropriately constructed neural network.

2. NEURAL CLASSIFICATION

One of the fundamental features of feedforward, multi-layer neural networks is their ability to map any ***u*** vector entered at the network input in a ***y*** output vector [3]. The form of representation is formulated by the topology of network and the values of weight coefficients at the input of individual neurons. The learning process of the

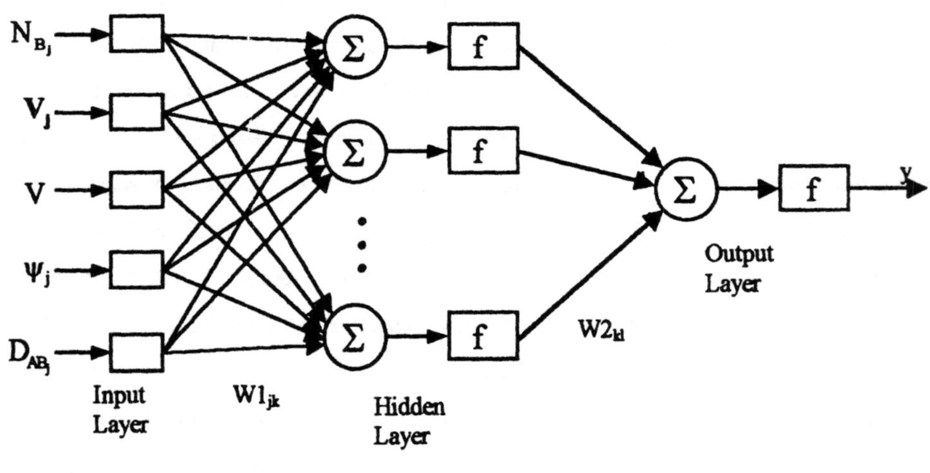

Fig. 1. Neural network for classification of collision situations - $W1_{jk}$ and $W2_{kl}$ describe the weights of neurons of the first and second layer, respectively, f describes the function for neuron activation.

network consists of such a tuning of the values of weight coefficients so that - for a given input vector (also referred to as image or an input pattern) the response at the network output is as close to the input pattern presented by the teacher as possible.

This is, therefore, a hetero-assocciative type of learning. Mapping of $y = F(u)$ demonstrated by a trained up network need not, then, be given through a known analytical relationship. It may be a heuristic representation of the operator's knowledge who prepares appropriate sets of input images and "reasonable" answers corresponding to them.

Let's consider a network with five inputs and one output whose task is to divide a set of possible navigational situations - meeting own vessel with foreign vessel - into two sub-sets described as "safe" and "unsafe" situations.

$$y = F\,(N_{BJ},\, V_j,\, V,\, \Psi_j,\, D_{ABJ})$$

where: y - network responses; F - mapping demonstrated by the network, N_{BJ} - bearing to foreign vessel B_j, V_j - foreign vessel's speed; V - own vessel's speed; Ψ_j - foreign vessel's relative course, D_{ABJ} - distance between own and foreign vessel. Such a network is shown in Fig. 1.

3. PREPARATION OF TEACHING PATTERNS

One of the basic problems in learning of a network is the preparation of input data sets and required output data corresponding to them. The data should be coherent and evenly cover the space of possible network inputs. In order to gather appropriate learning patterns a script was written in MATLAB environment which randomly presents a selected navigational situation to an experienced navigator. The parameters of the vessels' movements are represented on the screen in a form of vectors in a manner similar to the representation applied on the ARPA screen. It was assumed that the radius of the largest circle is 5 Nm. The responder may express his assessment of the situation presented by clicking the mouse onto one of the proposed response

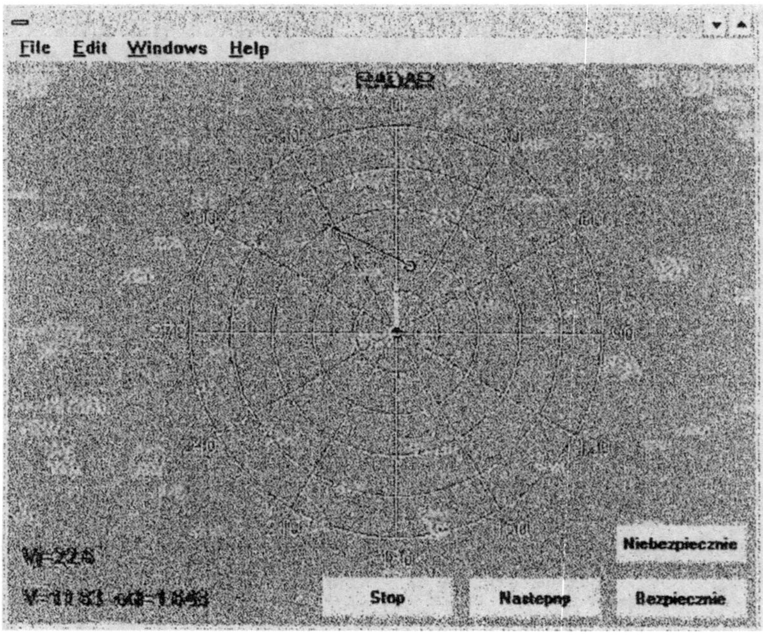

Fig. 2. Navigational situation presented to navigator for response.

buttons. In the version presented here the programme gathers two-state responses: 0 – safe situation; 1 - collision situation. Fig. 2. shows one of the patterns presented to navigator. The data collected are recorded in sets in a format which allows to directly use them by the functions of Neural Network Toolbox of the MATLAB package.

4. RESULTS OF SIMULATION

In the process of learning of the network a MATLAB package was used together with its Neural Network Toolbox. Various architectures of networks of different number of neurons in the hidden layer and for various functions of neuron activation in particular layers were tested. The results of the study indicate that for a five-element vector of input data three neurons in the hidden layer are enough for the network to operate correctly. Appropriate neuron activation functions had a sigmoidal and linear nature. During the process of learning of the network a backpropagation algorithm was used. [4].

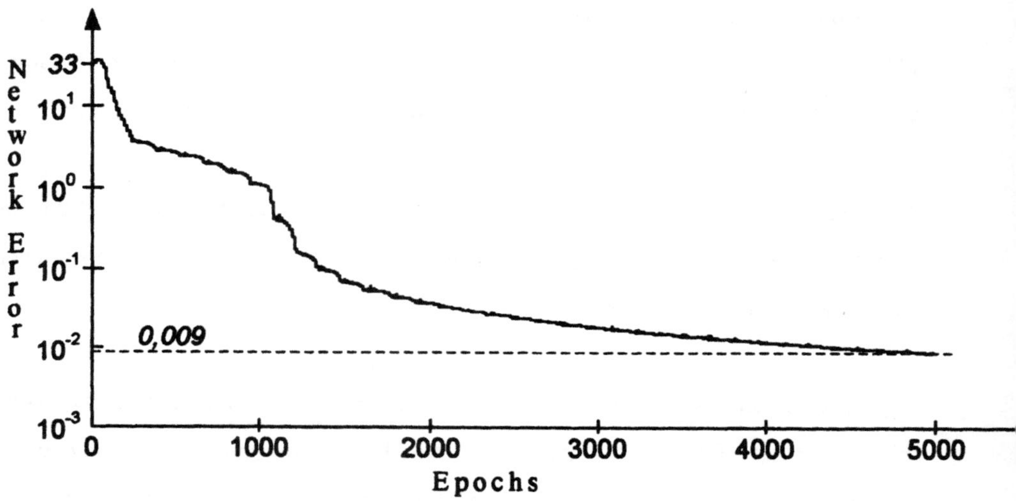

Fig. 3. Network sum squared error in learning process.

Fig. 3 shows an example of a record for an error function of the network during its learning process. The achievement of a required error level mainly depends on the coherence of data used at the learning stage. While preparing the learning patterns navigator, guided by his own subjective impressions, may assess two very similar navigational situations in a different manner. The network then may generate incorrect responses if the input signal vectors from the testing pattern are close enough to the learning vector with which an answer not identical in meaning is associated. The coherence of data may be improved by presenting navigator a few times the same data in a different sequence and rejecting responses which do not repeat. The results of experiments also indicate the necessity to interfere in dissipating of the patterns used during the learning process of the network in the space of all possible navigational situations. Condensation of input vectors considered in the vicinity of the hyper plane dividing this space into the "safe" and "unsafe" parts fundamentally reduces the number of incorrect responses of the network.

Fig. 4 shows responses of the network which was trained using a set of input data with patterns evenly distributed with the use of random number generator. The testing set contains one hundred patterns corresponding to specific navigational situations.

They were not presented to the network during learning. Ten of these situations were decided by navigator as unsafe (O). The network gave the same responses in nine case while one response was incorrect.

Fig 5 shows responses of the network to the testing set mentioned above, however, in this case a set of learning patterns was modified as to remove inconsistencies in responses and to increase the number of "uncertain" patterns in the vicinity of the division of the space of possible input vectors into two parts. This time the network gives the same responses as had previously been given by navigator.

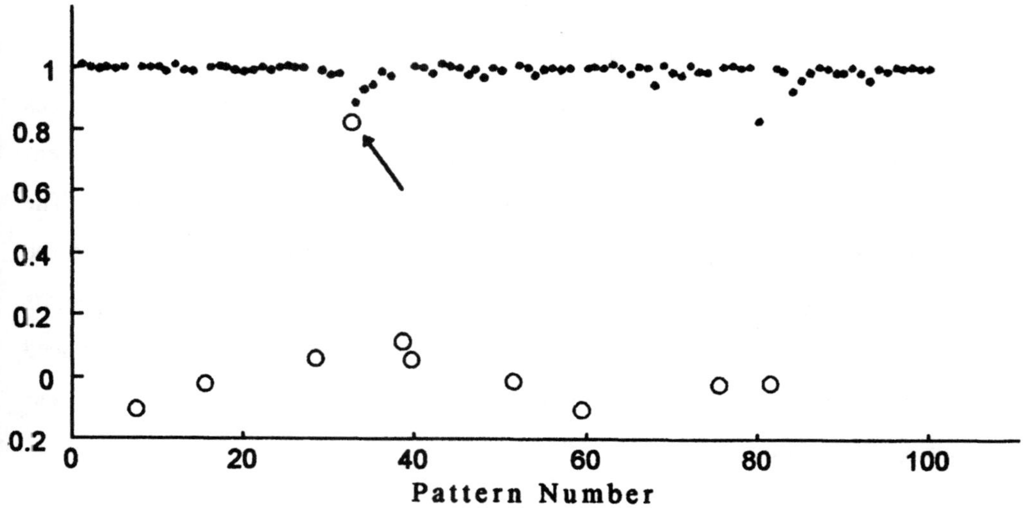

g 4. Classification of the testing pattern - the network with errors in the process of learning.

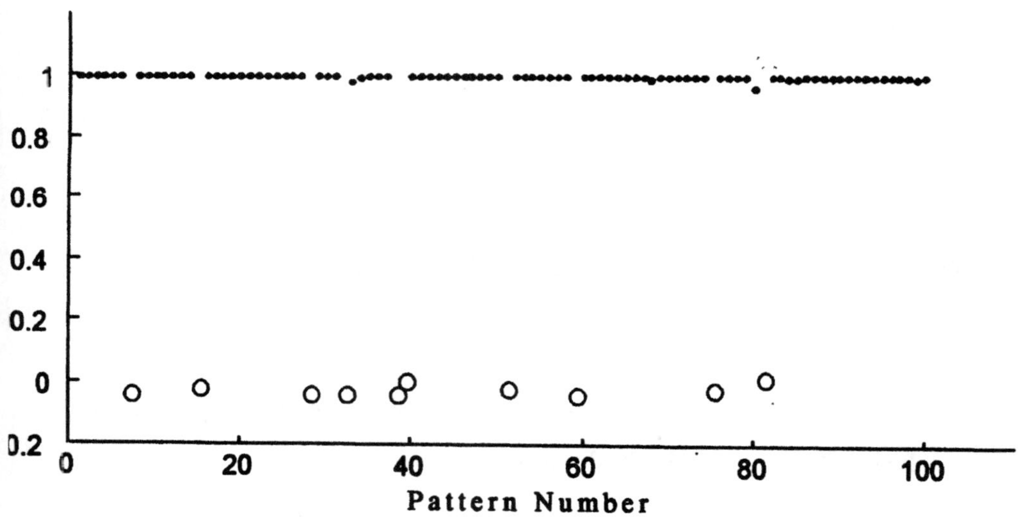

5. Classification of the testing set - correct process of the learning of the network.

5. CONCLUSIONS

Neural networks presented in this chapter may be used as elements of the systems applied to assess the safety of passing vessels at sea. They are able to represent heuristic knowledge similar to that demonstrated by an experienced navigator. The correctness of the assessment of the safety of passing vessels with the use of the network depends to a critical degree on the correctness of patterns used in the process of learning by the network. The use of the knowledge demonstrated by experienced navigators may lead to a situation when the network starts to demonstrate their "averaged" knowledge.

The introduction of an element of an artificial intelligence in a form of an appropriately prepared neural network into the systems of defining safe trajectory of a vessel under a collision situation may accelerate the process of selecting collision avoiding manoeuvre.

6. REFERENCES

[1] IMO *Preference Standards for Automatic Radar Plotting Aids (ARPA).* Resolution A.422 (XI), Nov. 1979

[2] Lisowski J., *Ship's Anti - Collision Systems*; Wydawnictwo Morskie Gdańsk, 1981. (In Polish)

[3] Tadeusiewicz R., *Neural Networks*, Akademicka Oficyna Wydawnicza RM, Warszawa 1993. (In Polish)

[4] Hertz J., Krogh A., Palmer R.G., *Introduction to the Theory of Neural Computation*, Addison-Wesley Publ., 1991

24

NEURAL-COMPUTING IN FABRICS INSPECTION SYSTEM

H. LIU, C.-H CHAN, T. KWAN and G. PANG
Dept. of Electrical and Electronic Engineering
The University of Hong Kong
Pokfulam Road, Hong Kong

1. Introduction

In the textile industry, in order to insure the quality of the textile fabrics, textile manufacturers utilize human inspectors to manually inspect fabrics before shipment to the customers. 100% inspection is generally needed to guarantee 100% defect-free production. When defects appear in the manufacturing process, 100% inspection will ensure that the occurrences of all the defects are recorded. A detailed inspection report is useful both to the manufacturer and the customer.

The efficiency of an inspector declines over time due to the dull and repetitive nature of the work. The replacement of the inspector with a cost-effective automated visual inspection system is an attractive scheme. This could leads to saving of labor costs, and the reduction in demand for highly skilled inspectors. Such an automated visual inspection system could also keep records of the defects for statistical analysis and for optimizing management decisions.

This chapter describes the development of an *Automated Visual Inspection (AVI) system* for weaving defect detection based on image processing and recognition algorithms. The purpose of the developed system is for the quality control of fabrics and weaving products from a textile factory. In addition to the use of standard image processing functions for enhancing and modifying the digital image, the chapter will describe the techniques from neural network for fabrics inspection and recognition. The aim is to obtain saving of manpower and time as well as increased accuracy in the inspection process.

Han and Yi [1] presented the approach for using linescan camera for the multicameras system. This system designed and implemented for high speed and high resolution

S.G. Tzafestas (ed.), Advances in Intelligent Systems, 255–266.

surface defect image acquisition in a large moving area. The cameras are used to obtain more than twenty thousand pixels per line. D. A. Lehotsky [2] described the approach for an intelligent camera technology. The system is modular. For multi-camera system the cameras are connected to personnel computer via an intelligent microprocessor based interconnection hub. Miettinen and Harkonen [3] proposed a systematic procedure for image acquisition. The need for a systematic method and tools in assisting the design process of illumination and camera arrangements is obvious. Fernandez et al. [4] developed general solutions for on-line surface inspection system for aluminum strip. This system achieves high performance, flexibility, possibility of multi-sensor integration, modularity. It may also be applied to wood surface, metal surface, plastic surface.

Laitinen [5] [6] presented the relation between the performance of an imaging unit of a web inspection system and the final image quality. The contrast of certain type of defects can be improved before segmentation by averaging the input data line by line. Norton-wayne and his colleagues [7] presented the general idea for the detection of defects in automated inspection. A method is presented for design of a system to achieve quantitative performance requirements in terms of probability of defect detection versus probability of false alarm. Macaire and Postaire [8] proposed the adaptive thresholding for detecting the defect in the real time. The specific hardware architecture was used for running it in the real time. Mlynarczuk [9] described the approach for segmentation of fabric direction by using the morphological operation. This algorithm helps to segment and outline the defect in the fabric inspection.

Lawson and Parker [10] have reported an artificial neural networks approach in industrial ultrasound image. This method has assisted to classify the defect. I. M. Dar et al. [11] developed the solution to detect and classify the pilling in textile fabric by using fuzzy logic. This approach presented a methodology that counts the number of pillings on textile fabric samples automatically and classifies them into one of the pre-defined classes with repeatable accuracy while accounting for the human judgement. Branca and his colleagues [12] have shown that the texture analysis had proved to be effective tool for detecting and classifying defects on natural material, like leather. Moreover, the multiresolution analysis was also proposed to deal with the risk of missing small details of potential significance in single resolution approaches. Abril et al. [13] established a sequential method for objective measurement. The method involves operation in the frequency domain. Ciamberlini et al. [14] presented the basic working principle for detecting the weaving defect by using fourier imaging. Well-defined spots corresponding to spatial frequencies of the tissue is shown, when good fabric passes in front of the optical system the fourier image, captured by camera. If the defect occurs in this inspection, the pattern changes significantly and a defect is easily detected in real time. Bayro-corrochano [15][16] in a review on automated visual inspection 1983-1993, gave a lot of general ideas in conventional approaches and intelligent systems approaches.

2. Requirements

Although significant advances has been made in the field of computer vision, image processing and pattern recognition in recent years, it is still not possible to design a flexible, cost effective, off-the-shelf visual inspection system suitable for all types of industrial applications. This is due to the need to consider the individual requirements of each application. Regarding an AVI system on weaving defects, the requirements are as follows: (1) Detection of defects on fabrics product as it moves through the image acquisition unit. The irregularities are detected as defects. (2) The image data must be processed at a sufficiently high rate. (3) The defects are grouped according to size, direction, shape and graded in terms of severity. It should be mentioned that the defects can come in complicated and irregular shapes. (4) The image acquired may be noisy. (5) Defect locations and their images can be displayed on monitor in real time. (6) Defect information is stored and used for subsequent analysis. (7) Marking unit and alarm can be activated. (8) Inspection report is prepared and printed.

Digital images are obtained as a high resolution camera scanned through the fabrics product. Image cleaning and modification is carried out for enhancement. Segmentation is then performed and the resultant image is used for analysis. Geometric features are obtained which is then used for pattern matching and classification. In this chapter, a **multilayer feedforward** neural network is used.

3. Different Levels of Digital Image Processing on Defects

Below is the different level of processing on the images of weaving defects. The low-level processing involves thresholding, edge detection, line and region finding. The following figure shows the captured image of a piece of fabric with defects on it.

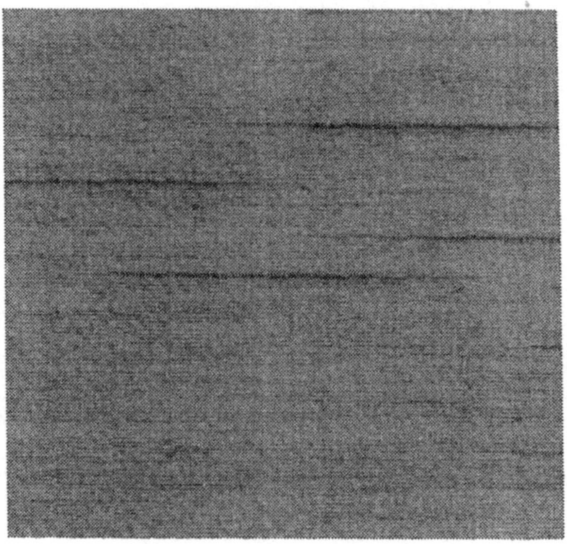

Fig.3.1 Image of a piece of fabric

After thresholding, the image would appear as the following figure.

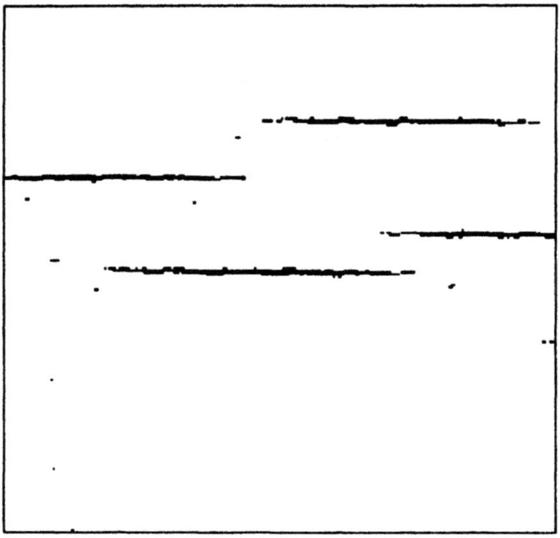

Fig.3.2 Fabric image after thresholding

The regions of the defects are found which is shown in the following figure.

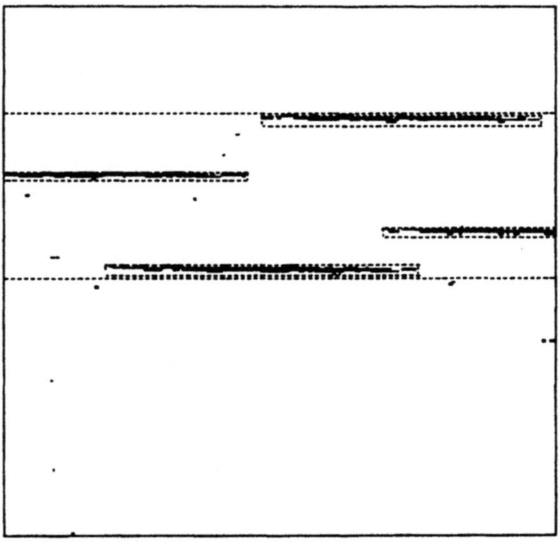

Fig.3.3 Fabric image after finding the region of defect

Individual defects and the overall region covered by all the defects are windowed.

The intermediate-level of image processing involves feature extraction. Five features are used in our tests. The first one the height of the overall defects window. The second feature is the width of the overall defect window. The third feature is the ratio of total defect area to the overall window area. The forth feature is the number of defects occurs in the overall defects window. The fifth feature is the ratio of the smallest defect area over the largest defect area.

Finally, the high-level processing consists of recognition and classification. For this part, neural network was used to conduct a test which will be described in a later section.

4. Analysis of the Defects

Twenty-four fabrics image with different defect were captured and processed. The following table shows the information about the fabric color, lighting, process before thresholding, threshold values and some comments

	Fabrics Color	Lighting	Process	Thres hold	Filter	Comment
Defect 1 (Uneven Yarn)	Yellow	Back Light	nil	158	nil	nil
Defect 2 (Thick Yarn)	Deep Blue	Direct light	Motion blur	114	nil	For each row, count white
Defect 3 (Slubs)	Yellow	Back Light	nil	165	nil	nil
Defect 4 (Trash)	Yellow	Back Light	nil	162	nil	nil
Defect 5 (Coloured Yarn)	White	Back Light	nil	127	nil	Reduce light intensity
Defect 6 (Dirty Yarn)	White	Back Light	nil	177	nil	Reduce light intensity
Defect 7 (Hessian Contamination)	Yellow	Back Light	nil	158	nil	nil
Defect 8a (Wood Pattern)	Yellow	Back Light	nil	154	nil	nil
Defect 9 (Nispick)	Yellow	Direct light	nil	174	nil	Grab image at 45 degree
Defect 10 (Double Weft)	Yellow	Back Light	Motion blur	154	nil	nil
Defect 11 (Thin Bar)	Yellow	Back Light	nil	201	Remove noise 4 times	Reduce light intensity
Defect 12 (Thick Bar)	Yellow	Back Light	nil	155	Remove noise 1 time	nil
Defect 13 (Kink)	Yellow	Back Light	nil	144	nil	nil
Defect 14 (Mixed Weft)	Yellow	Back Light	nil	157	nil	nil
Defect 15 (Wrong Draw)	Yellow	Back Light	nil	nil	nil	Defect color similar to background
Defect 16 (Float)	Yellow	Back Light	nil	152		

Defect 17 (Loom Fly)	Yellow	Back Light	nil	150	nil	nil
Defect 18 (Broken End)	Yellow	Back Light	nil	112		
Defect 19 (Bad Selvage)	Yellow	Back Light	nil	160	nil	nil
Defect 20 (Double, Tripewarp)	Yellow	Back Light	Guassian Blur	169	nil	nil
Defect 21 (Oil Stain)	Yellow	Room Light	nil	177	nil	nil
Defect 22 (Netting Multipies)	Yellow	Back Light	nil	210	Remove noise 4 times	nil
Defect 23a	Light Blue	Direct light	Band Pass Filtering	151	nil	nil
Defect 23b (Yarn Tails)	Yellow	Direct light	nil	152	Median noise removal 1 time	Use shadow of the defect for detection

Table 4.1 Fabric Image Processing information

The characteristics of defect, fabrics color, lighting used to see the defect and the noise characteristics are shown in the following table.

	Fabrics Color	Lighting	Defect Shape & Color	Noise shape & Color
Defect 1 (Uneven Yarn)	Yellow	Back Light	Thin, short, horizontal black lines	Few black lines of cloth texture
Defect 2 (Thick Yarn)	Deep Blue	Direct light	A long, white, horizontal line broken into segments	Many Short white line segments due to cloth texture
Defect 3 (Slubs)	Yellow	Back Light	A thicker, longer, horizontal black lines compared to defect1	Few short black lines due to cloth texture
Defect 4 (Trash)	Yellow	Back Light	Long, black, horizontal lines but broken	Few short black lines due to cloth texture
Defect 5 (Coloured Yarn)	White	Back Light	A thick, short, horizontal black line	nil
Defect 6 (Dirty Yarn)	White	Back Light	A long, black, horizontal line broken into many segments	Black dots and lines due to cloth texture
Defect 7 (Hessian Contamination)	Yellow	Back Light	Short, horizontal black lines	Black dots due to cloth texture
Defect 8a (Wood Pattern)	Yellow	Back Light	Thin, short, black lines formed into groups	Noise and pattern difficult to distinguish
Defect 9 (Nispick)	Yellow	Direct light	Very short, thin black horizontal lines forming a vertical line	Black dots and shadow due to uneven light

Defect 10 (Double Weft)	Yellow	Back Light	A long horizontal black line	Many thin & short black lines
Defect 11 (Thin Bar)	Yellow	Back Light	White thick line segments forming few parallel horizontal lines	Few white dots
Defect 12 (Thick Bar)	Yellow	Back Light	Black, thick segments forming a horizontal bar	Thin short black lines
Defect 13 (Kink)	Yellow	Back Light	Short black horizontal lines	Black dots
Defect 14 (Mixed Weft)	Yellow	Back Light	A thick, horizontal black line	Short black lines
Defect 15 (Wrong Draw)	Yellow	Back Light	A long vertical line	Short black lines
Defect 16 (Float)	Yellow	Back Light	A short, thin, horizontal black line	Short thin black lines
Defect 17 (Loom Fly)	Yellow	Back Light	Thick black short horizontal lines	Few black dots.
Defect 18 (Broken End)	Yellow	Back Light	A thin, white, long vertical line	White short lines
Defect 19 (Bad Selvage)	Yellow	Back Light	Black curves and lines forming a roots pattern	Many black dots
Defect 20 (Double, Tripewarp)	Yellow	Back Light	Black areas forming a vertical line	Quite big black areas
Defect 21 (Oil Stain)	Yellow	Room Light	Few black areas distributed over the cloth	Big black dots
Defect 22 (Netting Multipies)	Yellow	Back Light	White areas concentrated in a portion	Almost none
Defect 23a	Light Blue	Direct light	Black separated dots forming two parallel vertical lines	Black dots
Defect 23b (Yarn Tails)	Yellow	Direct light	Think horizontal black line segments	Black dots

Table 4.2 Defect and noise characteristics

5. Approaches to developing an Automated Visual Inspection System

The approaches to developing an Automated Visual Inspection System could be divided into two kinds. One is conventional approach and the other is intelligent approach. For the convention approach, pattern classification technique includes template-matching, deterministic and statistical classifier and structural classification. For template matching, a set of templates has been stored in a computer. The unknown input is compared with the template of each class to match the unknown to a class. For deterministic and statistical method, the most commonly used is Bayes classifier which shows a minimal probability of misclassification. Structural classification utilizes a relational structure of the pattern in form of a graph with labeled arcs. The unknown is matched with the graphs of each class by method such as graph isomorphism.

For the intelligent approach, the most common methods are knowledge-base systems, fuzzy logic and neural network. A knowledge-base is essentially an intelligent computer program which uses knowledge and inference procedures to solve problems which usually require human expertise for the solution. Example of application of knowledge-based system includes inspection of textile, printed circuit patterns, solder joints, automobile engine valves and dimension analysis of glass containers. Fuzzy logic is used at handling imprecise data. It provides a mechanism for representing attributes such as 'little', 'sometimes ' and 'often'. This enables the measurement of the degree to which a type of pattern is present. For the neural network, it can be used for pattern segmentation, classification, recognition, image analysis and restoration of noisy pictures.

5.1 Neural Network for Defect Classification

The architecture of a neural network for defect classification is shown in the figure below. A multi-layer perceptrons, using the back-propagation method of Rumelhart et al. [17] is used in this chapter.

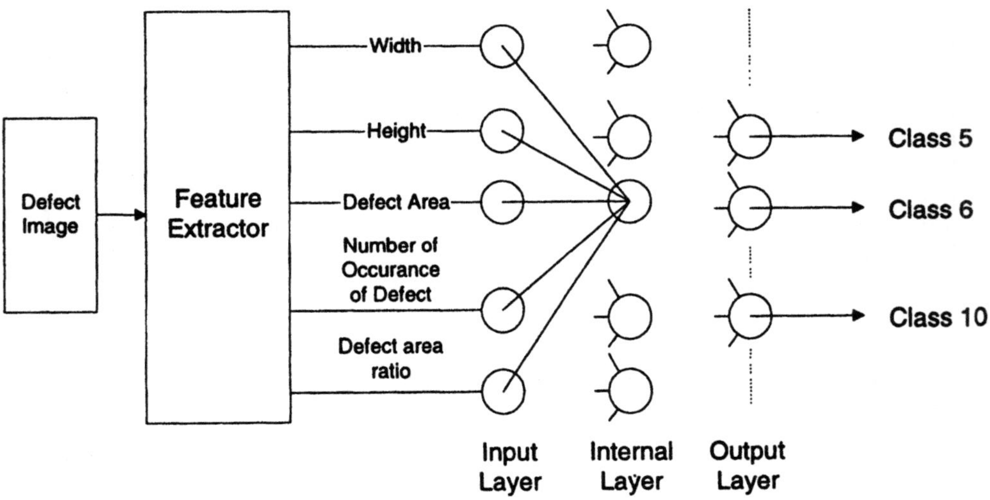

Fig. 5.1 Neural Network architecture for defect classification

The neural net was trained to classify fifteen defects. Backpropagation learning rules were used to adjust the weighs and biases of networks so as to minimize the sum squared error of the neural networks. Several parameters of the network were adjusted to yield the best result. The configuration of the neural network with a good performance is as follows: two layer network, twenty neurons in the hidden layer, 0.001 error goal and learning rate equals 1. It has been found that there is some improvement when the number of hidden unit is increased. However, the performance would drop if the learning rate is either too large or too small. The neural network is then tested using some test samples and an accuracy of 93% is obtained. With the use of a neural network, the time needed for defect classification is very fast and is well-suited for a real-time inspection system.

6. Hardware and Software Issue

6.1 Hardware

The hardware structure of the Fabric inspection system is shown below.

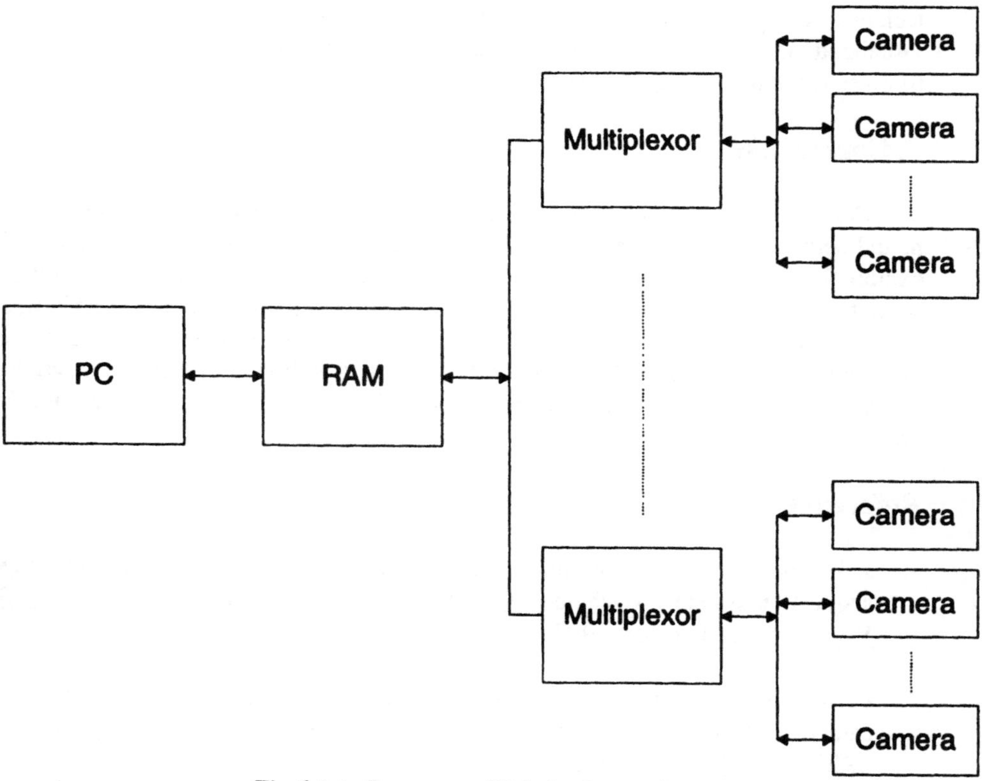

Fig.6.1.1 Structure of Fabrics inspection system

The image of the fabric is captured by the CCD element inside each camera. Multiplexors are used to provide channels for the image data to go from the cameras to the memory card. The image data is transferred from the RAM in the memory card to the PC for the different levels of image processing.

6.1.1 Omni Vision Camera Chip

The OV5017 digital camera chips produced by OmniVision are used to capture the image of the fabrics. This camera chip can provides high-quality video image at low cost. It has 384x288 pixels and it can provide up to fifty frames of image per second. It is on-chip programmable and enables the adjustment of the frame rate, exposure setting and image size required. All the camera features are register based which makes accessing of the video data and configuring the chip as easy as reading or writing of a static memory. The image core is a complete analog video camera which output CCIR standard video signal. The camera chip contains an on-chip 8-bit analogue to digital converter which convert the video signal at 50 frames per second.

6.2 Software

MATLAB was used in our test for weaving defects classification. MATLAB is a technical computing environment for high-performance numeric computation and visualization. The Neural Network Toolbox of MATLAB contains many useful routines for implementing the neural network used for our test.

7. Conclusion

An AVI system for weaving defect detection requires speed and noise tolerance. The neural network approach seems to be an effective tool for coping with such needs as it exhibit features of associative memory, pattern matching and learning. In addition, once the network is trained, the step of defect classification involves just a feedforward operation which is very fast. One drawback of neural network is that the weights do not give any physical meaning to the problem at hand. A hybrid approach of integrating neural network with symbolic processing such as fuzzy logic could compensate for the drawbacks of each separately.

References

1. J. H. Han, K. Yi, "Design and implementation of a metal surface defect image acquisition system". Machine vision applications in industrial inspection V. SPIE Vol. 3029. pp. 101-108. 1997.

2. D. A. Lehotsky, "Developments in high speed inspection using intelligent CCD cameras". Machine vision applications, architectures, and system integration V. SPIE Vol. 2908. pp. 2-13. 1996.

3. J. Miettinen, A. K. Harkonen, "Procedure for image acquisition design". Computer vision for industry. SPIE Vol. 1989. pp. 70-80. 1993.

4. C. Fernandez, P. Campoy, C. Platero, J. M. Sebastian, R. Aracil, "On-line surface inspection for continuous cast aluminum strip". Computer vision for industry. SPIE Vol. 1989. pp. 26-37. 1993.

5. J. Laitinen, "Image quality in automated visual web inspection". Machine vision applications in industrial inspection V. SPIE Vol. 3029. pp. 78-89. 1997.

6. J. Laitinen, "Evaluation Procedure for imaging in web inspection". Imaging and illumination for metrology and inspection. SPIE Vol. 2348. pp.14-25. 1994.

7. L. Norton-Wayne, M. Bradshaw, C. Sanby, "Machine vision for the automated inspection of web materials". Computer vision for industry. SPIE Vol. 1989. pp. 2-13. 1993.

8. L. Macaire, J. G. Postaire, "Flaw detection on galvanized metallic strips in real-time by adaptive thresholding". Computer vision for industry. SPIE Vol. 1989. pp. 14-25. 1993.

9. M. Mlynarczuk, "Directional fibers analysis", Vision system: applications. SPIE Vol. 2786. pp. 146-156. 1996.

10. S. W. Lawson, G. A. Parker, "Automatic detection of defects in industrial ultrasound images using a neural network". Vision system: applications. SPIE Vol. 2786. pp. 37-47. 1996.

11. I. M. Dar, W. Mahmood, G. Vachtsevanos, "Automated pilling detection and fuzzy classification of textile fabrics". Machine vision applications in industrial inspection V. SPIE Vol. 3029. pp. 26-36. 1997.

12. A. Branca, G. Attoloco, A. Distante, "Multiscale data analysis for leather defect detection". Machine vision applications, architectures, and system integration V. SPIE Vol. 2908. pp. 97-108. 1996.

13. H. C. Abril, M. S. Millan, R. Navarro, " Pilling evaluation in fabrics by digital image processing". Vision system: applications. SPIE Vol. 2786. pp. 19-28. 1996.

14. C. Ciamberlini, F. Francini, G. Longobardi, P. Poggi, P. Sansoni, B. Tiribilli, "Weaving defect detection by fourier imaging". Vision system: applications. SPIE Vol. 2786. pp. 9-18. 1996.

15. E. J. Bayro-corrochano, "Review of automated visual inspection 1983-1993, Part I: Conventional approaches". Intelligent Robots and Computer Vision XII: Algorithms and Techniques. SPIE Vol. 2055. pp. 128-158. 1993

16. E. J. Bayro-corrochano, "Review of automated visual inspection 1983-1993, Part II: approaches to intelligent systems". Intelligent Robots and Computer Vision XII: Algorithms and Techniques. SPIE Vol. 2055. pp. 159-172. 1993

17. D. E. Rumelhart, G. E. Hinton, R. J. Williams, "Learning internal representation by error propagation". Parallel distributed processing: Explorations in the microstructures of cognition, MIT Press, Cambridge, M.A. Vol. 1. pp 318-362. 1986.

25

CONTROL OF A FED-BATCH BIOPROCESS BY USING NEURAL NETWORK OBSERVERS

Y. S. BOUTALIS and O. I. KOSMIDOU
Automatic Control Systems Laboratory
Department of Electrical & Computer Engineering
Democritus University of Thrace
GR-67100 Xanthi, Greece

1. Introduction

Bio-technological systems have drawn considerable attention in the last few years due to their very precise industrial applications. They are processes characterized by the presence of bacteria whose behavior is highly nonlinear and time varying. Moreover, many of the process parameters are uncertain or naturally varying during the process operation. In addition, external disturbances are often present at the system input. Under these conditions, it is very important to ensure the process operation (named «stability») and the desired productivity. Thus, appropriate control laws have to be designed.

The major limitation in fed-batch bioprocesses is due to the lack of appropriate sensors providing direct and reliable measures of biological or biochemical variables, in real-time. Thus, estimated values of these variables have to be used [1].

The present chapter deals with the production of *Saccharomyces Cerevisiae* (baker's yeast). The control objective is to ensure the process stability in the presence of disturbances. The model of this system has been analyzed in a recent work [4], where two forms of nonlinear observers, namely a high-gain observer and an Extended Kalman type observer [2], [3], have been used for the state estimation. In the present chapter, a nonlinear observer is designed in terms of an Artificial Neural Network (ANN). Simulation results are given to show the ability of multilayer ANNs to act as

S.G. Tzafestas (ed.), Advances in Intelligent Systems, 267–276.

reliable emulators of the system model dynamics. Moreover, based on these emulators, an adaptive configuration is proposed, which could act as a nonlinear adaptive observer.

2. Description of the process

Saccharomyces Cerevisiae (SC) is produced from the fermentation of glucose. Due to the presence of bacteria, specific conditions, such as temperature, pH, and concentrations of the different substances, are required. The fermentation process takes lieu in a reactor whose input and output are glucose and SC, respectively. Measured values of different quantities (ethanol, glucose, acetate) can also be considered as outputs. Three physiological modes are possible:

- glucose fermentation produces SC, ethanol and acetate (mode 1)
- glucose respiration produces only SC (mode 2)
- ethanol and acetate respiration produce SC (mode 3)

Transition from one mode to another is made according to activation and/or inhibition laws and depends on the milieu's composition. Since there is no limitation from the oxygen, the whole process works in an aerobic way. Thus, the corresponding kinetic expressions are used to describe growth rates of the different substances. On the other hand, the matter balance laws allow to determine their concentrations at the different modes. These concentrations are the state variables of the process:

X_1 : SC concentration in mode 1
X_2 : SC concentration in mode 2
X_3 : SC concentration in mode 3
E : ethanol concentration
S : glucose concentration
A : acetate concentration

Denoting as X the total concentration of SC one has

$$X = X_1 + X_2 + X_3 \tag{1.a}$$

$$\dot{X} = R_1 + R_2 + R_3 - DX \tag{1.b}$$

where R_i, i=1,2,3 are the growth rates of SC in modes 1,2,3, respectively, and D is the dilution rate of glucose at the system input.

The following state-space model has been developed in [4] and is used subsequently:

$$\dot{X}_1 = R_1 + R_5 + R_9 - R_4 - R_8 - DX_1 \tag{2.a}$$

$$\dot{X}_2 = R_2 + R_4 + R_7 - R_5 - R_6 - DX_2 \tag{2.b}$$

$$\dot{X}_3 = R_3 + R_8 + R_6 - R_9 - R_7 - DX_3 \tag{2.c}$$

$$\dot{E} = \frac{45}{14}R_1 - \frac{20}{11}R_3 - DE \tag{2.d}$$

$$\dot{S} = -\frac{50}{7}R_1 - 2R_2 + D(S_{in} - S) \tag{2.e}$$

$$\dot{A} = \frac{1}{14}R_1 - DA \tag{2.f}$$

where S_{in} is the glucose concentration at the input and $R_1, R_2 \ldots, R_9$ are the growth rates at the transitions from one physiological mode to another; these kinetics can be represented by the graph of figure 1 and they are given by the following relations.

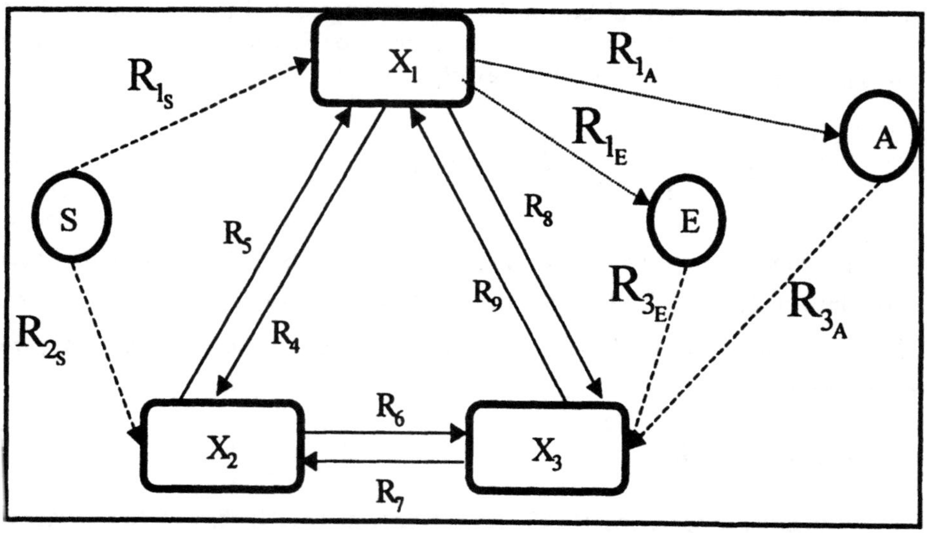

Figure 1. Model with three physiological modes

$$R_1 = \frac{12}{5}\frac{SX_1}{(1+2S)(2+5A)} \qquad R_2 = \frac{145}{2}\frac{SX_2}{1+250S}$$

$$R_3 = \frac{25}{2}\frac{EX_3}{(1+50E)(1+5A)} \qquad R_4 = 2\frac{X_1}{1+10S}$$

$$R_5 = \frac{SX_2}{1+10S} \qquad R_6 = \frac{4}{5}\frac{EX_2}{1+2E}$$

$$R_7 = 40\frac{SX_3}{1+20S} \qquad R_8 = \frac{EX_1}{(1+5E)(1+10S)}$$

$$R_9 = 50\frac{SX_3}{1+100S}$$

3. Control law

The control objective is to regulate the glucose concentration at the input in order to optimize, in some sense, the production of SC with respect to the glucose consumption. It has been found that this is achieved whenever the glucose concentration takes the value $S \equiv S_c = 0.071 g/l$. Thus, a set-point regulation problem has to be solved. Since the process is described by a nonlinear model, a possible control law consists in using pole placement for a linearizing control law [5]. Obviously, the only output to be regulated is the glucose concentration S. Thus, the linearizing control law is given by means of the following expression:

$$D = -\frac{1}{S_{in} - S}(-\frac{50}{7}R_1 - 2R_2 + \beta_s(S - S_c)) \tag{3}$$

The input should always be positive. In practice, one starts applying the control action at the instant t_0 when $E < 0.05 g/l$ and $S < 0.05 g/l$. Then, $S(t_0) < S_c$ and hence $D > 0$. One has $\dot{S} = -\beta_s(S - S_c) > 0$, and thus S exponentially tends to S_c, from inferior values.

Discrete-time model

In order to apply an Artificial Neural Network (ANN) observer to provide real-time estimates of the system variables, one has to use a discrete-time form of the system model. This can be obtained by using a first order Euler approximation. The discrete-time state equations are,

$$\begin{aligned} X_1(n+1) = X_1(n) + T(R_1(n) + R_5(n) + R_9(n) \\ - R_4(n) - R_8(n) - D(n)X_1(n)) + e_1(n) \end{aligned} \tag{4.a}$$

$$X_2(n+1) = X_2(n) + T(R_2(n) + R_4(n) + R_7(n)$$
$$- R_5(n) - R_6(n) - D(n)X_2(n)) + e_2(n) \quad (4.b)$$
$$X_3(n+1) = X_3(n) + T(R_3(n) + R_8(n) + R_6(n)$$
$$- R_9(n) - R_7(n) - D(n)X_3(n) + e_3(n) \quad (4.c)$$

$$E(n+1) = E(n) + T(\frac{45}{14}R_1(n) - \frac{20}{11}R_3(n) - D(n)E(n)) + e_4(n) \quad (4.d)$$

$$S(n+1) = S(n) + T(-\frac{50}{7}R_1(n) - 2R_2(n) + D(n)(S_{in}(n) - S(n))) + e_5(n) \quad (4.e)$$

$$A(n+1) = A(n) + T(\frac{1}{14}R_1(n) - D(n)A(n)) + e_6(n) \quad (4.f)$$

where n denotes the discrete-time index, (n=0,1, ...), T the sampling interval and $e_i(n)$ (i=1,2, ... , 6) denotes the combined effects of noise, discretization and modeling errors. Equivalently, for the growth rates one has

$$R_1(n) = \frac{12}{5}\frac{S(n)X_1(n)}{(1+2S(n))(2+5A(n))},\ R_2(n) = \frac{145}{2}\frac{S(n)X_2(n)}{1+250S(n)},\ \text{etc.}$$

4. Observation problem

In general the nonlinear state space model is of the form:

$$x_{k+1} = f(x_k, u_k)$$
$$y_k = h(x_k, u_k)$$

The objective is to control the nonlinear plant by a controller of known structure that uses the values of the state variables. However, the dynamic behavior of bioprocess systems is often characterized by state variables for which reliable measurements are difficult to be obtained. In these cases, it is preferable to estimate the state variables. Due to the variation of system parameters, an adaptive observer should be constructed that estimates the unmeasurable states subject to parameter variations. The typical adaptive observer configuration is shown in Figure 2. The difference between the output of the plant and that of the observer is the only information available and is used to formulate a suitable adaptation mechanism which will result in the states of the observer tracking closely the states of the plant. For linear systems, a great variety of design procedures have been developed providing stable and robust adaptation mechanisms, which under certain conditions ensure asymptotic convergence of the observation states to those of the plant. However this is not the case for nonlinear systems.

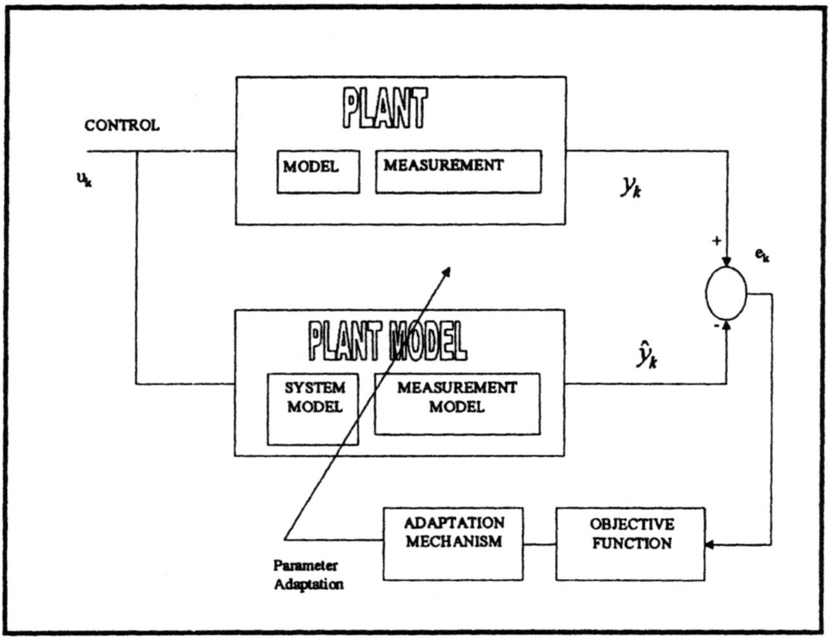

Figure 2. General configuration of an adaptive observer

Neural networks offer a promising alternative way of handling this problem. Artificial Neural Networks (ANNs) possess very interesting function approximation capabilities making them a very powerful tool in many scientific disciplines. More specifically ANNs have the theoretical ability to approximate arbitrary nonlinear mappings [6, 7]; there is also the possibility that such an ANN approximation is more parsimonious, i.e. it requires less parameters, than other competitive techniques such as orthogonal polynomials, splines, or Fourier series. Moreover since ANNs can have multi-inputs and multi-outputs, they can be naturally used for control of multivariable systems; they can be trained off-line using past data records of the system to be controlled or they can be adapted on-line in order to compensate for changes in the controlled system. It has been recognized that a controller based on multilayer neural network architectures offers a promising way of handling complex control problems.

5. Neural network observer

In the procedure described in section 2 of the chapter, the concentrations of ethanol, glucose and acetate can be measured, although the measurements are not extremely accurate. On the other hand the SC concentrations in modes 1,2,3 are not possible to be measured. Thus, the states X_1, X_2, X_3 have to be estimated. This is done by means of appropriately trained ANNs.

Three ANNs are trained to provide the open-loop state estimates for the three unmeasured concentrations. These ANN emulators are off-line trained to learn the dynamics of X_1, X_2, X_3, based on equations (2.a) - (2.c) of the model. The training configuration of the neural emulators is shown in figure 3. The same configuration applies for all of the three SC concentration modes dynamic models, but the outputs of the model and of the neural model emulator are $(X_1, \hat{X}_1)$, $(X_2, \hat{X}_2)$ and $(X_3, \hat{X}_3)$, respectively.

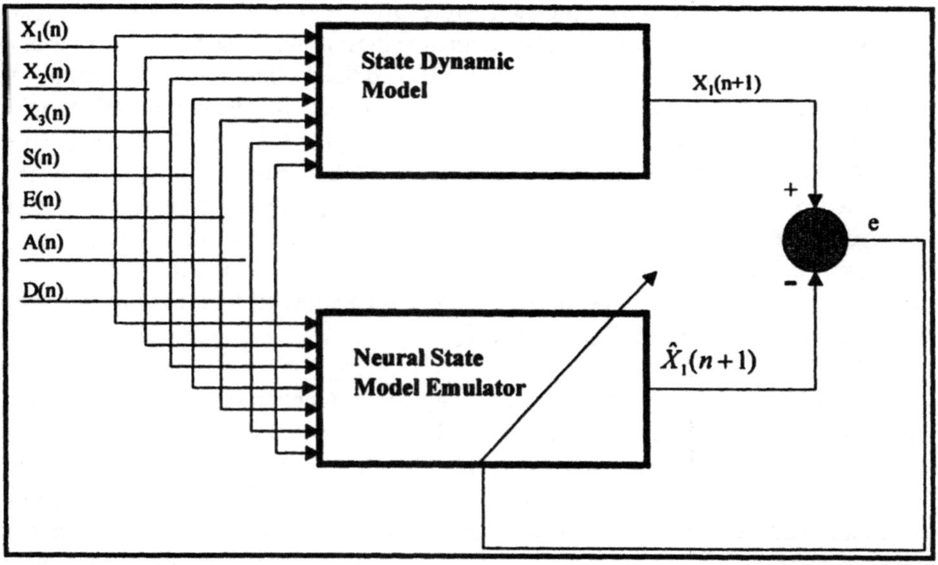

Figure 3. The training configuration of the SC concentration model (mode 1).

Estimated values produced by the open-loop neural state estimators, are used in the sequel by the control law of eqn (3). In the next section, the ability of a multilayer ANN to learn the system dynamics is examined and some further improvements for the training configuration are suggested.

However, even if the unmeasured states are estimated by a very accurate open loop emulator they may more or less seriously deviate from the real ones due to external disturbances or modeling errors (e.g. the very procedure of measuring E, S and A may disturb the hole procedure). Therefore, an adaptive configuration should be adopted to compensate for the changing plant conditions. It is suggested that this configuration comprises two stages: (i) The off-line training of six ANNs to act as state model emulators and (ii) the on-line adaptation of the three ANNs, which predict the unmeasured states. In the second stage the errors between the measured states (E, S, A) and their estimates by the corresponding emulators are used to adapt the ANNs that predict the unmeasured states. Since measurements are not considered very accurate, care should be taken so that the way by which the above mentioned errors influence the adaptation mechanism reflects the real need for adaptation. Although the details and experimental studies of this configuration are not given in this chapter, preliminary experiments have shown that it is a promising tool, which, if fine-tuned, could act as an adaptive nonlinear observer.

6. Training of the neural networks

The neural network emulators were off-line trained by using the nominal values of the quantities involved. The nominal values of the six states and the input (D) were produced by the model equations (4a – 4f) and the discrete form of the control law (3). The sampling period T were chosen to be 0.0009 h. The initial values of the procedure were $[X_1(0), X_2(0), X_3(0), E(0), S(0), A(0)] = [14.7, 0, 0, 1.8, 2.5, 0]$ g/l, Sin=300 g/l. The input D (which is the relative dilution rate of the input glucose) is computed by eqn (3) with $\beta_s = 1$, but it is always kept greater than $D_{min} = 0.02\ h^{-1}$. Figures 4 – 6 show 15000 samples of the nominal state and input values. This is equivalent to 13.5 hours of continuous operation. In figure 6 the effect of the control law is evident. The training of the ANNs were accomplished by means of the widely known back-propagation (BP) algorithm [8].

It has to be noticed that due to the sigmoidal function considered at the output neurons, the neural network outputs are restricted to vary within the range [0, 1] or [-1, +1] depending on the particular sigmoidal function used. This range is certainly different from the actual output variables range. A normalization procedure transforms each input and output range to the appropriate range of variation. Normalization keeps the connection weights at low level values thus avoiding saturation of neurons which might cause several convergence and stability problems during the learning process.

The training configuration of the neural emulators is shown in figure 3. The chosen neural networks were four layer networks 7-7-10-1 (2 hidden units). The choice of that particular structure was based on simulation experiments. Samples of selected areas of the nominal values presented in figures 4 – 6 were used to train the networks by a cross-validation procedure. In this procedure about 60% of the selected samples were used randomly as test samples and the rest of them were used for validation purposes. The same procedure is repeated for different random test and validation sample selection from the same initial sample areas. The network giving the best validation results was chosen to be the ANN model emulator of each state. In the sequel the whole test sample of the nominal values were used to test the generalization capabilities of each network. Figure 7 shows the nominal and the estimated values of state X_1. It is evident that the network is a good approximator of the model (4.a), but there are areas of the I/O sample space that need a more specialized learning. The same network was used to predict state X_1 based on a different set of input data. This set was produced by the same equations (4a – 4f, 3), but operating under the initial conditions $[X_1(0), X_2(0), X_3(0), E(0), S(0), A(0)] = [10, 4, 0.2, 2, 2, 0.02]$ g/l. The network showed good approximation capabilities and on this input set but some retraining is necessary in order to present optimal behavior. This phenomenon appears to be common in complex control tasks and it can be attributed to the difficulties of the associated learning problems. Thus, although MLPs are, theoretically, universal function approximators, practically, it is not easy to train them to approximate arbitrarily complex continuous functions, as the ones that might be encountered when dealing with the uncertainties met in control tasks, as previously depicted. A methodology that could lead to an improvement of the MLP function approximation capabilities is proposed in [9] and its application on the problem we are dealing with in this chapter is now under investigation

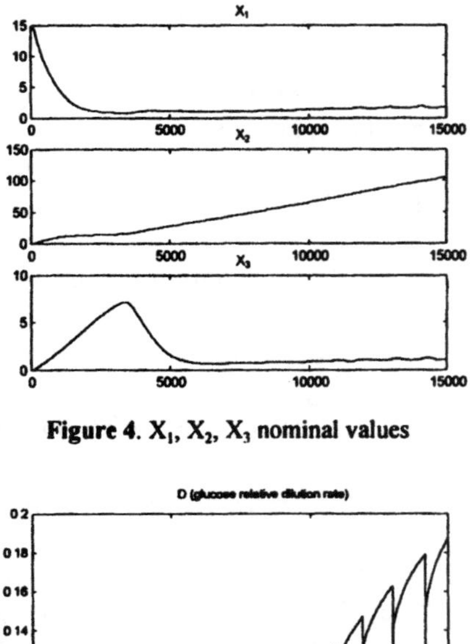

Figure 4. X_1, X_2, X_3 nominal values

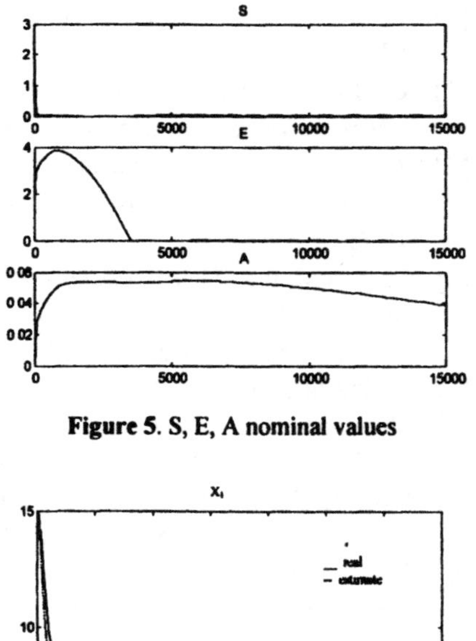

Figure 5. S, E, A nominal values

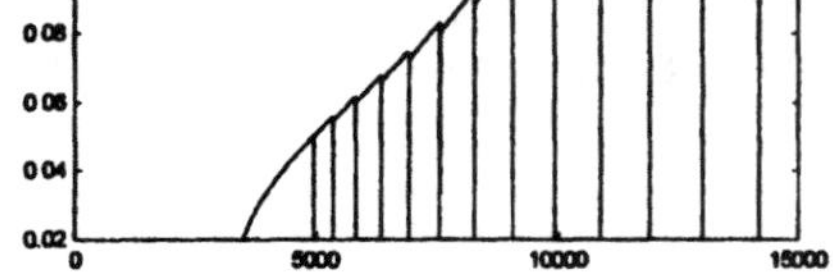

Figure 6. D (the relative dilution rate of glucose)

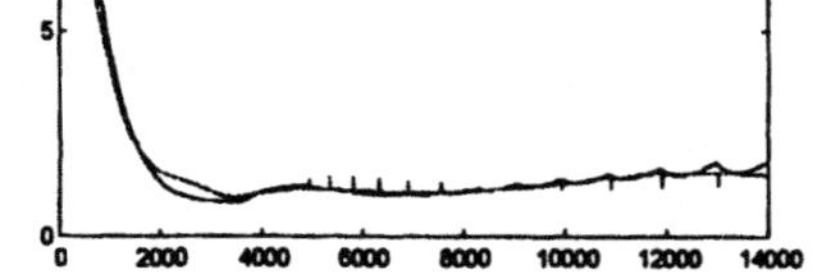

Figure 7.Nominal and estimated X_1 values

7. Conclusions

The control of the production of *Saccharomyces Cerevisiae* (baker's yeast) is used as an example of controlling fed-batch bioprocessess using ANN observers. The control objective is to ensure the process "stability" in the presence of disturbances. The model of this system has been analyzed in a recent work [4]. In the present chapter, a nonlinear observer is designed in terms of an Artificial Neural Network (ANN). Simulation results are given to show the ability of multilayer ANNs to act as reliable emulators of the system model dynamics. It is shown that, although ANNs are good approximators for this kind of models, their approximation abilities should be enhanced using more sophisticated techniques. Moreover, based on the ANN emulators, an adaptive configuration is proposed, which could act as a nonlinear adaptive observer.

References

[1] G. Bastin and D. Dochain, *On line estimation and adaptive control of bioreactors*, Elsevier, 1990.

[2] G. Bornard and H. Hammouri, *A high gain observer for a class of uniformly observable systems*, 30th IEEE CDC, December 1991, pp. 1494-1496.

[3] J.-P. Gauthier, H. Hammouri and S. Othman, *A simple observer for nonlinear systems: application to bioreactors*, IEEE Trans. Autom. Control, vol. 36, no 6, 1992.

[4] M. Pengov, *Deux observateurs non-linéaires: application à un procédé biotechnologique,* in «Théorie et Applications des Observateurs Non Linéaires», GR-Automatique CNRS, Paris, France, 8-9 Feb. 1996, pp. 40-49.

[5] A. Isidori, *Nolinear Control Systems,* Springer-Verlag, 1989.

[6] Cybenco G., *Approximation by superposition of a sigmoidal function*", Mathematics of Control, Signals and Systems, **2**, pp. 303-314, 1989.

[7] Hornik K., M. Stinchombe, and H. White, *Multilayer feedforward network are universal approximators*", Neural Networks **2**, pp. 359-366, 1989.

[8] D. E. Rumelhart, G. E. Hinton and R. J. Williams, *Learning internal representations by error propagation*, in Parallel Distributed Processing, D. E. Rumelhart and J. L. McLelland, Eds., vol. 1. Cambridge, MA: MIT Press, 1986.

[9] D. A. Karras, Y. S. Boutalis and B. G. Mertzios, *A two-stage methodology for approximating functions based on neural networks with applications to intelligent dynamic robot control*, Proceedings IWSSIP'98, pp. 139-142, Zagreb, Croatia, June 1998.

26

NEURAL NETWORKS FOR THE MODEL IDENTIFICATION OF NAVAL TURBOCHARGERS

N. G. PANTELELIS, A. E. KANARACHOS
and N.GOTZIAS
Mechanical Design and Control Systems Division
Department of Mechanical Engineering
National Technical University of Athens
Zografou Campus,P.O. Box 64078, 15710 Athens, Greece

1. Introduction

The condition monitoring of complex machines is still a very open research topic as until recently experience and signal processing techniques were the only tools for the fault diagnosis of such machines. Attempts to solve the predictive maintenance problem using numerical simulations have been presented extensively for example in [1, 2, 3]. Furthermore, analytical and Neural network methods have been used extensively in the identification of dynamic models [4, 5, 6, 7, 8]. With respect to the modelling of the nonlinear and the transient behaviour of the journal bearings a very helpful approximate analytical procedure can be found in [9]. The nonlinear interaction between the hydrodynamic bearings and the mass unbalance response has been studied in [10].

On the other hand, the very popular Artificial Neural Networks (ANN) have been employed almost everywhere to solve problems such as control, modelling, system identification [11, 12], pattern recognition etc. The advantages of the Neural Networks over other numerical methods is the parallel and massive processing using simple functions, the inherent property of the adaptation, learning and training and the potential to operate in real time processing. Because of these properties ANN are the ideal choice for real time applications where large amount of data are being processed and analysed. ANN have been employed in the machine condition monitoring mainly for fault detection via the training of ANN with healthy and faulty signals resulted either from test rigs and/ or theoretical simulations [13, 14, 15,16]. A basic feature of these systems is the assumption of accurate simulations (experimental or theoretical) and the accurate production and/or recording of massive simulated data with various potential faults [17]. Of course the accuracy of the simulated data even in the case of test rig data is an open issue.

S.G. Tzafestas (ed.), Advances in Intelligent Systems, 277–288.

The condition monitoring of naval turbochargers is a challenge as apart from the high rotational speed of the flexible rotor and the journal bearings' non-linear behaviour, the system is excited with large displacements from the main engines of the ship where the turbochargers are founded on. The main target of this work is the automatic diagnosis of potential faults of a turbocharger using a series of acceleration measurements from certain points and directions of the machine. This will be possible combining the measurements with simple Finite Element models and Artificial Neural Networks, in an attempt to combine only the advantages of each of the these methods. In order to diagnose potential faults of the turbochargers by just monitoring, requires considerable experience and guessing allowing for considerable errors due to the complexity of the problem. However, measurements contains primary information which should be taken into consideration. On the other hand, using only Finite Element (FE) models to simulate this complex system is almost impossible to achieve fast and accurate results due to the difficulty to approximate all the unknown characteristics of the system. However, FE models, and especially the simple ones, should be also considered as a powerful tool to guide the fault diagnosis procedure through the physics of the problem. Finally, Artificial Neural Networks (ANN) are introduced as a powerful tool for the coupling of the above methods in a real time tool that will diagnose possible faults and their severity. The principal of the proposed method is that by feeding the system with acceleration measurements, the ANN will identify several parameters of the FE model in order to match the simulated output to the measured data. These tuneable parameters of the FE model refer to certain possible faults of the machine (mass imbalance, cracks, oil whirl etc.). So, the output of the proposed system will be the fault diagnosis of the machine together with additional information on the severity of the corresponding faults. Using this approach, the possibilities for unnatural outputs, likely to result from a simple Neural Networks Fault Diagnosis system due to several possible causes (a fault that is out of range, a fault not taught correctly, non-linear behaviour of the machine etc.), are eliminated without requiring significant computational costs.

The rest of this chapter is organised as follows: In section 2, the models that has been used to simulate accurately the behaviour of the turbocharger will be explained. The proposed strategy will be demonstrated in section 3 and the ANN method that is adopted will be presented in section 4. In section 5, the performance of the system will be demonstrated for a number of simplified test cases and finally, conclusions will be drawn and our future plans will be presented in section 6.

2. System Modelling

For the simulation of the general case of a flexible rotor with disks which is supported in journal bearings, the classic finite element type is used, having two nodes with two displacements (v and w) and two slopes (θ and ψ) at each node:

$$q_{fb} = \{v_1 \quad \theta_1 \quad w_1 \quad \psi_1 \; v_2 \quad \theta_2 \quad w_2 \quad \psi_2\}^T \tag{1}$$

Introducing the bending behaviour of the shaft the Element Mass Matrix is given as:

$$\mathbf{M}_{fb} = \begin{bmatrix} M_{yy} & 0 & M_{yz} & 0 & -M_{yy} & 0 & -M_{yz} & 0 \\ 0 & 0 & 0 & 0 & 0 & 0 & 0 & 0 \\ M_{yy} & 0 & M_{zz} & 0 & -M_{zy} & 0 & -M_{zz} & 0 \\ 0 & 0 & 0 & 0 & 0 & 0 & 0 & 0 \\ -M_{yy} & 0 & -M_{yz} & 0 & M_{yy}+M_h & 0 & M_{yz} & 0 \\ 0 & 0 & 0 & 0 & 0 & 0 & 0 & 0 \\ -M_{yz} & 0 & -M_{zz} & 0 & M_{zy} & 0 & M_{zz}+M_h & 0 \\ 0 & 0 & 0 & 0 & 0 & 0 & 0 & 0 \end{bmatrix} \quad (2)$$

The corresponding Element Damping Matrix by:

$$\mathbf{C}_{fb} = \begin{bmatrix} C_{yy} & 0 & C_{yz} & 0 & -C_{yy} & 0 & -C_{yz} & 0 \\ 0 & 0 & 0 & 0 & 0 & 0 & 0 & 0 \\ C_{yy} & 0 & C_{zz} & 0 & -C_{zy} & 0 & -C_{zz} & 0 \\ 0 & 0 & 0 & 0 & 0 & 0 & 0 & 0 \\ -C_{yy} & 0 & -C_{yz} & 0 & C_{yy}+C_{yys} & 0 & C_{yz} & 0 \\ 0 & 0 & 0 & 0 & 0 & 0 & 0 & 0 \\ -C_{yz} & 0 & -C_{zz} & 0 & C_{zy} & 0 & C_{zz}+C_{zzs} & 0 \\ 0 & 0 & 0 & 0 & 0 & 0 & 0 & 0 \end{bmatrix} \quad (3)$$

and the Element Stiffness Matrix as:

$$K_{fb} = \begin{bmatrix} K_{yy} & 0 & K_{yz} & 0 & -K_{yy} & 0 & -K_{yz} & 0 \\ 0 & 0 & 0 & 0 & 0 & 0 & 0 & 0 \\ K_{yy} & 0 & K_{zz} & 0 & -K_{zy} & 0 & -K_{zz} & 0 \\ 0 & 0 & 0 & 0 & 0 & 0 & 0 & 0 \\ -K_{yy} & 0 & -K_{zy} & 0 & K_{yy}+K_{yys} & 0 & K_{yz} & 0 \\ 0 & 0 & 0 & 0 & 0 & 0 & 0 & 0 \\ -K_{yz} & 0 & -K_{zz} & 0 & K_{zy} & 0 & K_{zz}+K_{zzs} & 0 \\ 0 & 0 & 0 & 0 & 0 & 0 & 0 & 0 \end{bmatrix} \quad (4)$$

To simplify the analysis, the unbalance force vector is introduced corresponding to the force resulted from the mass unbalance m_i at a distance e_i at running speed Ω rad/s:

$$F_i = \begin{Bmatrix} F_{yi} \\ M_{zi} \\ F_{zi} \\ M_{yi} \end{Bmatrix} = \begin{Bmatrix} m_i \cdot e_i \cdot \Omega^2 \cdot cos(\Omega t) \\ 0 \\ m_i \cdot e_i \cdot \Omega^2 \cdot sin(\Omega t) \\ 0 \end{Bmatrix} \quad (5)$$

The corresponding inertia and damping matrices for a disk (concentrated element) are:

$$\begin{bmatrix} M_D & 0 & 0 & 0 \\ 0 & M_D & 0 & 0 \\ 0 & 0 & I_{Dy} & 0 \\ 0 & 0 & 0 & I_{Dy} \end{bmatrix} \cdot \begin{bmatrix} u^{\circ\circ} \\ w^{\circ\circ} \\ \theta^{\circ\circ} \\ \psi^{\circ\circ} \end{bmatrix} + \Omega \cdot \begin{bmatrix} 0 & 0 & 0 & 0 \\ 0 & 0 & 0 & 0 \\ 0 & 0 & 0 & -I_{Dx} \\ 0 & 0 & I_{Dx} & 0 \end{bmatrix} \cdot \begin{bmatrix} u^{\circ} \\ w^{\circ} \\ \theta^{\circ} \\ \psi^{\circ} \end{bmatrix} \qquad (6)$$

where the second matrix of term (6) is the gyroscopic (Coriolis) matrix.

Assuming that the journal bearings present a linear behaviour, the mass unbalance at the disks would allow the journals to rotate in an elliptical orbit around the equilibrium (no mass unbalance force) position, at synchronous speed. Similarly, the corresponding matrices for the bearings are given as:

$$-\begin{bmatrix} k_{yy} & k_{yz} & 0 & 0 \\ k_{zy} & k_{zz} & 0 & 0 \\ 0 & 0 & 0 & 0 \\ 0 & 0 & 0 & 0 \end{bmatrix} \cdot \begin{bmatrix} v \\ w \\ \theta \\ \psi \end{bmatrix} - \begin{bmatrix} c_{yy} & c_{yz} & 0 & 0 \\ c_{zy} & c_{zz} & 0 & 0 \\ 0 & 0 & 0 & 0 \\ 0 & 0 & 0 & 0 \end{bmatrix} \cdot \begin{bmatrix} v^{\circ} \\ w^{\circ} \\ \theta^{\circ} \\ \psi^{\circ} \end{bmatrix} \qquad (7)$$

The first matrix is a stiffness matrix and the second a viscous damping matrix. These matrices are generally asymmetric ($k_{yz} \neq k_{zy}$ and $c_{yz} \neq c_{zy}$) and the terms of the matrices can vary significantly as a function of the running speed. However, at the presence of non-linearities, especially for highly unbalanced rotors, the journal motion produces harmonics one order higher than the synchronous frequency. Hence, it is essential that the full non-linear hydrodynamic equations [9] should be studied in order to attain a realistic model of the transient journal bearing behaviour (fig.1).

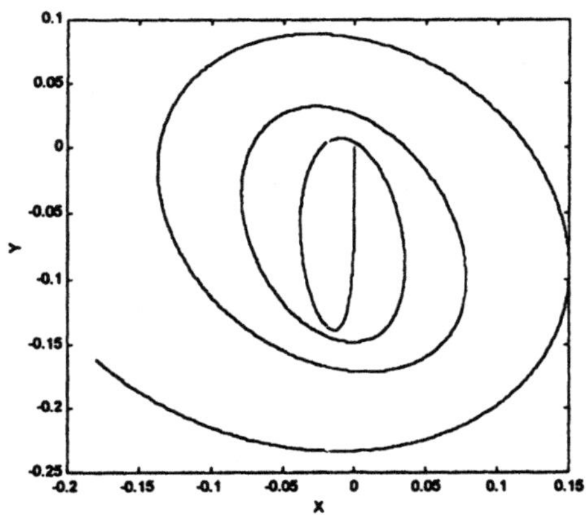

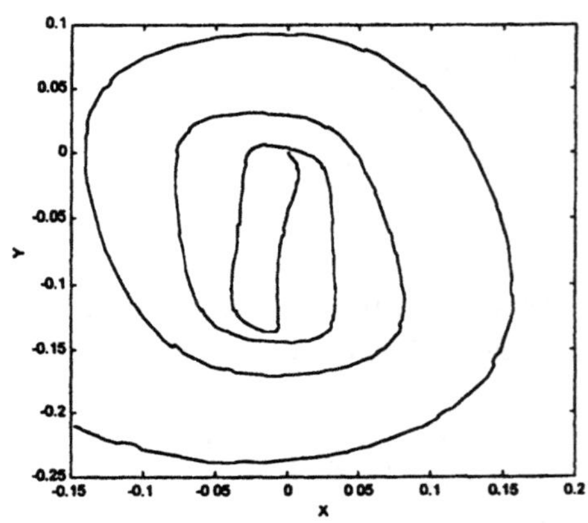

Figure 1. The locus of the center of a journal in a start up simulation for a balanced (left) and an unbalanced nonlinear case.

Combining equations (1) through (7) the following dynamic system occurs:

$$\mathbf{M} \cdot \mathbf{q}^{\cdot\cdot} + \mathbf{C}(\Omega) \cdot \mathbf{q}^{\cdot} + \mathbf{K} \cdot \mathbf{q} = \mathbf{f}(t) \quad (8)$$

where q is the vector that contains all the nodal displacements and slopes, M is the symmetric mass matrix, C the asymmetric matrix, including an antisymmetric gyroscopic matrix function of the running circular speed, Ω, K the stiffness matrix, frequently asymmetric owing to the characteristics of the bearings and f(t) the external force vector that includes bearing forces, imbalance forces etc.

Apart from the simulation of the mass unbalance fault several other potential faults such as oil whirl, cracked shaft, can be introduced in our model. As the turbochargers are always founded on the main engine low, with respect to its running speed, frequency vibrations are transferred from the main engine to the impeller through the hydrodynamic bearings. Even though these low frequency vibrations (around 500 rpm in the present case) are far away from the impeller' s running speed (12.000 rpm) they do affect its operation in a nonlinear way through the clearance variation due to the vibrations of the bearing itself.

Moreover, the ability to measure only the bearings' vibrations and not the journal, and consequently, the impeller itself complicates further the problem. To tackle this problem, thanks to the significant distance between the two basic frequencies filters can be introduced to separate the lower and the higher frequency regions. Thus, using a lowpass and a highpass filters with cut-off frequency at 6 times the running speed of the main engine the bearings and the journal vibrations, respectively can be identified.

3. The Fault Diagnosis System

The automatic fault diagnosis of complex machines with strongly nonlinear behaviour is a task that requires highly expertise engineers, involving many parameters that small deviations can lead to erroneous results. For these reasons intelligent systems are necessary to prevent expensive machines from sudden breakdowns which, consequently, may cause serious problems to the operation of a whole industry, ship or plane. The approach followed at the present work combines the, as simple as possible, modelling of all the possible faults of a turbocharger with the Neural Networks having a dual role into the system. First, to learn from several faults that can be simulated by the turbocharger's finite element models as well as from previous collected measured data and second, to combine both modelling and Neural Networks in a real time identification procedure with reference the current measurements of the machine. This way we can take full advantage of the learning characteristics and the simple nature of the Neural Networks. However, due to the strong nonlinearities resulted from the journal bearings and the low frequency main engine vibrations from the turbocharger's foundations it is very difficult to teach correctly and accurately the Neural Networks. This is where the simulation model is required.

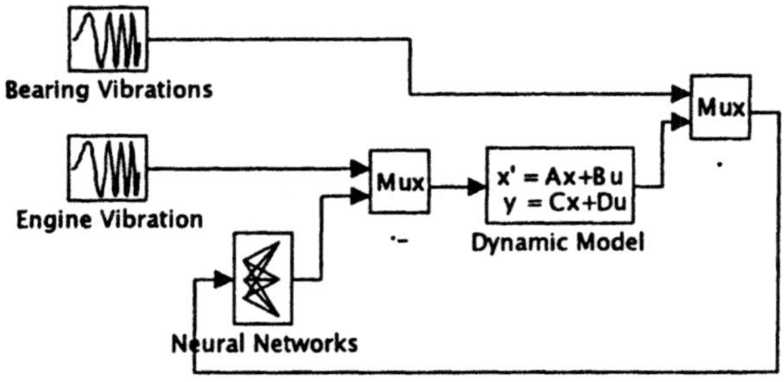

Figure 2. The real time faults diagnosis system.

Within the present approach the machine is monitored at certain points and directions. Feeding these signals into the proposed system, the model together with the Neural Networks try to produce similar signals by tuning the parameters- faults of the model and consequently to identify which are the faults of the machine as well as which is the severity of the corresponding faults. The procedure that the system follows to identify the potential faults of the model is depicted in fig. 2. The ANN identification procedure iterates to identify the possible faults of the dynamic model using as an input vector the vibrations from the foundation (engine vibrations box in fig. 2) together with the simulation error of the model and the measured data (bearings vibrations box in fig. 2).

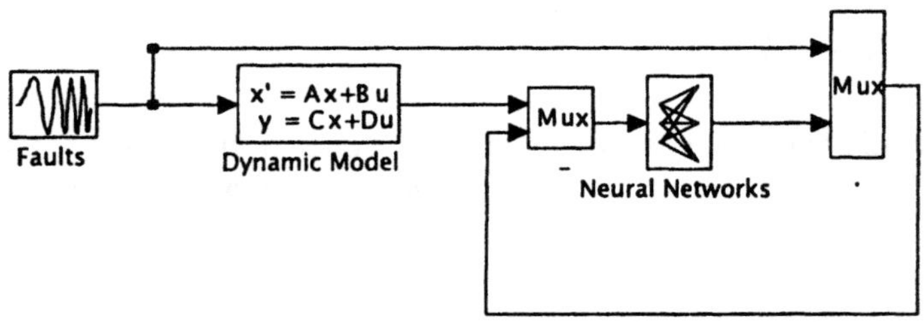

Figure 3. The simplified faults diagnosis system.

As a first approach towards the development of the proposed system, a simpler system has been developed to focus mainly on the Neural Networks models. In this model certain simplifications to the original Finite Element model have been introduced. The problem will be studied as a steady state problem and not as a transient one, thus no nonlinearities from the bearings motion are considered. Furthermore, the position of the ANN will be after the dynamic model and not before as in the original system. This will accelerate considerably the learning time, as the dynamic behaviour is not calculated at each NN identification step but it is taken from already stored data.

4. Neural Networks

In general for the identification of a dynamic system several optimisation methods could be used. The neural networks were chosen because of the dual role that can play as discussed in section 3. Thus, the identification procedure consists of the neural networks training phase and the validation phase. During the first phase the ANN is fed with a sequence of input vectors together with the corresponding output vectors. At the present study and the simple model as input vector the power spectral density components of the vertical deflection of the bearings are considered. To avoid including the individual frequencies in the input vector the frequency axis is scaled with respect to the running speed of the turbocharger, as the running speed fully characterises the faults' behaviour. Using the same dynamic analysis data the identification procedure starts from an approximate initial vector of the weights and biases of the ANN and the resulted output vector is compared to the desired one. As the output vector is the quantified components of the potential faults, in the current study the unbalanced masses at the disks is the output vector of the system. An objective function based on the error between the current and the desired output is calculated to guide the learning procedure to the direction where weights and biases will decrease it. When the iterative procedure is converged i.e. the error between the model and the 'known' learning output is small enough the ANN is considered that it has identified the parameters-faults of the model.

For the learning procedure the back propagation algorithm with optimal learning rate has been employed. The input vector consists of 30 power spectral components of the velocity at the journal bearing that corresponds to circular speeds from half to 15 times the running speed of the impeller. One hidden layer was introduced with 20 neurons and for the output vector the mass unbalance was considered. The logistic sigmoid transfer function ($\frac{1}{1+e^{-x}}$) was employed both for the input vector to the hidden layer as well as a from the hidden layer to the output vector. Using the feedforward algorithm and the optimal weights and biases found from the learning step unbalanced masses can be identified from an unknown deflection signal with acceptable accuracy as it will be shown in section 5.

Of course, few parameters have been studied with respect to the NN set up and many more remains to be studied especially in conjunction with the increase of the model complexity and the nonlinearities introduction.

5. Results

As the problem that we deal with is a very complicated problem a step by step approach will be considered. At the present study, the mass imbalance fault will be studied for a real turbocharger, shown in fig. 4 (left). Its finite element model with the two disks and the two bearings are depicted in fig. 4 (right).

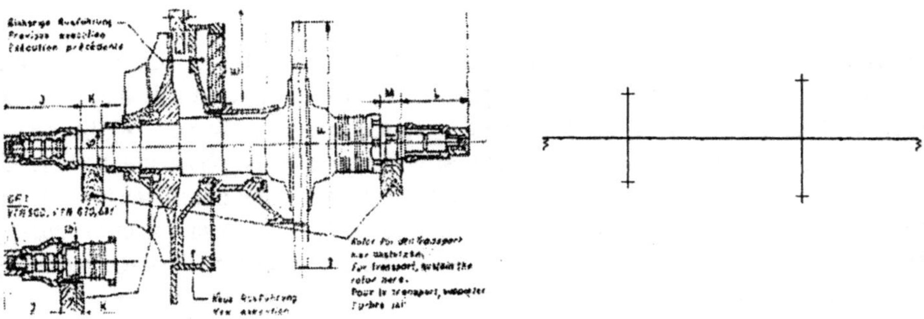

Figure 4. Drawings of the real impeller (left) and the corresponding FE model (right).

The simulated dynamic response (deflections at a journal bearing) of the model in the case of two unbalanced masses and vibrations from the main engine (the impeller runs at 12.000 rpm and the main engine at 500 rpm) is shown in fig. (5). The higher frequency component is due to the mass unbalance in one of the disks at the running speed of the impeller and the lower frequency contribution is due to the ships' large diesel main engine applied at the foundations (bearings) of the rotor.

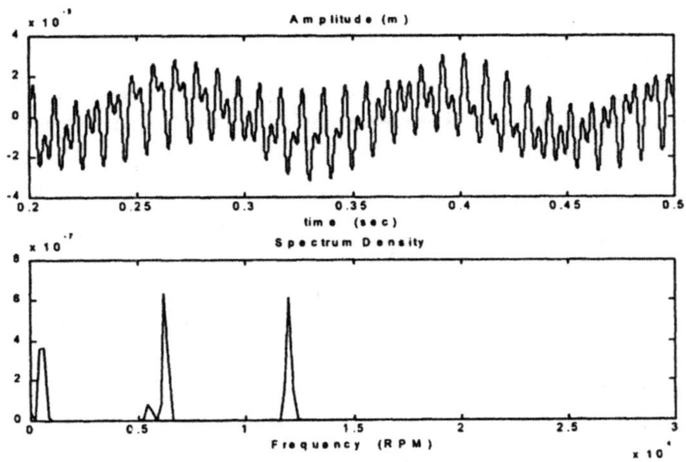

Figure 5. Time history and power spectral density of the vertical deflections at one of the journals due to mass imbalance at both disks and vibrations from the main engine.

With respect to the Neural Networks set up parameters after several trials the optimal performance was achieved employing one hidden layer with 20 neurons. For the evaluation of the learning performance two tests were performed. In the first one, the case of a single unbalanced mass at one of the disks was considered. The convergence of the learning procedure was terminated at 10^{-5} error of the unbalanced mass after 40.000 iterations. In order to validate the solution unknown input signals, different from those fed for learning purposes, were used as input to the model and the output result was compared to the correct one. As shown in fig. 6 the «unknown» mass unbalances (circles in fig. 6) are identified with good accuracy with respect to the correct output (solid line).

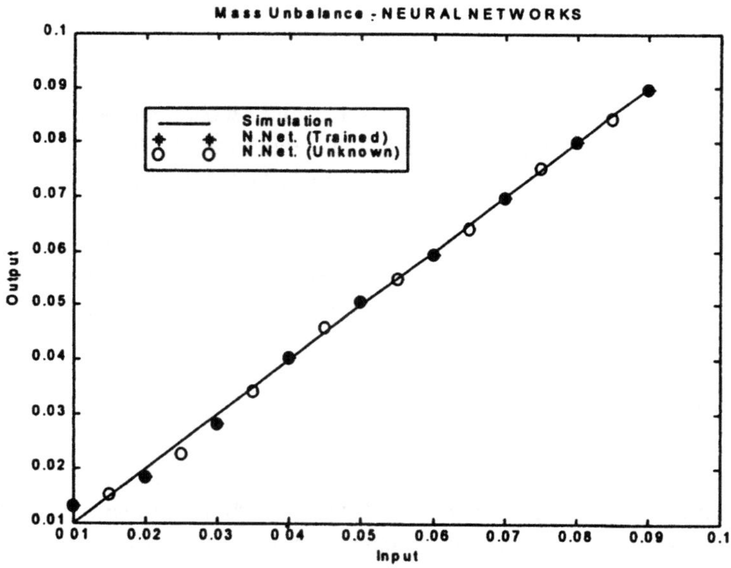

Figure 6. Learning performance of the ANN in the first test case. The expected outputs (solid line) are compared to fed outputs (star+ circle) and «unknown» outputs (circle).

In the second test a more complicated fault case was introduced: both of the disks are unbalanced. Only three unbalanced masses for each disk (only 9 cases) were fed into the ANN identification system for the learning procedure and the converged accuracy is shown in fig. 7 (left). The performance of the ANN with respect to «unknown» mass unbalances can be depicted in fig. 7 (right). The accuracy is very satisfactory considering that from the power spectral density the system does not find only the type of fault but also the severity of the fault in some adequate accuracy. Of course, further experimentation is required in order to explore all the possible improvements. For example, the scaling of the input vectors which consists of the power spectral density components of the vibrations with respect to their maximum value [17], may raise considerably the neural networks' performance.

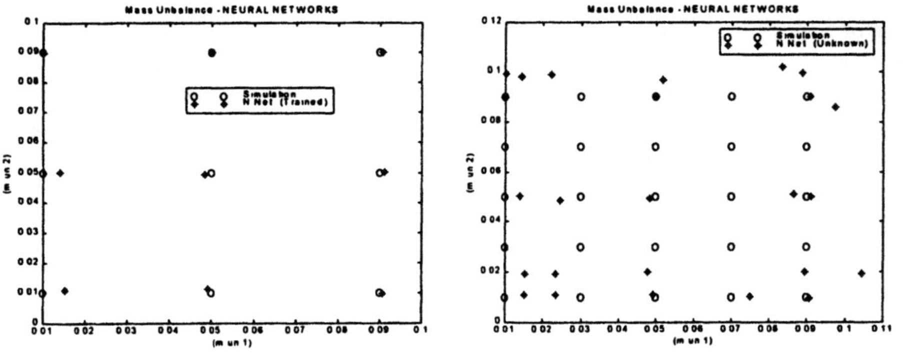

Figure 7. Learning performance of the ANN in the second test case. On the left the correct outputs (circle) are compared with converged outputs (star). On the right the «unknown» outputs (star) are compared to the correct ones (circle).

Alongside four naval turbochargers are being monitored for six months time and significant amount of data is being collected. The incorporation of these data into the under development system either for learning purposes or for evaluation purposes is one of our next steps.

6. Conclusions

A new method for the fault diagnosis of naval turbochargers was presented. Trying to combine only the advantages of very popular tools such as the theoretical simulations, measurements and signal processing together with the Neural Networks, the proposed method promises better physics- oriented and more generalised diagnosis, with quantification of the corresponding faults. Many difficulties to set up the Neural Networks and the development of theoretical models have been encountered due to complexity of the problem.

Real time accurate fault diagnosis is a real challenge but also it is the ultimate target. Our next steps towards the achievement of this target are the incorporation of all possible faults in the theoretical model, the improvement of the performance of the NN, the coupling of the measured data from the user site and a possible inclusion of data from a test rig which is operating under another CE project.

ACKNOWLEDGEMENTS

This research was funded by EU under the Brite-Euram VISION Project, Contract No. BREU-CT96-1313. The co-operation with the partners of the consortium and especially with Maritime Company of Lesvos is also gratefully acknowledged.

REFERENCES

1. Bettig, B.P. and R.P.S. Han (1998). Predictive Maintenance using the Rotordynamic Model of a Hydraulic Turbine- Generator rotor, Transactions of ASME, Journal of Vibration and Acoustics, **120**, 441-448.

2. Lee, A-C. and Y-P Shih (1996). Identification of the unbalance distribution and dynamic characteristics of bearings in flexible rotors, Proc. Instn. Mech. Engrs., **210**, 409-432

3. Russo, M. and R.Russo (1993). Parametric excitation instability of a rigid unbalanced rotor in short turbulent journal bearings, Proc. Instn. Mech. Engrs., **207**, 149-160.

4. Lee, C-W. and S-W Hong (1989). Identification of bearing dynamic coefficients by unbalance response measurements, Proc. Instn. Mech. Engrs., **203**, 93-101.

5. Zhu, C. and F.W. Paul (1995). A Fourier series Neural Network and its application to system identification, Transactions of ASME, Journal of Dynamic systems, Measurement, and Control, **117**, 253-261

6. Chen, S. and S.A. Billings (1992). Neural networks for nonlinear dynamic system modelling and identification, Int. J. Control, **56**, 319-346.

7. Masri, S.F., A.G. Chassiakos and T.K. Caughey (1993). Identification of Nonlinear Dynamic Systems Using Neural Networks, Journal of Applied Mechanics, **60**, 123-133.

8. Narendra, K., K. Parthasarathy (1990). Identification and Control of Dynamical Systems Using Neural Networks, IEEE Transactions on Neural Networks, **1**, 4-27.

9. Childs, D. (1993). Turbomachinery rotordynamics. Phenomena, Modelling and Analysis, J. Wiley and Sons, Inc, NY, 1993.

10. Zhao, J.Y., I.W. Linnett and L.J. McLean (1998). Unbalance response of a flexible rotor supported by a squeeze film damper, Transactions of ASME, Journal of Vibration and Acoustics, **120**, 32-38.

11. Annaswany, A.M. and S-H. Yu (1996). θ - Adaptive Neural Networks: A new approach to parameter estimation, IEEE Transactions on Neural Networks, 7, 907-918.

12. Chen, S., S.A. Billings and P.M.Grant (1990). Non-linear system identification using neural networks, Int. J. Control, **51**, 1191-1214.

13. Chen, Y. And R.Du (1996). Artificial Neural Network architecture design using similarity measure with applications in engineering monitoring and diagnosis, Transactions of ASME, Journal of Dynamic systems, Measurement, and Control, **118**, 635-639.

14. Emmanouilidis, C., J. MacIntyre, C. Cox (1998). An integrated, soft computing approach for machine condition diagnosis, (to appear in EUFIT' 98).

15. Wang, X.F. and P.D. McFadden (1996). Simulation models for bearing vibration generation and fault detection via neural networks, in: IMechE Conference Transactions, Sixth Intl Conference on Vibrations in Rotating Machinery, MEP, London, 1996.

16. McCormick, A.C. and A.K. Nandi (1997). Classification of the rotating machine condition using artificial neural networks, Proc. Instn. Mech. Engrs., **211**, 93-101.

17. Jantunen, E. and K. Vaha-Pietila (1997). Simulation of Faults in Rotating Machines Proc. of COMADEM 97, Vol 1, pp283-292.

27

AN OPTIMUM SUSPENSION SYSTEM FOR VEHICLES USING FUZZY REASONING

A. E. KANARACHOS, D. KOULOCHERIS
and TH.GION
Mechanical Design and Control Systems Division
Department of Mechanical Engineering
National Technical University of Athens
Zografou Campus, P.O. Box 64078, 15710 Athens, Greece

1. INTRODUCTION

Advanced suspension systems play a vital role in the performance of modern vehicles. They must support the vehicle body, keep the rider's comfort within permissible allowances, retain the vehicle stability during various handling actions, control body and wheel attitude, and minimize the vertical force variation of the road-to-tire contact. Another trade-off exists between the rider's comfort and safety and the economics of producing advanced suspensions. Presently, advanced suspensions implemented in modern vehicles are often described in confusing and conflicting ways. Therefore a great effort is made nowadays to develop or perfect adaptive or active suspension systems for vehicles [1-3]. These systems, compared to the passive ones [4-8], have a superior performance, but are very expensive, technically very complicated, much less reliable, require regular service and some of them consume non negligible quantities of energy. The semi-active system achieves some performance capabilities of fully active systems with components close to passive ones in terms of cost and complexity. The idea was to employ a spring to support the isolated mass in parallel with an adjustable damper whose force-velocity relationship could be modulated. Investigation of active and semi-active suspensions of ground vehicles in transportation is recently increasing [9-24]. However the semi-active suspensions which are denoted as less expensive alternatives to the active suspensions replace the active force generators by adjustable suspension parts according to the dynamic response of the vehicles.

The scope of this paper is to present an optimised semi-active suspension system for ground vehicles, using fuzzy reasoning. The vehicle system is described by linear differential equations subject to excitations from road surfaces. The fuzzy logic rules are determined such that the maximum value of vertical and rotary acceleration of vehicle body at the passengers seats are minimized from the view point of ride comfort under the constraint on the suspension travels and tyre deflections.

S.G. Tzafestas (ed.), Advances in Intelligent Systems, 289–300.

2. SYSTEM DESCRIPTION

2.1 Notation

damping coefficient	c
frequency	f
stiffness coefficient of springs	k
time	t
acceleration vector	**A**
damping matrix	**C**
vertical displacement vector	**x**
stiffness matrix	**K**
excitation force	**F**
excitation displacement	x_0
mass matrix	**M**
forward velocity	u
velocity vector	**U**

2.2 Vehicle model

In a first approach, we considered a quarter car model Fig.1 (passive), Fig. 2 (semi-active with fuzzy controller) and also a half car model Fig. 3 (passive), Fig. 4 (semi-active with fuzzy controller). The elastic coefficient for tyre is k_1=155900 [N/m] and the damping coefficient is c_1=2500 [Ns/m]. The sprung mass is m_2=290 [kg], the mass inertia J=400[kg.m^2] and the unsprung mass m_1=29 [kg]. In the general case we considered linear as well as non linear springs of the hardening type having a cubic characteristic equation:

$$F = K_{2l}\, x + K_{2nl} x^3$$

The system of differential equations of motion for both linear models can be brought in the usual general form:

$$\mathbf{M} * \mathbf{A} + \mathbf{C} * \mathbf{U} + \mathbf{K} * \mathbf{x} = \mathbf{F}\,(t)$$

The excitation vector **F** resumes the action of the environment on the vehicle.

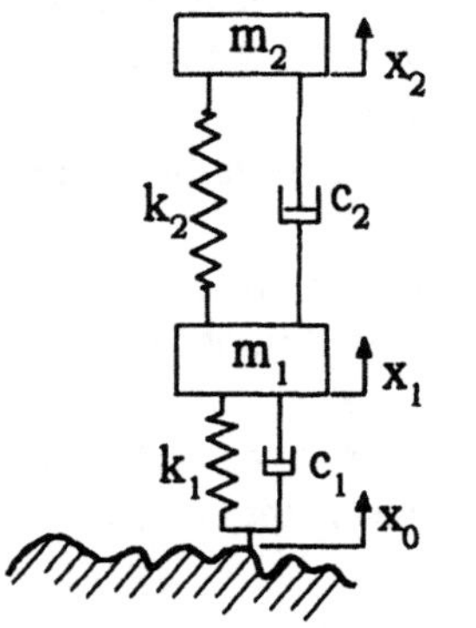

Fig. 1 Quarter car model

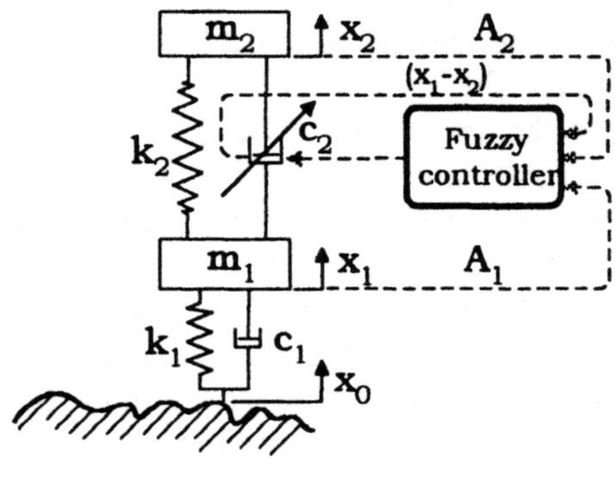

Fig. 2 Quarter car model with fuzzy controller

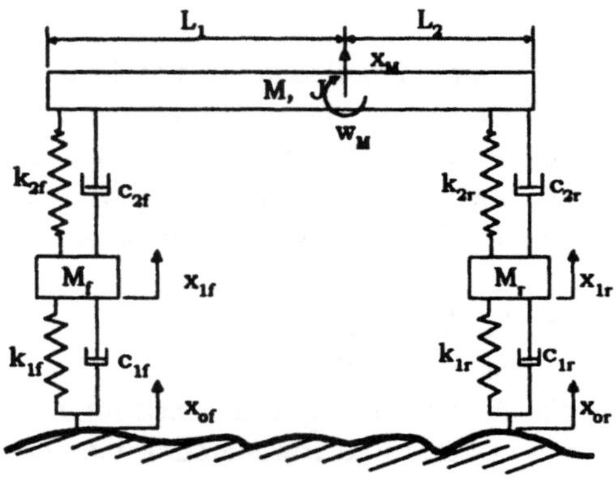

Fig. 3 Half car model

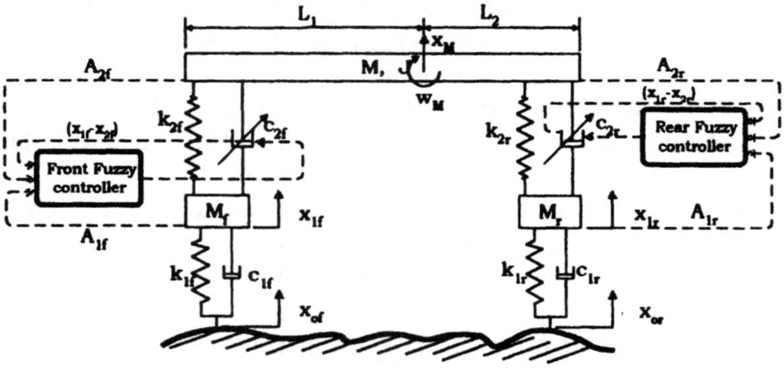

Fig. 4 Half car model with fuzzy controller

2.3 Road excitations

Excitations by road irregularities can be considered, either deterministic or random. In the following, we focused our interest to two of them corresponding to the transition of obstacles: step function excitation and harmonic excitation.

The step function excitation that we considered is:

$h(t) = u * t$	for	$0 < t < 0.06 / u$	[s]
$h(t) = 0.06$	for	$0.06 / u < t < 0.19 / u$	[s]
$h(t) = 0.06 - (u * t - 0.19)$	for	$0.19 / u < t < 0.25 / u$	[s]
$h(t) = 0$	for	$0.25 / u < t$	[s]

The harmonic excitation due to road irregularities is:

$$h_i(t) = 0.02 * [\, 1 - \cos(2 \pi f_i t) \,]$$

which is a cosine shaped bump of height 0.04m (peak to peak), with a wavelength proportional to the forward velocity of the car.

2.4 Solution of equations

The system of differential equations of motion of the model were solved numerically on a PC-Pentium computer. Since it is necessary to solve them at each step of the optimisation process, it is important to select an appropriate and fast solution method. Our preference was for the Runge - Kutta fourth order method with a variable time step. Convergence to the solution did not present any problem.

3. OPTIMISATION STRATEGY

The optimisation problem has been solved by use of the Box method and we determined the optimum elastic and damping characteristics of the passive suspension system. The objective function takes into consideration the peak vertical and rotary acceleration of the sprung mass. Moreover all the design parameters (such as the suspension working space, the elastic characteristics of the springs and the damping characteristics of the dampers) have been limited to acceptable values from the engineering point of view by use of constraints.

Before considering the specific optimisation problem we will resume three very important points which reflect our experience gained in our computation laboratory while dealing with optimisation problems for several years:

1. Optimisation algorithms are mathematical tools that can not take into consideration the physical or engineering aspects of the problem and so they give us optimum solutions that are mathematically correct, but most of the time are of no practical interest to the engineer.
2. Physical or engineering aspects of the problem must be taken into consideration by incorporating appropriate constraints in the formulation of the optimisation problem.
3. Optimisation algorithms may be fouled very easily by the existence of local minima. If we are not aware of this situation, we may finish by considering a local minimum very distant from the absolute minimum as being the optimum solution. This, hopefully, will not be the case if successive uses of the optimisation algorithm from different initial conditions every time render the same optimum.

With these points in mind we may select a suitable optimisation algorithm. Our experience with different algorithms over the past few years is in favour of the Box algorithm, which has the additional advantage of taking directly into consideration explicit and implicit constraints. The optimisation algorithm that we used is the original one developed by Box [25], slightly modified in our computation laboratory after several years of use, in order to achieve faster convergence.

Moreover, we consider important to point out that the solution of a non linear problem always depends on the prehistory of this system. This prehistory can not be known in all cases, because of the random character of excitations imposed by the road irregularities to the vehicle. Consequently, the optimised results obtained by this method will not be representative of the real situation in all cases of excitation.

All the design parameters were constrained to values that do have a meaning and correspond to real and industrially produced springs and dampers.

The minimum and maximum values of the feasible domain for the design parameters in the case of the model are:

Parameter		min. value	max. value
k_{21}	[N/m]	10000	50000
k_{2nl}	[N/m^3]	200000	3000000
c_2	[Ns/m]	1000	5000

Geometrical constraints were applied, regarding the relative displacements of the different masses of the model, in order not to have design incompatibilities in the working space of the suspension:

$$|x_1 - x_2| \leq 0.150 \qquad \text{[m]}$$

4. FUZZY LOGIC CONTROL

A semi-active suspension to be proposed here is realized by changing the damping coefficients of the front and rear dampers.

Figures 2 & 4 show the closed loop systems for the quarter and half car models respectively. The fuzzy controller makes use of three inputs from equal sensors: 1. unsprung mass acceleration A_1, 2. sprung mass acceleration A_2 and 3. suspension travel $x_{12}=(x_1-x_2)$. It generates a fourth one, the change-rate of the suspension travel by discrete derivation of the third input:

$$u_{12}{}^{i} = \frac{x_{12}{}^{i} - x_{12}{}^{i-1}}{t^{i} - t^{i-1}}.$$

The output of the fuzzy controller is the optimal damping coefficient. The damper adjusting valve is modeled as a first order system with transfer function:

$$G(s) = \frac{1}{s+160}$$

Its rise time is approximately $20m\sec$.

Figures 5 & 6 show the membership functions for the four inputs and the output. The abbreviations at the input membership functions stand respectively for : large negative (L-), small negative (S-), very small (VS), small positive (S+) and large positive (L+). The use of two-sided Gaussian membership functions instead of standard trapezoidal ones achieved a smoother response and eliminated a problem of slight unsteadiness in the region of 'small' excitations.

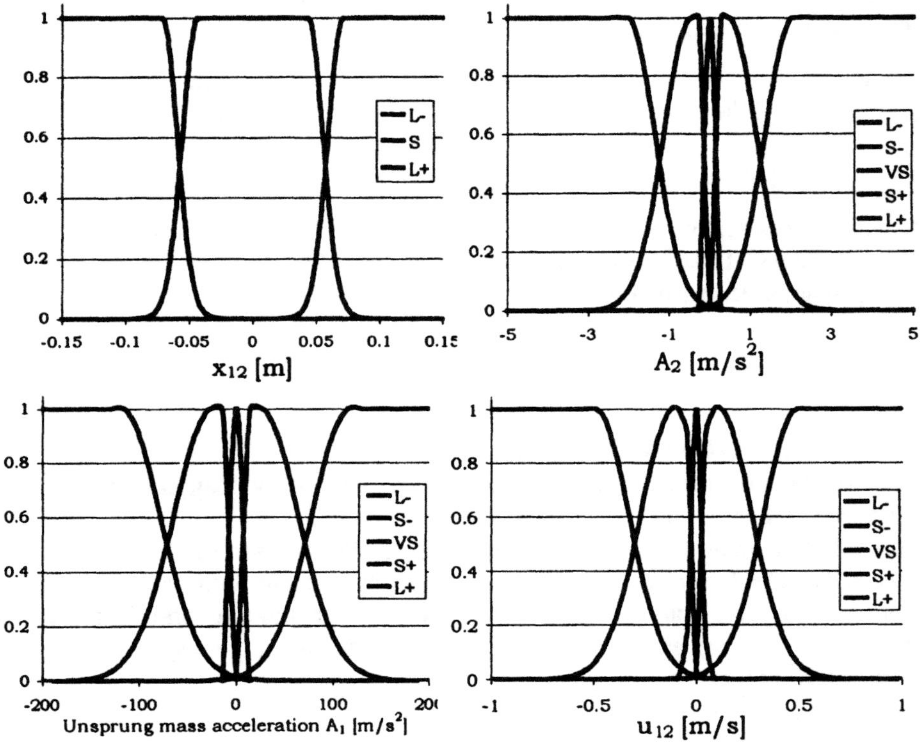

Fig. 5 Input membership functions

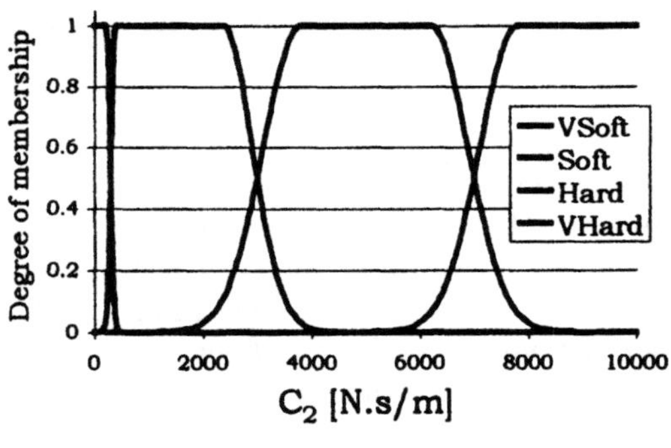

Fig. 6 Output membership functions

The main objective in designing and tuning the fuzzy controller was optimizing passenger comfort i.e. minimizing the sprung mass acceleration and settling time. The goal is to absorb the energy due to the road excitation in the minimum time and with the minimum sprung mass acceleration. The fuzzy rules for the quarter-car controller, the front half–car controller and the rear half-car controller, shown in Table 1, 2 & 3, describe strategies to handle each case.

- When the damping force is of opposite direction to the sprung mass acceleration the damping is '*very hard*' in order to absorb as much energy as possible and reduce the acceleration (rules 1,2).
- When the damping force is of the same direction to the sprung mass acceleration the damping is '*very soft*' in order not to contribute to further acceleration (rules 3,4). Especially for the rear half car model controller the damping is '*soft*' to help the pitching motion settle.
- When the suspension travel is '*large*', the sprung mass acceleration '*very small*' and the change-rate of the suspension travel '*very small*', the elastic forces are large and are about to accelerate the sprung mass. The damping is '*very hard*' to oppose to that before rule 1 or 2 is activated (rule 5).
- When all inputs are '*very small*', the damping is '*very soft*' (rule 6).
- When a hard excitation occurs, i.e. a bump, ramp or step, the unsprung mass excitation is '*large*' before the excitation affects the sprung mass. Then the damping is '*very soft*', in order to feed the least energy to the sprung mass. In the case of an active suspension the wheel would pull up. In the case of the semi-active suspension there is no actuator to pull up the wheel. The appropriate action towards this goal is to cut off the damping force that opposes to the lifting of the unsprung mass. (rule 7).
- When an excitation occurs to another wheel, then the sprung mass acceleration is '*large*' before the unsprung mass is excited. Same principle here, the damping is '*very soft*' to accept the displacement without feeding much energy to the unsprung mass (rule 8).
- When the suspension travel tends to reach its geometrical limit the damping is '*hard*' to brake it down (rules 9, 10).

All the relative weighting factors of the rules were initially set to 1, except the last two rules which could interfere with the 'low sprung mass acceleration' strategy. They were set at 0.1. In order to achieve significant damping of the very small residual motions, the damping for all inputs '*very small*' should be tuned to around 1000N.s/m. This was accomplished by tuning the weighting factor of rule 5 to 0.2.

A_1	A_2	X_{12}	U_{12}	C_2	Weight
	S+ or L+		S- or L-	VHard	1
	S- or L-		S+ or L+	VHard	1
	S+ or L+		S+ or L+	VSoft	1
	S- or L-		S- or L-	VSoft	1
	VS	L+ or L-	VS	VHard	1
VS	VS	S	VS	VSoft	0.2
L+ or L-	VS			VSoft	1
		L-	L-	Hard	0.1
		L+	L+	Hard	0.1

Table 1 : Fuzzy rules for quarter car model

A_1	A_2	X_{12}	U_{12}	C_2	Weight
	S+ or L+		S- or L-	VHard	1
	S- or L-		S+ or L+	VHard	1
	S+ or L+		S+ or L+	VSoft	1
	S- or L-		S- or L-	VSoft	1
	VS	L+ or L-	VS	VHard	1
VS	VS	S	VS	VSoft	0.2
L+ or L-	VS			VSoft	1
VS	L-			VSoft	1
		L-	L-	Hard	0.1
		L+	L+	Hard	0.1

Table 2 : Fuzzy rules for half car model (front)

A_1	A_2	X_{12}	U_{12}	C_2	Weight
	S+ or L+		S- or L-	VHard	1
	S- or L-		S+ or L+	VHard	1
	S+ or L+		S+ or L+	VSoft	1
	S- or L-		S- or L-	Soft	1
	VS	L+ or L-	VS	VHard	1
VS	VS	S	VS	VSoft	0.2
L+ or L-	VS			VSoft	1
VS	L-			VSoft	1
		L-	L-	Hard	0.1
		L+	L+	Hard	0.1

Table 3 : Fuzzy rules for half car model (rear)

5. NUMERICAL RESULTS

In Fig. 7 to 14 we present numerical results to step excitation obtained by considering linear/non-linear springs and linear dampers. In Fig. 15 to 22 we present numerical results to sine excitation. As can be seen, the maximum value of the suspended mass acceleration is neatly improved in the case of the quarter car as well as the half car model with fuzzy, compared to the passive ones. The semi-active models with fuzzy controller achieve better results in all other performance criteria, such as maximum overshoot, settling time, maximum pitching angle and road holding ability.

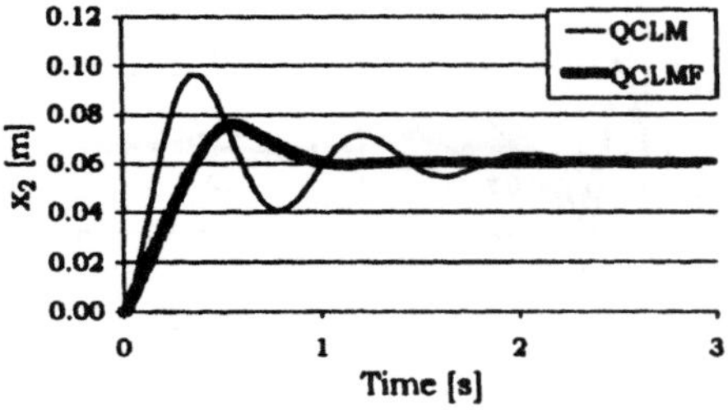

Fig. 7 Sprung mass displacement

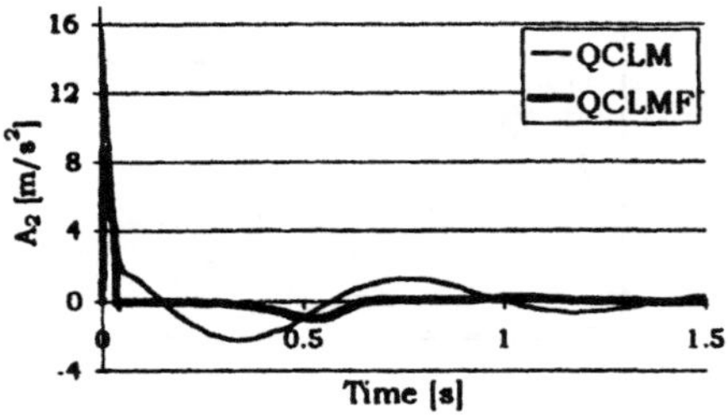

Fig. 8 Sprung mass acceleration

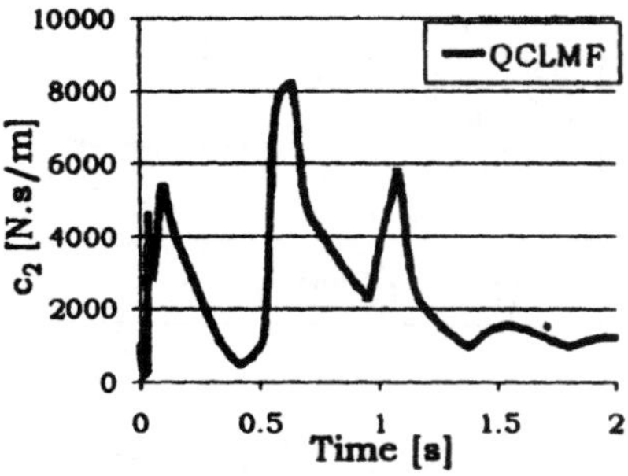

Fig. 9 Damping coefficient

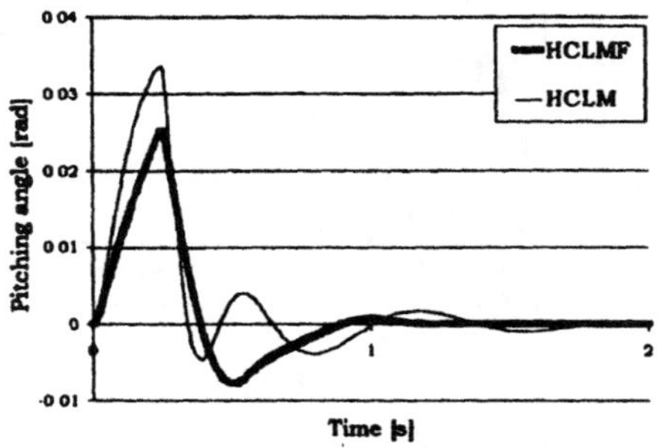

Fig. 10 Sprung mass pitching angle

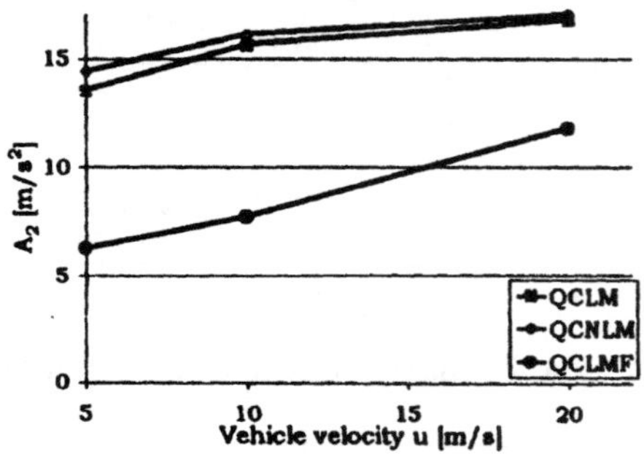

Fig. 11 Max sprung mass acceleration

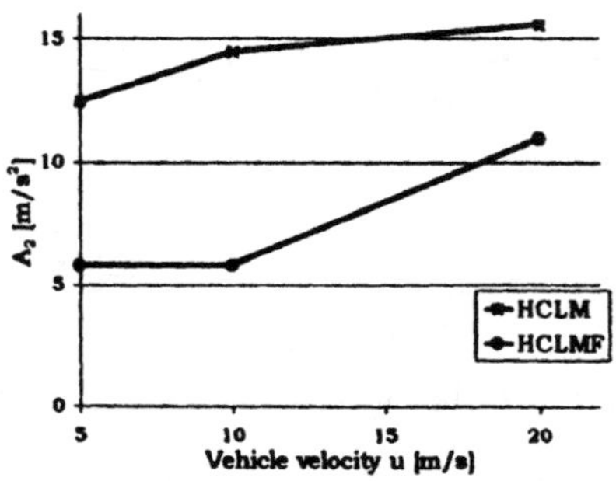

Fig. 12 Max sprung mass acceleration

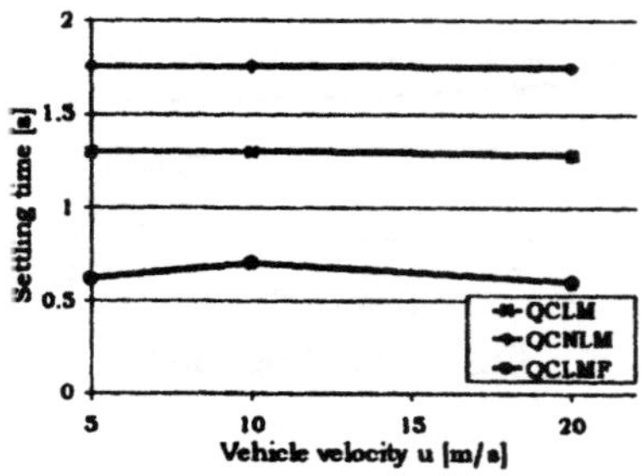

Fig. 13 Settling time

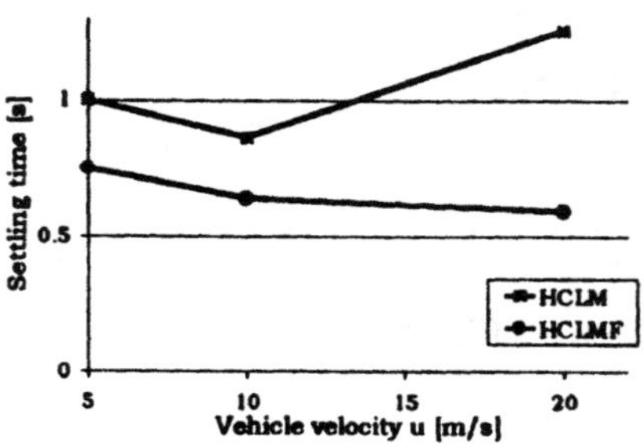

Fig. 14 Settling time

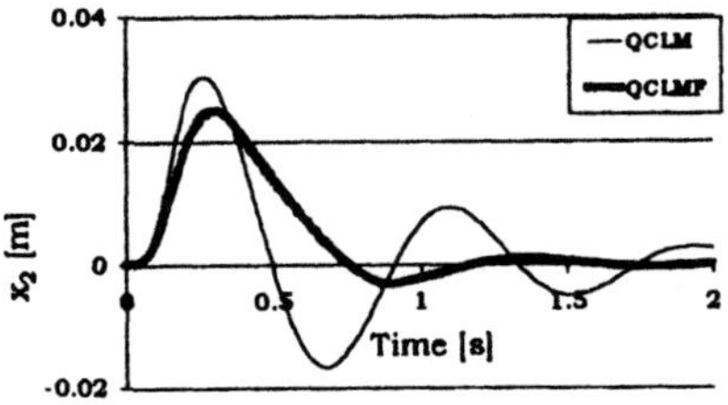

Fig.15 Sprung mass displacement

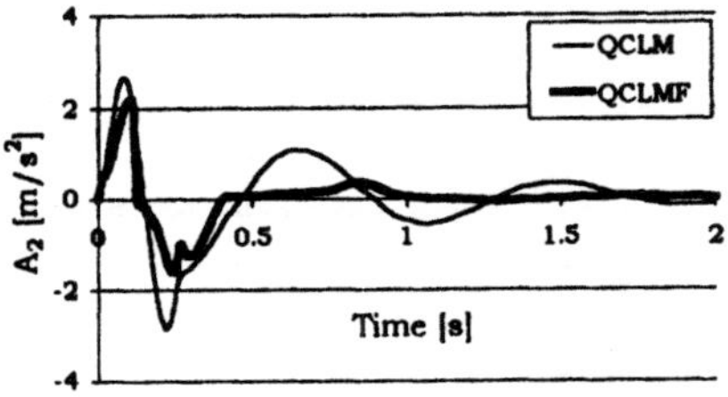

Fig. 16 Sprung mass acceleration

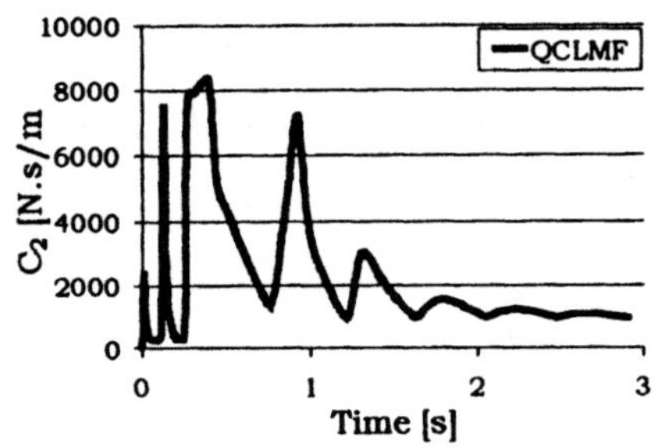

Fig. 17 Damping coefficient

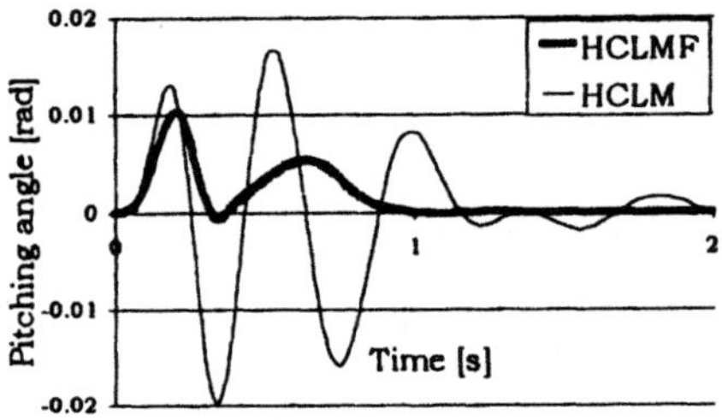

Fig. 18 Sprung mass pitching angle

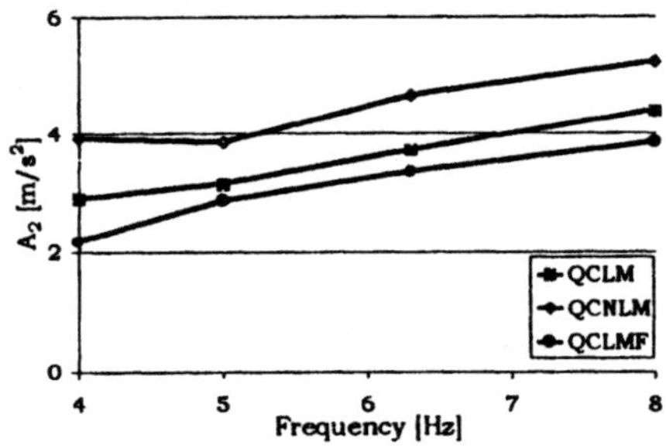

Fig. 19 Max sprung mass acceleration

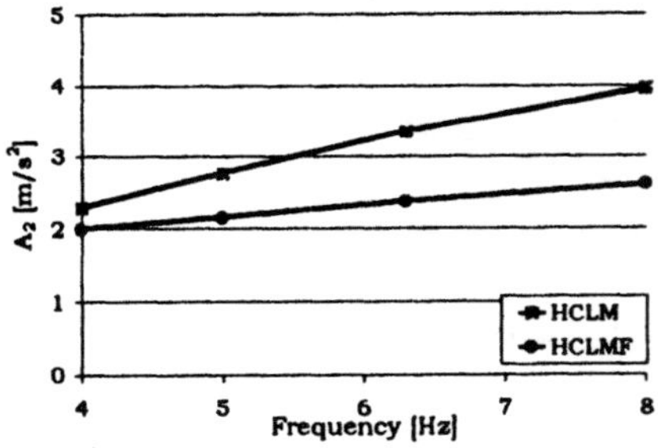

Fig. 20 Max sprung mass acceleration

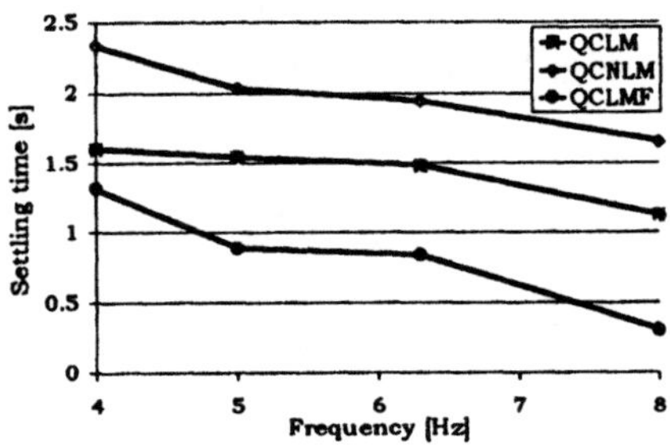

Fig. 21 Settling time

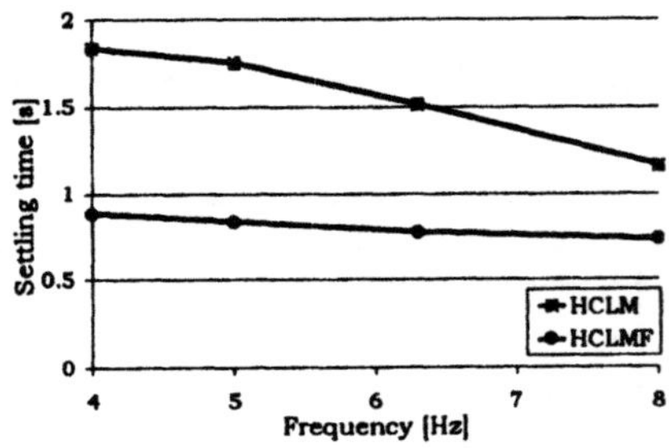

Fig. 22 Settling time

6. CONCLUSIONS

This paper proposed the construction of a semi-active suspension of passenger cars using fuzzy reasoning. The sprung and unsprung mass accelerations, the suspension travel and its time derivative were treated as the input variables and the damping coefficient of the suspension was adjusted as the output variable in the fuzzy control rules at every suspension location.

The performance of this suspension system was evaluated by considering a quarter car model and a half car model and was proved to contribute greatly to the reduction of the peak acceleration of the suspended mass. The simulation results indicate that the proposed semi-active suspension is much improved in vertical and rotary acceleration. A future extension is planned by considering full car model with preview.

7. REFERENCES

1. Tsao, Y.J., Yeh, E.C., A Fuzzy preview control scheme of active suspension for rough road, Int. Journal of Vehicle Design, Vol. 15, Nos 1/2, pp. 166-180.
2. Asami, T., Optimum Design of Dynamic Absorbers for a System Subjected to Random Excitation, JSME series III, Vol. 34, No. 2,1991, pp. 218-226.
3. Yamamoto, M., Active Control Strategy for Improved Handling and Stability, SAE 911902, pp. 21-31.
4. Spentzas, C.N.; Koulocheris, D.B., (1995) OLAF, An Optimised Large Band Filter for Vehicles Passive Suspensions, AUTOTECH 95, Paper No. C498/25/023.
5. Koulocheris, D.B; Spentzas, C.N. An optimal passive suspension system for ground vehicles, 4th Intern. Conference ATA 1997, Paper No.97A3047.
6. Pintado, N.; et al., Optimisation for Vehicle Suspension Part I: Time Domain. Vehicle System Dynamics, 1990, 19, pp. 273-288.
7. Demie, M., Optimisation of characteristics of elasto-damping elements of cars from the aspect of comfort and handling. Int. Journal of Vehicle Design, 1992, vol. 13, no. 1, pp. 29-46.
8. Spentzas, C. N., Optimisation of vehicle ride characteristics by means of the Box method, 1993, vol. 14, nos. 5/6, pp. 539-551.
9. Crawford, I.I., Semi-Active Suspension System, pp. 602-619.
10. Crolla, D.A., Abdel-Hady, M.B.A., Semi-Active Suspension Control for a Full Vehicle Model, SAE 911904, pp. 45-51.
11. Daver, R., et al, Nouvellestrategie de suspension semi-active, pp. 57-62.
12. Hedrick, J.K., Yi, K. Active and Semi-Active Heavy Truck Suspensions to Reduce Pavement Damage, SAE 892486 pp. 29-36.
13. Ivers, D.E., Miller, L.R., Experimental Comparison of Passive, Semi-Active On/Off, and Semi-Active Continuous Suspensions, SAE 892484, pp.1-7.
14. Li, W., Active and Semi-Active Systems for Optimisation of Bogie Vehicle Primary and Secondary Suspensions in the Lateral Plane, Proceedings of 13th IAVSD Symposium, pp. 297-307,1993.

15. Sharp, R.S., et al, Second Generation Approaches to Active and Semi-Active Suspension Control System Design, Proceedings of 13th IAVSD Symposium, pp. 158- 171,1993.
16. Sharp, R.S., Pilbeam, C., On the Ride Comfort Benefits available from Road Preview with Slow-active Car Suspensions, Proceedings of 13th IAVSD Symposium, pp. 437-448,1993.
17. Venhovens, P.J.Th., et al, Semi-Active Control of Vibration and Attitude of Vehicles, Proceedings of 13th IAVSD Symposium, pp. 522-540,1993.
18. Venhovens, R.J.Th., The Development and Implementation of Adaptive Semi-Active Suspension Control, Vehicle System Dynamics, 23 (1994), pp. 211-235.
19. Chou, J.H., et al, Grey-fuzzy control for active suspension design, Int. Journal of Vehicle Design, 1998, vol. 19, No.1, pp. 65-77.
20. Lin, Y. J.; Y. O. Lu; Padovan, J., Fuzzy logic control of vehicle suspension systems. Int. Journal of Vehicle Design, 1993, vol. 14, nos. 5/6, pp. 457-470.
21. Yuech-Jaw LIn, et al, Fuzzy logic control of vehicle suspension systems.
22. Yoshimura T. , et al, A semi-active suspension with dynamic absorbers of ground vehicles using fuzzy reasoning, Int. J. of Vehicle Design, Vol. 18, No. 1,1997.
23. Yoshimura T. , A semi-active suspension of passenger cars using fuzzy reasoning and the field testing, Int. J. of Vehicle Design, Vol. 19, No. 2,1998.
24. Woodard, S.E., Garg, D.P., A numerical optimisation approach for tuning fuzzy logic controllers.
25. Box, M. J., A method of constrained optimisation and comparison with other methods, Computer Journal, 1965, vol. 8, pp. 42-45.

28

FUZZY MARKOV SYSTEMS FOR THE DESCRIPTION AND CONTROL OF POPULATION DYNAMICS

M. A. SYMEONAKI, G.B. STAMOU
and S.G. TZAFESTAS
Intelligent Robotics and Automation Laboratory
Department of Electrical and Computer Engineering
National Technical University of Athens (NTUA)
Zografou Campus, Zografou, 15773, Athens, Greece

1. Introduction

The theory of nonhomogeneous Markov systems (NHMSs) was introduced in Vassiliou [1982] in order to determine a more inclusive background for a number of Markov chain population models (manpower planning, ecology, etc.). A great assortment of applied probability population models can be adapted in this extensive framework. Consider a population which is stratified into classes (states) according to distinguishing properties and let $S = \{1,2,\ldots,n\}$ be the set of states that are assumed to be exclusive and exhaustive, i.e. each member of the system may be in one and only one state at any given time. The state of the system at any given t is represented by the row vector $\mathbf{N}(t) = [N_1(t), N_2(t),\ldots,N_n(t)]$, where $N_i(t)$ is the expected number of members of the system in state i at time t, $T(t)$ denotes the total number of members in the system at time t and $\Delta T(t) = T(t+1) - T(t)$. The sequence of the transition matrices is represented

S.G. Tzafestas (ed.), Advances in Intelligent Systems, 301–310.

by $\{\mathbf{P}(t)\}_{t=0}^{\infty}$ and $\{\mathbf{p}_o(t)\}_{t=0}^{\infty}$, $\{\mathbf{p}_{n+1}(t)\}_{t=0}^{\infty}$ are the sequences of the input and loss probability vectors respectively. It is proved (Vassiliou [1982]) that the sequence of the matrices $\{\mathbf{Q}(t)\}_{t=0}^{\infty}$, where

$$\mathbf{Q}(t) = \mathbf{P}(t) + \mathbf{p}'_{n+1}(t)\mathbf{p}_o(t) = \{q_{ij}(t)\}_{i,j\in S}$$

defines uniquely a nonhomogeneous Markov chain, which is called the embedded nonhomogeneous Markov chain ($(\cdot)'$ denotes the transpose of the respective vector). The (i,j)-element of $\mathbf{Q}(t)$ represents the total transition probability, i.e. the probability that either a member which is in state i moves to state j, or a member which is in state i leaves the system and a new member enters state j, in the time interval $(t-1,t]$. The expected number of members in the various states at time t is given by the following equation:

$$\mathbf{N}(t) = \mathbf{N}(t-1)\mathbf{Q}(t-1) + \Delta T(t-1)\mathbf{p}_o(t-1) \tag{1}$$

One of the basic problems in the theory of NHMS is the uncertainty that is inherented in the estimation of the transition probabilities. More specifically, in order to apply the model mentioned above and estimate the population structure of the system the elements of the matrices $\mathbf{Q}(t)$ and of the vectors $\mathbf{p}_o(t)$ must be determined. This estimation is based on statistical data analysis, whenever enough data is available. The above process obviously introduces uncertainty due to measurement errors and lack of data. In this chapter we study this problem using the theory of fuzzy logic and fuzzy reasoning. A different approach is given in Stanford [1982], where it is assumed that the probabilities $p_{ij}(t)$ lie in the closed interval $[a_i, b_i]$ and that these values are such that $\sum_{j\in S} p_{ij}(t) = 1$.

Whenever this method is applied though, it is extremely difficult to maintain the stochastic property of the matrices $\mathbf{Q}(t)$. Another approach can be found in Symeonaki [1998], where the perturbation theory is applied in the theory of NHMSs.

2. The fuzzy nonhomogeneous Markov system (F-NHMS)

In this section we introduce and define for the first time the concept of a fuzzy nonhomogeneous Markov system (F-NHMS). Consider now a population system. In each of these systems there exists a set of m parameters $(\text{bp1}, \text{bp2}, \ldots, \text{bp}m)$, which we call the basic parameters of the system and which determine the population parameters of the system $(\text{pp1}, \text{pp2}, \ldots, \text{pp}l)$. These population parameters are used in order to estimate the transition probabilities, in the sense that each transition probability $q_{ij}(t)$ is a function of the population parameters of the system:

$$q_{ij}(t) = f_{ij}(\text{pp1}, \text{pp2}, \ldots, \text{pp}\, l)$$

where

$$\sum_{j \in S} f_{ij}(\text{pp1}, \text{pp2}, \ldots, \text{pp}\, l) = 1,$$

for any value of the parameters $\text{pp1}, \text{pp2}, \ldots, \text{pp}l$.

We use a Fuzzy Inference System (FIS) in order to determine the population parameters of the system from the basic parameters of the system (Figure 2.1).

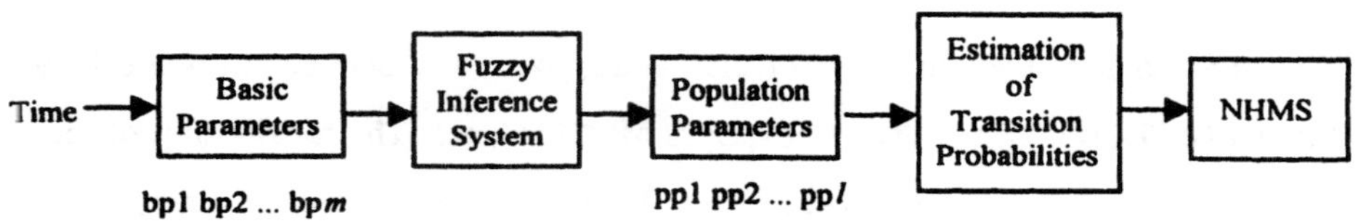

Figure 2.1. The structure of the fuzzy NHMS.

For example, in the case of an ecological population there are two basic parameters *Temperature* and *Pollution* and three population parameters: *Fecundity, F, Mortality, M,* and *Longevity, L.* In this case the transition matrix is given by the well known Leslie matrix (Leslie [1945]):

$$\mathbf{A}(t)=\begin{bmatrix} S_1(t)+F_1(t) & F_2(t) & F_3(t) & \cdots & F_{n-1}(t) & F_n(t) \\ P_1(t) & S_2(t) & 0 & \cdots & 0 & 0 \\ 0 & P_2(t) & S_3(t) & \cdots & 0 & 0 \\ \vdots & \vdots & \vdots & \cdots & \vdots & \vdots \\ 0 & 0 & 0 & \cdots & P_{n-1}(t) & S_n(t)+P_n(t) \end{bmatrix}$$

where $S_i=\left(1-M_i(t)\right)\frac{L_i(t)-1}{L_i(t)}$, $P_i(t)=\frac{1-M_i(t)}{L_i(t)}$ and $i=1,2,\ldots,n$. Notice that each transition probability is a function of the population parameters of the system, which themselves depend upon the values of the basic parameters. This application of fuzzy reasoning in ecological systems has been studied by Stamou [1998].

In this chapter we study the problem of estimating the elements of the transition matrices, using a Fuzzy Inference System (FIS) for each parameter and therefore, taking advantage of the heuristic knowledge that the experts of the system posses.

Assume that the values of the $i-th$ basic parameter of the system range between two values a_i and b_i, i.e. the values of the bp*i* parameter belong to the closed interval $[a_i,b_i]$. We define a fuzzy partition A_i of order d_i on the domain $[a_i,b_i]$. A possible partition is shown in Figure 2.2. A fuzzy partition *B* is also defined on the universe of discourse of the population parameters. The fuzzy partitions A_i and *B* are linguistic representations of their universe of discourses, therefore their elements are linguistic terms like "LOW", "BIG", "QUITE OR LESS MEDIUM", etc. The relationship of the crisp universe of discourses is represented using linguistic rules that define a mapping of the fuzzy partition A_i to the fuzzy partition *B*. This mapping is said to be a *fuzzy association* and represents the empirical, linguistic rules.

As far as the elements of A_i and the elements of *B* have a linguistic meaning, heuristic or empirical linguistic rules can be used in order to describe the input-output relationship. Combining these rules with the well-known generalised modus ponens (GMP) rule of inference [Klir 1996], the multiconditional approximate reasoning schema can be formulated as follows:

<u>RULE 1</u>: IF $(a_1, a_2, ..., a_m)$ IS $(A_1^{(1)}, A_1^{(2)}, ..., A_1^{(d)})$ THEN b IS B_1

<u>RULE 2</u>: IF $(a_1, a_2, ..., a_m)$ IS $(A_2^{(1)}, A_2^{(2)}, ..., A_2^{(d)})$ THEN b IS B_2

....

<u>RULE k</u>: IF $(a_1, a_2, ..., a_m)$ IS $(A_k^{(1)}, A_k^{(2)}, ..., A_k^{(d)})$ THEN b IS B_k

<u>ANTECEDENT</u>: $(a_1, a_2, ..., a_m)$ IS $(A_1, A_2, ..., A_m)$

<u>CONSEQUENSE</u>: b IS B

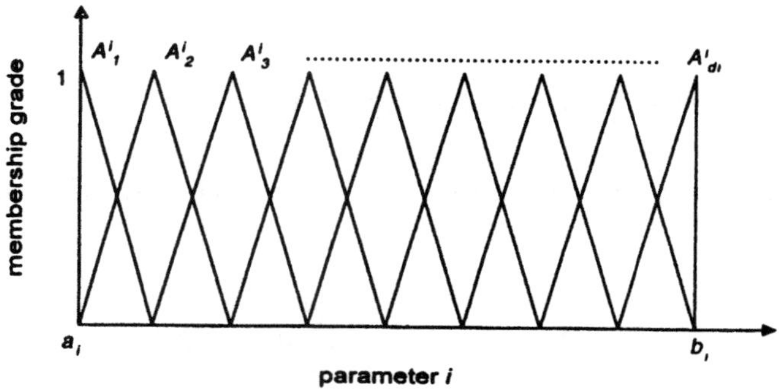

Figure 2.2. A fuzzy partition of the i-th parameter

The number of all different rules in the system is denoted by k and it is obviously given by $k = d_1 d_2 \cdots d_m$. We denote by $w_i(t)$ the degree in which the rule i fires at time t. Each rule corresponds to a matrix $\mathbf{Q}_i$ and it can be proved by induction (the prove is omitted) that if we use as t-norm the *product*, then $\sum_{i=1}^{k} w_i(t) = 1$.

Therefore for each t, the matrix of the embedded nonhomogeneous Markov chain $\{\mathbf{Q}(t)\}_{t=0}^{\infty}$ is of the form:

$$\mathbf{Q}(t) = \sum_{i=1}^{k} w_i(t) \mathbf{Q}_i \tag{2}$$

with

$$\mathbf{Q}_i \mathbf{1}' = \mathbf{1}' \text{ and } \sum_{i=1}^{k} w_i(t) = 1 \text{ for each } t = 0,1,2,...$$

where $\mathbf{1}' = [1,1,\ldots,1]'$.

We assume at this point that the matrices $\mathbf{Q}_i$ $(i = 1,2,\ldots,k)$ are irreducible, regular, stochastic matrices.

Following the same reasoning for the sequence of input probability vectors we have that, for each t the vector $\mathbf{p}_o(t)$ is of the form:

$$\mathbf{p}_o(t) = \sum_{i=1}^{m} u_i(t)\mathbf{p}_{o_i} \tag{3}$$

where

$$\mathbf{p}_{o_i}\mathbf{1}' = \mathbf{1}' \text{ and } \sum_{i=1}^{m} u_i(t) = 1 \text{ for each } t = 0,1,2,\ldots .$$

Therefore, the expected number of members in the various states of the system at time t, is given by:

$$\mathbf{N}(t) = \mathbf{N}(t-1)\sum_{i=1}^{k} w_i(t-1)\mathbf{Q}_i + \Delta T(t-1)\sum_{i=1}^{m} u_i(t-1)\mathbf{p}_{o_i}$$

which recursively can be written in the following form:

$$\mathbf{N}(t) = \mathbf{N}(0)\prod_{\tau=0}^{t-1}\sum_{i=1}^{k} w_i(\tau)\mathbf{Q}_i + \sum_{\tau=1}^{t}\Delta T(\tau-1)\sum_{i=1}^{m} u_i(\tau-1)\mathbf{p}_{o_i}\prod_{j=\tau}^{t-1}\sum_{i=1}^{k} w_i(j)\mathbf{Q}_i \tag{4}$$

We call the population system that we have just described a *fuzzy nonhomogeneous Markov system (F-NHMS).*

3. Asymptotic behaviour of the F-NHMS

As in the theory of Markov chains (homogeneous and nonhomogeneous), the asymptotic behaviour is one of the most significant and engaging problems in the theory of NHMS. The answer to this problem lies in finding the limit of the population structure, $\mathbf{N}(t)$, and the limit of the matrix of variances and covariances of the state population, $\mathbf{V}(t)$, as $t \to \infty$, in a closed analytic form, under conditions, which are well met in real problems. More specifically, the above mentioned problem is converted into finding the equation, which evaluates the population structure, $\mathbf{N}(t)$, as a function of the initial population structure of the system, $\mathbf{N}(0)$ (which is known), and then determining the necessary conditions under which the limit, $\lim_{t\to\infty}\mathbf{N}(t)=\mathbf{N}(\infty)$, exists. The asymptotic behaviour of a NHMS was studied in Vassiliou [1982].

We will now study the asymptotic behaviour of a F-NHMS.

It can easily be proved that $\mathbf{p}_o(t)$ is a stochastic vector, for each t (since $\mathbf{p}_{o_i}\mathbf{1}'=\mathbf{1}'$). Moreover, we prove by employing the method of induction, that the matrix $\mathbf{Q}(t)=\sum_{i=1}^{k}w_i(t)\mathbf{Q}_i$ is an irreducible, regular, stochastic matrix for each t.

We now assume that $\lim_{t\to\infty}w_i(t)=w_i$ and consequently $\lim_{t\to\infty}w_i(t)\mathbf{Q}_i=w_i\mathbf{Q}_i$, which implies that $\lim_{t\to\infty}\mathbf{Q}(t)=\lim_{t\to\infty}\sum_{i=1}^{k}w_i(t)\mathbf{Q}_i=\sum_{i=1}^{k}w_i\mathbf{Q}_i$. Obviously, $\sum_{i=1}^{k}w_i\mathbf{Q}_i$ is also an irreducible, regular, stochastic matrix. Assume, moreover, that $\lim_{t\to\infty}u_i(t)=u_i$ and therefore $\lim_{t\to\infty}\mathbf{p}_o(t)=\sum_{i=1}^{m}u_i\mathbf{p}_{o_i}$, which is a stochastic vector. Then, the asymptotic behaviour of a F-NHMS is given by the following theorem (the proof is omitted).

THEOREM 1.

Let a F-NHMS as defined above and let that $\lim_{t\to\infty}T(t)=T$. Then, the limit of the expected population structure exists and is given in a closed analytic form by the following equation:

$$\lim_{t\to\infty}\mathbf{N}(t)=\mathbf{N}(\infty)=T\mathbf{e}_i\left[\mathbf{I}-\left(\mathbf{I}-\sum_{i=1}^{k}w_i\mathbf{Q}_i\right)\left(\mathbf{I}-\sum_{i=1}^{k}w_i\mathbf{Q}_i\right)^{\#}\right],$$

where $(\cdot)^{\#}$ represents the generalised group inverse (Meyer [1975]) of the respective matrix. ■

The question that rises immediately from the above analysis is, which are the variances and covariances of the state sizes, at any given time t. Following the same reasoning we can easily prove the following theorem concerning the matrix $\mathbf{V}(t)$, which denotes the matrix of the variances and covariances of the state sizes, i.e.

$$\mathbf{V}(t)=\begin{bmatrix} Var(N_1(t)) & \operatorname{cov}(N_1(t),N_2(t)) & \cdots & \operatorname{cov}(N_1(t),N_n(t)) \\ \operatorname{cov}(N_2(t),N_1(t)) & Var(N_2(t)) & \cdots & \operatorname{cov}(N_2(t),N_n(t)) \\ \vdots & \vdots & \vdots & \vdots \\ \operatorname{cov}(N_n(t),N_1(t)) & \operatorname{cov}(N_n(t),N_2(t)) & \cdots & Var(N_n(t)) \end{bmatrix}$$

THEOREM 2.

In a F-NHMS where $\lim_{t\to\infty} T(t)=T$, the limiting matrix $\mathbf{V}(\infty)=\lim_{t\to\infty}\mathbf{V}(t)$, is given by:

$$\mathbf{V}(\infty)=T\left[\mathbf{I}-\left(\mathbf{I}-\sum_{i=1}^{k}w_i\mathbf{Q}_i\right)\left(\mathbf{I}-\sum_{i=1}^{k}w_i\mathbf{Q}_i\right)^{\#}\right]'\left[\mathbf{I}-\left(\mathbf{I}-\sum_{i=1}^{k}w_i\mathbf{Q}_i\right)\left(\mathbf{I}-\sum_{i=1}^{k}w_i\mathbf{Q}_i\right)^{\#}\right] \quad (5)$$

where $(\cdot)'$ represents the transpose of the respective matrix. ■

Furthermore, it can be proved that the rate of convergence in both cases is geometrical. This is an considerable result, since in most real problems, we are not only interested in whether the system converges but also in how fast it converges.

4. Asymptotically attainable structures of the F-NHMS

In this section we study the problem of finding which structures are possible as limiting ones in a F-NHMS, provided that we control the sequence of input probability vectors and more precisely the limit of the sequence, $\{\mathbf{p}_o(t)\}_{t=0}^{\infty}$. In the case of a NHMS the problem was studied in Vassiliou and Georgiou [1990]. It is known (Vassiliou and Georgiou [1990]) that a NHMS has an asymptotically attainable structure $\mathbf{q}^*$ under asymptotic recruitment control, if there exists a sequence of input probabilities vectors $\{\mathbf{p}_o(t)\}_{t=0}^{\infty}$ with $\lim_{t\to\infty}\mathbf{p}_o(t)=\mathbf{p}_o$ such that $\lim_{t\to\infty}\mathbf{q}(t)=\mathbf{q}^*$, where $\mathbf{q}(t)$ is the stochastic vector of the relative population structure given by $\mathbf{q}(t)=\dfrac{\mathbf{N}(t)}{T(t)}$. We denote the set of all asymptotically attainable structures by $\mathbf{A}^{\infty}$.

THEOREM 3.

In a F-NHMS where $\lim_{t\to\infty} T(t)=T_{*}$, the set of all asymptotically attainable structures, $\mathbf{A}^{\infty}$, is the convex hull of the points:

$$\mathbf{z}_i=\lambda_i^{-1}\mathbf{e}_i\left[\mathbf{I}-\sum_{i=1}^{k} w_i\mathbf{P}_i\right]^{-1} \tag{6}$$

where λ_i is the sum of the $i-$th row of the matrix $\left[\mathbf{I}-\sum_{i=1}^{k} w_i\mathbf{P}_i\right]^{-1}$. ■

Therefore, we have proved that all asymptotically attainable structures in a F-NHMS have a specific form and that they are a convex combination of k points (Equation (6)), where k is the number of all different rules in the system. Thus, there is no point in applying a certain asymptotic input control, in order to direct the system to a relative population structure, if this structure does not belong to the set $\mathbf{A}^{\infty}$.

5. Conclusions

In this chapter, we have introduced for the first time the concept of a fuzzy nonhomogeneous Markov system and studied its asymptotic behaviour. More precisely, we have applied fuzzy reasoning to the theory of NHMSs and presented a new method to determine the transition probabilities of the system. The proposed method takes into consideration the heuristic knowledge that the experts of the system posses, without altering the stochastic property of the embedded transition matrices. Further work needs to be done, in order to estimate the fuzzy input control for maintaining the population structure of the F-NHMS between a given upper and lower bound.

References

1. Klir G. J. and B. Yuan, Fuzzy sets and fuzzy logic: theory and applications, *Prentice Hall PTR*, USA, (1995).
2. Leslie, P. H., On the use of matrices in certain population mathematics, *Biometrica*, **33**, 183-212, (1945).
3. Meyer, C. D., The role of the group generalised inverse in the theory of finite Markov chains, *Siam Review*, **17**, 443-464, (1975).
4. Stamou, G. B., Fuzzy reasoning and neural networks in system modelling and decision making: robotic and environmental applications, *PhD thesis*, Athens, (1998).
5. Stanford, R. E., The set of limiting distributions for a Markov chain with fuzzy transition probabilities, *Fuzzy Sets and Systems*, **7**, 71-78, (1982).
6. Symeonaki, M. A., Theory of perturbed Nonhomogeneous Markov Systems, *PhD thesis*, Thessaloniki, (1998).
7. Vassiliou, P.-C. G., Asymptotic behaviour of Markov systems, *J. Appl. Prob.*, **19**, 851-857, (1982).
8. Vassiliou, P.-C. G. and A. C. Georgiou, Asymptotically attainable structures in nonhomogeneous Markov systems, *Oper. Res.*, **3**, 537-545, (1990).

29

CONCERNING HOPFIELD NETWORKS: AN OVERVIEW WITH APPLICATION TO SYSTEM IDENTIFICATION AND CONTROL

S.G TZAFESTAS and D.VOGIATZIS
Intelligent Robotics and Automation Laboratory
Department of Electrical and Computer Engineering
National Technical University of Athens (NTUA)
Zografou Campus, Zografou, 15773, Athens, Greece

1. Introduction

Associative memory is defined to be a kind of a memory which is *content addressable.* More specifically, a series of patterns have been stored somehow, in that memory. These patterns form a partitioning of a multidimentional space or alternatively a set of distinct classes. Then a new pattern, which is a noisy version of a stored one, appears and the memory seeks among the stored patterns the closest match. Numerous are the applications of an associative memory: removing noise from incoming signals, given a set of stored prototypes; or seeking a bibliographic reference given partial information about the title of the book.

Supposing we are concerned with binary patterns, then a kind of associative memory can be implemented by considering the Hamming distance between the input pattern and the stored patterns:

$$\sum_i^N [\xi_i^\mu (1-\zeta_i) + (1-\xi_i^\mu)\zeta_i)]$$

where ζ_i is a dimension of a stored pattern, ξ_i a dimension of the input patterns, $i = 1, 2, \ldots, N$ the units of the network, and $\mu = 1, 2, \ldots, p$ is the number of the stored patterns.

A Hopfield network, which will be defined in the next section, is not only capable of implementing associativity, but also combinatorial optimization. A good example, is the solution of the Traveling Salesman Problem.

Hopfield published in 1982 [9] a seminal paper which introduced his original idea of storing patterns as attractors of a dynamical system. That is the stored patterns are at the bottom of an energy valley, and an input pattern traverses the valley until it reaches a stable state which is a stored pattern.

The basic model employs McCulloch-Pitts neurons, i.e. weighted summation of inputs and a threshold. These neurons are fully interconnected forming a complete bi-directional weighted graph, the weights being a representation of the stored

S.G. Tzafestas (ed.), Advances in Intelligent Systems, 311–334.

patterns. In addition, each neuron has an external input. The operation involves "feeding" the inputs with a specific pattern. Should this pattern be unstable, then the outputs of the neurons are updated either *asynchronously* or *synchronously* until a stable state is reached.

This paper provides an overview of the fundamental ideas and concepts involved in the theory of Hopfield networks (operation, energy function, weight adjustment, spurious states, capacity, Cohen-Grossberg theorem, nonmonotonic transfer functions) along with their applications to search, optimisation, graph bipartitioning and system identification and control problems. The type of results expected from the use of Hopfield networks are illustrated by three examples of robot identification and control.

2. Operating the Hopfield network

Consider N fully interconnected units and their connection weights w_{ij} ($i = 1 \ldots N$, $j = 1 \ldots N$). It is assumed that the units take spin values (i.e ± 1). The operation of the Hopfield network is as follows:

1. Insert a vector S into the inputs of the network, the *ith* dimension of S coincides with the *ith* unit of the network.

2. For each unit compute the weighted sum of input which reaches it, i.e. $\sum_{j=0}^{j=N} w_{ij} S_j$ and look at its sign. If the sign is positive then, update the unit's value to +1; if the sign is negative then update the units value to -1; finally, if the sign is zero then do nothing. The aforementioned operation is summarized in the following equation:

$$S_i(t+1) = sign\left(\sum_{j=1}^{N} w_{ij} S_j(t)\right), \quad i = 1 \ldots N \tag{1}$$

3. If the updated values of the units of the network equal the pre-adaptation values, then the network has reached a stable state.

4. Updating the values of the units can be performed either *synchronously* or *asynchronously*.

3. The energy function

Hopfield's original contribution [9] was to think in terms of energy as far as a fully recurrent network is concerned. More specifically, the network starts from a state of high energy and as the states of the network change the energy is reduced. Thus, in a sense the operation of a Hopfield network is "energy falling" until a local state of minimal energy is reached. It is possible to find such a state since the total number of states for a network of N spin units is 2^N. The form of the energy function is:

$$H = -\frac{1}{2} \sum_{ij} w_{ij} S_i S_j \tag{2}$$

where S_i, S_j represent the state of units i, j of the network. It will be shown that the operation of a Hopfield network results in a non increasing energy change. To this end, two assumptions are needed:

1. The weight strengths are symmetric, i.e. $w_{ij} = w_{ji}$.
2. The self-coupling terms are zero, i.e. $w_{ii} = 0$.

Proof of the energy reduction [1] Suppose that the network's state has changed, then one might say that is due to the change of unit k, where $1 \leq k \leq N$. Therefore, S_k has been turned into $-S_k$. Call this new state S'. Then the energy change is described as follows:

$$\begin{aligned}
\Delta H &= H(S') - H(S) \\
&= -\tfrac{1}{2}\left\{\textstyle\sum_{i,j} w_{ij} S_i' S_j' - w_{ij} S_i S_j\right\} \\
&= -\tfrac{1}{2}\textstyle\sum_{i,j} w_{ij}\{S_i' S_j' - S_i S_j\}
\end{aligned}$$

Now, only those terms in the sum which have either $i = k$ or $j = k$ are non-zero, since for all other values of i we know that $S_i' = S_i \; i \neq k$. Thus, the sum simplifies to

$$\begin{aligned}
\Delta H &= -\tfrac{1}{2}\left\{\textstyle\sum_{i\neq k} w_{ik}(S_i' S_k' - S_i S_k) + \sum_{j\neq k} w_{kj}(S_k' S_j' - S_k S_j)\right\} \\
&= -\textstyle\sum_{i\neq k} w_{ik}(S'_i S'_k - S_i S_k) \\
&= -\textstyle\sum_{i\neq k} w_{ik} S_i (S_k' - S_k) \\
&= -\Delta S_k \textstyle\sum_{i,j} w_{ij} S_i \\
&= -\Delta S_k h_k
\end{aligned}$$

where ΔS_k is $S_k' - S_k$, which equals to $-2S_k$ since the state of unit k was changed. Note that the symmetry of connections in the second simplification above was used in which subscipt j was replaced by i and the order of subscripts on w_{ij} was swapped to make the two sums identical.

Finally, the change in energy resulting from the update is equal to:

$$\Delta H = 2S_k * h_k$$

This ΔH is always negative: the unit was *unstable* so S_k has the opposite sign to h_k and their product is less than zero (note tha h_k is non-zero; this is guaranteed by the definition of stability for an unstable unit). Thus the energy decreases when an unstable unit is changed, $H(s)$ is the desired energy function and one can state the following result:

[1] From the lecture notes on connectionist computing, (University of Edinburgh, Dept of AI)

For a Hopfield Net with symmetric weigths and no-self connection an energy function E(S) exists and the update algorithm for the net decreases the energy; therefore such a net, if updated asynchronously, will always reach a stable state.

It can be shown that if a Hopfield network converges to a stable state is does so after at most N updates, where N is the number of units. Thus it is possible to terminate the update after N units have been updated, and test whether a stable state has been reached: if any unit is unstable at that point, the net will never settle, but will oscillate in some cyclic pattern of state, if further updates are done.

3.1. PROGRAMMING THE NETWORK

The network searches (locally) for a state of minimum energy. Thus, one way to store a pattern into the network is to devise an energy function which has a local minimum at the stored pattern. Let us consider the case of one stored pattern then :

$$H = -\frac{1}{2N}\left(\sum_{i=1}^{i=N} S_i \xi_i\right)^2 \tag{3}$$

where S_i and ξ_i are the ith component of the stored pattern and the input pattern respectively.

In essence, the above equation is the inner product of the stored and the input pattern. Hence, the above formula measures the overlap between these two vectors. Matching equations 2 and 3 if follows that:

$$w_{ij} = \frac{1}{N}\xi_i \xi_j \tag{4}$$

For the many pattern case, one gets:

$$H = -\frac{1}{2N}\sum_{\mu=1}^{p}\left(\sum_{i} S_i \xi_i^{\mu}\right)^2 \tag{5}$$

$$w_{ij} = \frac{1}{N}\sum_{\mu=1}^{\mu=p} \xi_i^{\mu} \xi_j^{\mu} \tag{6}$$

3.2. SPURIOUS STATES

The stored patterns are the attractors of the Hopfield network, however it is possible to have additional attractors, i.e attractors which have not been programmed. The following properties hold:

1. The energy function (equation 2) implies that the stored patterns and their negation have the same energy. Therefore, for any pattern its negation is also a stable state.
2. A linear combination of an *odd* number of stored patterns is a local minimum [1].
3. Finally, a large number of stored patterns creates local minima which cannot be correlated to any of the stored patterns (see [2]).

In cases 2 and 3 the local minima are called ***spurious states***. It should be noted that the basins of attraction for local minima are rather limited in size, thus it is not very easy to reach them.

4. Capacity of the discrete Hopfield network

Considering the Hopfield network as an associative memory, we shall try to estimate its capacity. There are many possible approaches to this; for example, the size of the basins of attraction of the stored memories or the number of the spurious states are certain issues. Here, our attention will be confined to the estimation of the stability properties of stored memories (see [8] for a more complete presentation).

Consider the stability of pattern ξ^ν. By equation 1, the stability condition is

$$\xi_i^\nu = sign\left(\sum_j w_{ij}\xi_j^\nu\right), \quad i = 1\dots N. \tag{7}$$

Then the weights are eliminated by using equation 6

$$\xi_i^\nu = sign\left(\frac{1}{N}\sum_j\sum_\mu \xi_i^\mu\xi_j^\mu\xi_j^\nu\right), \quad i = 1\dots N. \tag{8}$$

and the sum on μ is splitted into the particular term $\mu = \nu$ and all the all the rest terms, as:

$$\begin{aligned}\xi_i^\nu &= sign\left(\xi_i^\nu + \frac{1}{n}\sum_j\sum_{\mu\neq\nu} \xi_i^\mu\xi_j^\mu\xi_j^\nu\right)\\ &= sign(\xi_i^\nu(1 - C_i^\nu)), \quad i = 1\dots N.\end{aligned} \tag{9}$$

where

$$C_i^\nu = -\xi_i^\nu\frac{1}{N}\sum_j\sum_{\mu\neq\nu} \xi_i^\mu\xi_j^\mu\xi_j^\nu. \tag{10}$$

By observing that the function sign is multiplicative and that ξ_i^ν is equal to its own sign and non-zero, equation 9 may be further simplified to

$$sign(1 - C_i^\nu) = 1, \quad i = 1 \ldots N, \tag{11}$$

which in turn is equivalent to

$$C_i^\nu < 1, \quad i = 1 \ldots N. \tag{12}$$

The c_i^ν's just depend on the patterns that we attempt to store. Subsequently we consider independent random patterns, and try to estimate the probability

$$P_{error} = P[C_i^\nu > 1]. \tag{13}$$

From the form of C_i^ν one can observe that P_{error} increases with the number p of patterns that are to be stored. Choosing a criterion for acceptable performance (e.g., $P_{error} < 0.01$) lets us calculate the storage capacity $p_{max} = p_{max}(P_{error})$ of the network, i.e., the maximum number of patterns that can be stored without "unacceptable" error. Actually, there are several different expressions for p_{max}, depending on the type of criterion used for P_{error}.

Let us first calculate P_{error}. It depends on the number of units N and the number of patterns p. It is assumed that both N and p are large compared to 1, because this is typically the case and because it makes mathematics easier. Now C_i^ν is $1/N$ times the sum of about Np independent random numbers, each of which is 1 or -1. From the theory of random coin tosses it has therefore a binomial distribution with mean zero and variance $\sigma^2 = p/N$. But since Np is assumed large this can be approximated by a Gaussian distibution with the same mean and variance.

Thus, the probability P_{error} that C_i^ν exceeds 1 is:

$$\begin{aligned} P_{error} &= \frac{1}{\sqrt{2\pi}\sigma} \int_1^\infty e^{-x^2/2\sigma^2} dx \\ &= \frac{1}{2}\left[1 - erf(\frac{1}{\sqrt{2\sigma^2}})\right] = \frac{1}{2}\left[1 - erf(\sqrt{N/2p})\right] \end{aligned} \tag{14}$$

where the *error function* erf(x) is defined by

$$erf(x) \stackrel{def}{=} \frac{2}{sqrt(\pi)} \int_0^x e^{-u^2} du. \tag{15}$$

Table 1 shows the values of p/N required to obtain the values of P_{error}. Thus if one chooses the criterion $P_{error} < 0.01$ for example, he arrives at $p_{max} = 0.15N$.

This calculation only tells us about the *initial* stability of the patterns. If we choose $p < 0.185N$ for example, it tells us that no more than 1% of the pattern bits will be unstable initially. But if we start the system in a particular patter ξ^ν and about 1% of the bits flip, what happens next? It may be that the first few flips will cause more bits to flip. In the worst case there could be an avalanche phenomenon

Table 1: Capacities

P_{error}	p_{max}/N
0.001	0.105
0.0036	0.138
0.01	0.185
0.05	0.3
0.1	0.61

in which more and more bits flip until the final state bears no resemblance to the original pattern. Therefore the estimates of p_{max} are upper bounds; smaller values may be required to keep the final attractors close to the desired patterns. A more sophisticated calculation [8] shows that an avalanche occurs if $p > 0.138N$, making the whole memory useless. Thus $p_{max} = 0.138N$ if one is willing to accept the errors that occur up to that point. At $p = 0.138N$ table 1 shows that only 0.37% of the bits will be unstable initially, though it turns out that about 1.5% of them flip before a stable attractor is reached.

An alternative definition of the capacity insists that most of the memories be recalled *perfectly*. Since its pattern contains N bits, one needs $P_{error} < 0.01/N$ to get all N bits right with 99% probability[2] This clearly implies $p/N \to 0$ as $N \to \infty$, thus one can write the asymptotic expansion of the error function

$$1 - erf(x) \to e^{-x^2}/\sqrt{\pi}x \quad (as\ x \to \infty) \tag{16}$$

to obtain

$$\log(P_{error} \approx -\log 2 - N/2p - \frac{1}{2}log\pi - \frac{1}{2}\log(N/2p). \tag{17}$$

This turns the condition $P_{error} < 0.01/N$ into

$$-log2 - N/2p - \frac{1}{2}log\pi - \frac{1}{2}\log(N/2p) < \log 0.01 - logN \tag{18}$$

or, taking only the leading terms for large N,

$$N/2p > \log N \tag{19}$$

giving the capacity $p_{max} = N/2\log N$ for this case.

Even more strigently one could ask that all the patterns be recalled perfectly. This requires to get Np bits right with, say, 99% probability, and so needs $P_{error} < 0.01/pN$. It is easy to see that this changes equation 19 to

$$N/2p > \log(Np) \tag{20}$$

which gives $p_{max} = N/4\log N$ because $\log Np \rightsquigarrow \log N^2 = 2logN$ in leading order.

[2]Stricly speaking one should write $(1 - P_{error}^N > 0.99$ here, but $P_{error} < 0.01/N$ is a good approximation from the binomial expansion.

Note that it was assumed in the perfect recall cases that C_i^{ν}'s are independent of one another. Closer examination shows that this is justified. More detailed derivations of the $N/\log N$ results are available in [19, 15].

In summary the capacity p_{max} is proportional to N (but never higher than $0.138N$) if one is willing to accept a small percentage of errors in each pattern, but is is proportional to $N/\log N$ if he insists on most or all the patterns be recalled perfectly.

5. A stochastic extension

Now consider an extension of the update rule for the Hopfield network. The update is stochastic in accordance to the following equations:

$$Prob(S_i = +1) = \frac{1}{1 + exp(-\frac{2*h_i}{T})} \tag{21}$$

$$Prob(S_i = -1) = \frac{1}{1 + exp(+\frac{2*h_i}{T})} \tag{22}$$

where h_i is the weighted sum of all inputs to unit S_i, and T is the analogue of temperature in thermodynamic systems. The higher the temperature, the more time the system spends in higher energy states; if the temperature is lowered, the system looses energy and spends more time in lower energy states. However, high energy states are not impossible for the cool system, just improbable.

If the temperature is very high then $Prob(S_1 = \pm 1) = 1/2$ therefore the states are completely random. If the temperature is lowered, it becomes increasingly improbable for the system to move to higher energy states. Finally, if $T \to 0$ then the update rule reduces to the original Hopfield rule.

The reason for the introduction of the aforementioned extension is the existence of spurious states which are local minima. Usually local minima have high energy, and so the stochastic fluctuations of the system reduce the probability of being trapped in non optimal states.

6. The continuous Hopfiled Network

In this model the activity of the neurons varies continuously in (a suitably large interval of) $\Re$. Let $u_i(t)$ be the activation potential of analogue neuron i, and $x_i(t)$ the corresponding value of the neuron's output. These two quantities are related by

$$x_i = \phi(u_i), \quad i = 1 \ldots N. \tag{23}$$

where ϕ is the sigmoidal nonlinearity of the neurons. Both u_i and x_i are functions of time t. The state of the network may be described in terms of the vector of activation potentials $\mathbf{v(t)} = (\mathbf{u_i(t)})$ or, equivalently, via the vector of output signals $\mathbf{x(t)} = (\mathbf{x_i(t)})$. Here, the first representation will be used because it is better suited to the analysis that follows. The dynamics of the network can be described as a system of differential equations as follows (see [4, 10] for a more general treatment):

$$\tau \frac{du_i}{dt} = -u_i + \sum_{j \neq i} w_{ij}\phi(u_j), \quad i = 1 \ldots N. \tag{24}$$

Note that the diagonal elements of the weight matrix (self coupling weights) do not appear in the above equations, and so it will be safe to assume that they are equal to zero. With this assumption, the differential equations take the following simple vector form:

$$\tau \frac{d\mathbf{v}}{dt} = -\mathbf{v} + \mathbf{W}\dot{\phi}(\mathbf{v}). \tag{25}$$

It is not true that the network is always able to recall one of the stored patterns ξ^μ from an arbitrary initial condition. In fact one of the major disadvantages of this type of network is its particular sensitivity to "memory load": when the number μ of stored patterns exceeds the limit $N/2lnN$, the stored patterns become increasingly unstable and new stable configurations (spurious states) appear. Despite this apparent limitation however, the Hopfield network attracted a lot of interest in the previous decade due to the following powerful convergence theorem.

7. The Cohen-Crossberg Theorem

7.1. THE THEOREM

At about the same time that the Hopfield network was launched, M. A. Cohen and S. Grossberg stated a general result for assessing the stability of a broad class of recurrent networks, described by the following system of differential equations:

$$\frac{d}{dt}u_i = a_i(u_i)\left[b_i(u_i) - \sum_{j=1}^{N} c_{ij}\phi_j(u_j)\right], \quad i = 1 \ldots N. \tag{26}$$

Precicely, they proved the following

Theorem 1 *If the system of differential equations (26) satisfies the conditions:*

1. *The matrix of synaptic weights is symmetric:*

$$c_{ij} = c_{ji} \tag{27}$$

2. *The functions $\phi_i(u_i)$ are monotonic:*

$$\phi'(u_i) = \frac{d}{du_i}\phi_i(u_i) \geq 0, \tag{28}$$

then the function

$$E = \frac{1}{2}\sum_{i=1}^{N}\sum_{j=1}^{N} c_{ij}\phi_i(u_i)\phi_j u_j - \sum_{i=1}^{N}\int_0^{u_i} b_i(\lambda)\phi_i'(\lambda)d\lambda \tag{29}$$

satisfies the condition

$$\frac{dE}{dt} \leq 0 \tag{30}$$

i.e. it is a Lyapunov function of the system, and therefore the system is globally asymptotically stable.

The Cohen-Grossberg theorem is a general neurodynamical result with a wide range of applications.

7.2. RELEVANCE OF THE COHEN-GROSSBERG THEOREM FOR THE HOPFIELD NETWORK

Is is easily seen that the Hopfield network described above falls within the scope of the Cohen-Grossberg theorem by making the following substitutions:

$$\begin{aligned} u_i &= v_i \\ a_i(u_i) &= \tau^{-1} \\ b_i(u_i) &= -u_i \\ c_{ij} &= -w_{ij} \\ \phi_i &= \phi \end{aligned}$$

Hence, the function (29) takes the form

$$E = -\frac{1}{2}\sum_{i=1}^{N}\sum_{j=1}^{N} w_{ij}\phi(u_i)\phi(u_j) + \sum_{i=1}^{N}\int_0^{u_i} \lambda\phi'(\lambda)d\lambda \tag{31}$$

8. Nonmonotonic transfer functions

As mentioned in section 4., the Hopfield network suffers from a very low memory capacity and the existence of spurious states. In order to overcome these limitations, a variety of modifications have been proposed in the last few years. Of particular interest is the nonmonotonic approach proposed recently by several researchers in this area [20, 6, 16]

8.1. THE MORITA NETWORK

The transfer function suggested by Morita [16] is defined as

$$\phi(x) = \left(\frac{1 - exp(-gx)}{1 + exp(-gx)}\right)\left(\frac{1 + \kappa exp(g_0(|x| - h))}{1 + exp(g_0(|x| - h))}\right) \tag{32}$$

What is special about this function is that it is not *monotonic*. The advantages of this novelty are of great importance and are exposed in detail in [20]. First, the network does not recall any spurious states; if it is unable to recall a stored state, then it exhibits chaotic behaviour. Second, the storage capacity of the network of N nodes is $0.3N$, which is better than the $0.138N$,

Morita suggested the following values for the parameters:

$$g = 50 \quad , \quad g_0 = 15,$$
$$h = 0.5, \quad \kappa = -1.$$

8.2. THE PIECEWISE LINEAR MODEL

This model was studied in [20] .It employs a piecewise linear approximation to the transfer function of the Morita network, given by

$$\phi(u) = sign(u) - u/k, \tag{33}$$

where k is a suitably chosen parameter. This choice of function allowed the use of a simple geometrical argument to show that the absolute capacity of the network is equal to $0.4N$ if k is taken equal to the size N of the network.

9. Searching and Optimisation problems

Hopfield nets can be used, amongst other things, to solve optimisation and combinatorial problems such as the the Travelling Salesman problem. The net can't be programmed to solve these problems using the desired attractors: that begs the question. Thus it has to be programmed directly from the energy function of which the maximum is sought.

Consider a function of N variables: $f(x_1, \ldots, x_N)$ for which each x_i can take only two values and no products of more than two x variables occur (i.e. the function is a quadratic form of the x_i). To program a Hopfield net with such a function, the following steps are performed:

1. Write each x_i, using a suitable linear transformation, in terms of the output S_i of each unit i if the Hopfield network.
2. Introduce the transformations into $f(x_1, \ldots, x_N)$ so that it becomes a quadratic form of S_i.

3. Sort out the terms in $f()$ which are linear in some S_i. Make all such terms quadratic by replacing S_i with $S_i S_o$, where S_o is the invented unit with output frozen at +1.

4. Now $f()$ consists of quadratic and constant terms. Ignore all the constant terms. Read off the connection weight w_{ij}, by matching terms with the energy function (see equation 2), as twice the coefficient of $S_i S_j$ in the quadratic form $f()$.

10. System Identification and Control Using Recurrent Neural Networks

10.1. INVERSE MODEL APPROACH

Identifying and building a controller for a nonlinear process can be a daunting task. However, with the aid of neural network technology it is possible to simplify this process [12, 17, 11, 18, 5, 13]. Recently, a most interesting result has been proved (see [12]). That is, once a recurrent network, the Hopfield type being a suitable candidate, which models a nonlinear process has been trained, the Inverse Model (see [7]) of the same network can be used as a controller of the nonlinear process without any additional training being needed. The Inverse model of a network is the network itself assuming a different input; this notion will be formally explained in the following.

Before exposing the method, the reason for choosing Hopfield networks will be explained as reported by the authors:

- The Hopfield network can model non-linear mappings
- The dynamics of the inverse Hopfield network is the same as that of the original network. Therefore no additional training is needed.
- Stability can be easily established.

In section 6. we described the continuous Hopfield network in terms of the activation potentials. Now, for the purpose of control analysis it is better to expose it in terms of vector output signals:

$$\begin{aligned} \dot{x}_i &= -\beta_i x_i + \sum_{j=1}^{n} w_{i,j}\sigma(x_j) + \gamma_i u, \qquad i = 1, 2, \ldots, n \\ y &= x_i \end{aligned} \tag{34}$$

where u is the external input to the network, and x represents its state.

The above equations can be transcribed so as to produce a Single Input Single Output (SISO) affine system. Assuming that:

$$f_i(x) = -\beta_i * x_i + \sum_{j=1}^{n} w_{i,j}\sigma x_j$$

$$g_i(x) = \gamma_i \qquad (35)$$
$$h(x) = x_1$$

the folowing SISO affine system is produced:

$$\dot{x}_i = f_i(x) + g_i(x), \qquad i = 1, 2, \ldots, n \qquad (36)$$
$$y = h(x)$$

The network (36) can be used to model (i.e. identify the dynamics of) a non-linear process. The second, and more revolutionary step is that the inverse of this network can control the non-linear process. The way to achieve this without further training is through *Hirschorn's theorem of Inversion*. This theorem incorporates the concept of *relative order* which must be defined first.

Definition 1 *For the SISO system (36), the relative order (degree) α is defined as the smallest integer α such that $L_g L_f^{a-1} h(x) \neq 0$, where $L_g L_f^{a-1} h(x)$ is the Lie derivative of $h(x)$ in the directions of $f(x)$ and $g(x)$.*

In other words, the relative order a of a SISO system is equal to the number of times that the output $y(t)$ must be differentiated with respect to time in order to have the input $u(t)$ appearing explicitly. For example, if the system (36) has $a = 1$, then

$$\dot{y}(t) = L_f h(x) + [L_g h(x)]u$$

where $L_f h(x) = \sum_{i=1}^{n} \frac{\vartheta h(x)}{\vartheta x_i} f_i(x)$, $L_g h(x) = \sum_{i=1}^{n} \frac{\vartheta h(x)}{\vartheta x_i} g_i$, $L_g h(x) \neq 0$

Theorem 2 (Hirschorn) *For a nonlinear system to be invertible, it must have finite relative order, i.e. $\alpha < \infty$. The left inverse of a nonlinear system is another nonlinear system which, when driven by the appropriate derivative of y, produces u as its output.*

In Hopfield networks the presupposition of Hirschorn's theorem is true, because the relative order is bounded by the network states and the network states are equal to the number of nodes of the network. (this result is formally established in [12], section 4.1, where the weights of the network must satisfy certain conditions). Finaly, the left inverse of the system is

$$\dot{Z} = f(Z) + g(Z)\acute{u} \qquad (37)$$

where

$$\acute{u} = \frac{\frac{d^\alpha y}{dt^\alpha} - L_f^\alpha h(Z)}{L_g L_f^{a-1} h(Z)} \qquad (38)$$

Once a recurrent network has been trained to model (identify) a plant, it can be used for:

- calculating the linearising state feedback control, i.e. the control that cancels the nonlinear terms of the plant and yields any desired linear dynamics.
- to observe in open-loop the plant and generate the state vector x needed in the linearising feedback.

Then it is required to determine whether the trained neural network as an open-loop observer is stable, so that the linearising feedback will surely work properly. Stability of the neural network can be established through Lyapunov's first method and a suitable matrix measure [12, 5]

10.2. HIGH-ORDER RECURRENT NEURAL NETWORKS APPROACH

In a second-order recurrent neural network the total input to the neuron involves linear and quadratic combinations of the inputs. This concept has been extended in [13] by allowing input tripplets, quadruplets, etc, thus resulting in high-order recurrent neural network models (HORNNs).

Introducing an L-dimensional vector z as

$$z = \begin{bmatrix} z_1 \\ z_2 \\ \vdots \\ z_L \end{bmatrix}, \qquad z_i = \prod_{j \in I_i} y_j^{d_j(i)} \tag{39}$$

where y_j is the jth input to the i^{th}neuron, and $d_j(i)$ are nonnegative integers, $\{I_1, I_2, \ldots I_L\}$ is a collection of L not-ordered subsets of $\{1, 2, \ldots, m+n\}$, and

$$y = \begin{bmatrix} y_1 \\ \vdots \\ y_n \\ y_{n+1} \\ \vdots \\ y_{m+n} \end{bmatrix} = \begin{bmatrix} \sigma(x_1) \\ \vdots \\ \sigma(x_n) \\ u_1 \\ \vdots \\ u_m \end{bmatrix}, \qquad u = \begin{bmatrix} u_1 \\ u_2 \\ \vdots \\ u_n \end{bmatrix}$$

where u is the external input vector and $\sigma(x_i) = \lambda_0 \frac{1}{1+e^{-\mu x_i}} - \lambda_1$ is the sigmoid function, the genral HORNN model can be written as:

$$\dot{x}_i = -\beta_i x_i + b_i \sum_{k=1}^{L} w_{i,k} z_k \tag{40}$$

Now, defining the vector θ_i of adjustable weights as:

$$\theta_i = \begin{bmatrix} b_i w_{i1} \\ b_i w_{i2} \\ \vdots \\ b_i w_{iL} \end{bmatrix}, \qquad i = 1, 2, \ldots, n \tag{41}$$

the network model(40) takes the form:

$$\dot{x}_i = -\beta_i x_i + \theta_i^T z \tag{42}$$

The vectors $\theta_i, i = 1, 2, \ldots n$ represent adjustable weights of the network, whereas the coefficients $\beta_i, i = 1, 2, \ldots n$ are part of the underlying network structure and are fixed during training. To ensure bounded-input bounded-output (BIBO) stability of each neuron model, each β_i is assumed to be positive. In vector notation, equation (42) is written as:

$$\dot{x} = Ax + \Theta^T z, \tag{43}$$

where

$$x = \begin{bmatrix} x1 \\ x2 \\ \vdots \\ x_n \end{bmatrix}, \quad \Theta^T = \begin{bmatrix} \theta_1^T \\ \theta_2^T \\ \vdots \\ \theta_n^T \end{bmatrix} \in \Re^{L\times n}, \quad A = \begin{bmatrix} -\beta_1 & & & 0 \\ & -\beta_2 & & \\ & & \ddots & \\ 0 & & & -\beta_n \end{bmatrix}$$

Clearly, the vector z is a function of both the network state and the external input u. Now the following modelling (approximation) theorem holds [13].

Theorem 3 *Consider a nonlinear dynamic plant described by*

$$\dot{\xi} = F(\xi, u) \tag{44}$$

where $\xi \in \Re^n$, $u \in \Re^m$ *and* $F : \Re^{n+m} \longrightarrow \Re^n$ *is a smooth vector field defined on a compact set* $y \subset \Re^{n+m}$, *and assume that it is initially at the same state as the high-order neural model (43), i.e.* $x(0) = \xi(0)$. *Then, for any* $\epsilon > 0$ *and any finite time* $T > 0$, *there exists an integer* L *and a matrix* $\Theta^* \in \Re^{L\times n}$ *such that the HORNN's state* $x(t)$, *with* L *high-order connections and weight values* $\Theta = \Theta^*$ *satisfies the condition:*

$$\sup_{0\leq t\leq T} |x(t) - \xi(t)| \leq \epsilon \tag{45}$$

This theorem says that if a sufficiently large number of high-order connections are allowed in the recurrent network model then it is possible to approximate any dynamical system with any desired accuracy. Particular learning weight adjustment algorithms under the no-modelling-error assumption can be found in [13]. These algorithms can be extended to hold for the higher-order neuron model of the type (26), i.e. the model (26) with $\phi_j(u_j)$ being replaced by z_j (see (39)). The learning algorithms can also be modified so as to relax the "no-modeling error assumption" and obtain *robust learning algorithms*. For this case the following result holds [13]:

Theorem 4 *Consider the HORNN identifier (42) where* θ_i *is the estimate of the unknown optimal weight vector* θ_i^*. *The state vector* $e_i = x_i - \xi_i$ *is described by*

$\dot{e}_i = -\beta_i e_i + \phi_i^T z - v_i$ *where* $\phi_i = \theta_i - \theta_i^*$. *Now, assume that the weights are updated (adjusted) as*

$$\dot{\theta}_i = \begin{cases} -\Gamma_i z e_i & \text{if } |\theta_i| \leq M_i \\ -\Gamma_i z e_i - \sigma_i \Gamma_i \theta_i & \text{if } |\theta_i| > M_i \end{cases} \tag{46}$$

where σ_i *is a positive constant selected by the designer. Then, for* $i = 1, 2, \ldots n$:

1. $e_i, \phi_i \in L_\infty$

2. *there exist constants* λ *and* μ *such that*

$$\int_0^t |e(\tau)|^2 d\tau \leq \lambda + \mu \int_o^t |v(\tau)|^2 d\tau \tag{47}$$

11. Graph Bipartitioning

Graph Bipartitioning is an interesting problem in graph theory. The formal definition is as follows: *Given a graph with an even set of vertices, partition the graph into two sets such as the number of vertices in each set is exactly half of the vertices of the original graph.* Furthermore the connections between the two sets of vertices should be minimal.

In this case we shall consider a connectionist solution of the problem. Let us assume that the number of vertices is N and that C_{ij} denotes the strength of connection between the vertices of the partitioned graphs; i.e. $C_{ij} = 0$ if there is no connection, otherwise $C_{ij} = 1$. Then the problem is to minimize

$$L = -\sum_{ij} C_{ij} S_i S_j$$

where S_i denotes the vertex i of the graph, and $S_i = \pm 1$ depending on the partioned graph it belongs in. Thus:

$$\sum_i S_i = 0$$

The following cost function has the nice property that penalises solutions in which the number of edges in the subgraphs is not equal

$$H = -\sum_{ij} C_{ij} S_i S_j + \mu (\sum_i S_i)^2$$

12. Some Examples

To illustrate some of the reviewed concepts and methods we will give in this section three system identification and control examples.

Example 1: System Identification using a conventional RNN [5]

The system to be identified is a 1-link robot described by (see Fig 1):

$$ml^2 \ddot{\theta}(t) + f\dot{\theta}(t) + mgl \sin\theta(t) = u(t)$$

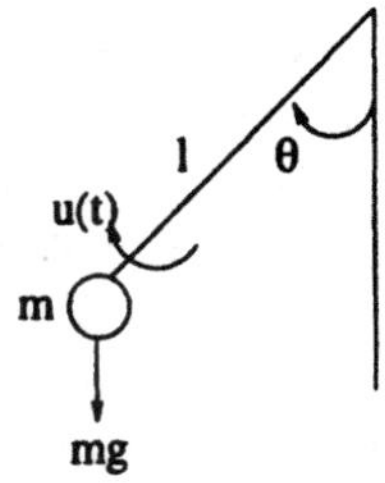

Figure 1: Single link robot

with parameters: $l = 1m, \quad m = 2kg, \quad f = 1kgm^2/sec.$

The state space model of this system is:

$$\begin{aligned} \dot{\xi}_1 &= \xi_1 \\ \dot{\xi}_2 &= -9.8\sin\xi_1 - 0.5\xi_2 + 0.5u \\ y &= \xi_1 \end{aligned} \tag{48}$$

with $\xi_1(0) = \xi_2(0) - 0$ and $y = \theta(t)$. This state-space model was identified with a RNN of the type

$$\begin{aligned} \dot{x}_i &= \hat{f}_i(x) + \hat{g}_i u, \quad i = 1, 2, \ldots N \\ y_{NN} &= \hat{h}(x) \end{aligned} \tag{49}$$

where

$$\begin{aligned} \hat{f}_i(x) &= -x_i + \sum_{j=1}^{N} w_{ij}\sigma(x_j) \\ \hat{g}_i(x) &= \gamma_i, \quad i = 1, 2, \ldots N; \hat{h}(x) = x_1 \end{aligned} \tag{50}$$

The actual neural network used has $N = 5$ neurons and $\sigma(x) = \tanh x$. The training was carried out repetitively over the time interval $[0, t_f]$ so as to minimise the error function:

$$J = \left(\frac{1}{t_f} \int_0^{t_f} e^2(t)dt \right)^{1/2}, \qquad t_f = 20sec$$

where $e(t) = y(t) - y_{NN}(t)$. The training input was random noise and the initial conditions for the neuron states were selected at random. The real parts of the eigenvalues of the linearised trained network lie in the interval [-5.8919, 4.0203]. The exact eigenvalues are: $-0.2260 \pm j3.1483, \quad -2.0288 \pm j0.1798$ and -0.4082. Thus the trained network is stable. To investigate the generalisation capability, the RNN was also fed with different inputs. The value of the error function for these inputs was found to be very low (~ 0.006 *to* 0.007). This means that the trained RNN has a good generalisation capability. The first component of the weight vector is $\gamma_1 \approx 0$ and so the relative degree of the trained network is $a = 2$. The system to be identified has the same relative degree. Thus we see that the RNN is able to identify both the dynamics of the plant and its relative degree.

Example 2: System Identification using a HORNN [13]

This example aims to show the effectiveness of using a HORNN identifier of the form (42) in the case of 2-DOF robotic manipulator with revolute joints, described by:

$$D(\theta(t), p)\ddot{\theta}(t) + C(\theta(t), \dot{\theta}(t), p)\dot{\theta}(t) + G(\theta, t) = \tau(t)$$

where
τ is the $n \times 1$ vector of joint torques
$\theta(t)$ is the $n \times 1$ vector of joint variables
$D(\theta(t), p)$ is the $n \times n$ inertia matrix
$C(\theta, \dot{\theta}, p)\dot{\theta}(t)$ is the vector of Coriolis/centripedal forces.
$G(\theta, t)$ is the vector of gravitational forces.
p is a parameter vector which depends on the robot structure an payload.

The parameter vector p has the following components [14]:

$$\begin{aligned} p_1 &= I_1 + I_2 + I_{3C} + L_3^2 M_2 + L_1^2(M_3 + M_4 + M_p) + p_2 \\ p_2 &= I_3 + I_4 + I_p + L_4^2 M_4 + L_2^2 M_p \\ p_3 &= L_1 L_4 M_4 + L_1 L_2 M_p \end{aligned}$$

where I_1 is the rotor 1 inertia, I_2 the arm 1 inertia about the centre of gravity, I_3 the motor 2 rotor inertia, I_{3C} the motor 2 stator inertia, I_4 the arm 2 inertia about the centre of gravity, I_p the payload ineritia, M_1 the motor 1 mass, M_2 the arm 1 mass, M_3 the motor 2 mass, M_4 the arm 2 mass, M_p the payload mass,L_1 the arm 1 length, L_2 the arm 1 radius of gyration, and L_3 the arm 2 radius of gyration. The parameter values used are the following:

$I_1 = 0.2675$	$I_p = 0.046$	$M_p = 6.81$
$I_2 = 0.360$	$M_1 = 73.0$	$L_1 = 0.36$
$I_3 = 0.0077$	$M_2 = 10.6$	$L_2 = 0.24$
$I_{3C} = 0.040$	$M_3 = 12.0$	$L_3 = 0.139$
$I_4 = 0.051$	$M_4 = 4.85$	$L_4 = 0.099$

The system matrices C and D are:

$$D = \begin{bmatrix} p & 0 & 2p\cos\theta_2 & 0 & p & p\cos\theta_2 \\ 0 & p & 2p\cos\theta_2 & 0 & 0 & 0 \end{bmatrix}$$

$$C = \begin{bmatrix} 0 & 0 & -p\dot{\theta}_2 \sin\theta_2 & 0 & 0 & -p(\dot{\theta}_1 + \dot{\theta}_2)\sin\theta_2 \\ 0 & 0 & -p\dot{\theta}_1 \sin\theta_2 & 0 & 0 & 0 \end{bmatrix}$$

The sampling period used is 0.001 seconds and the learning rate is 0.05. The parameters of the sigmoid function $\sigma(x)$ are: $\lambda_0 = 4$, $\lambda_1 = 2$ $\mu = 0.1$. During the

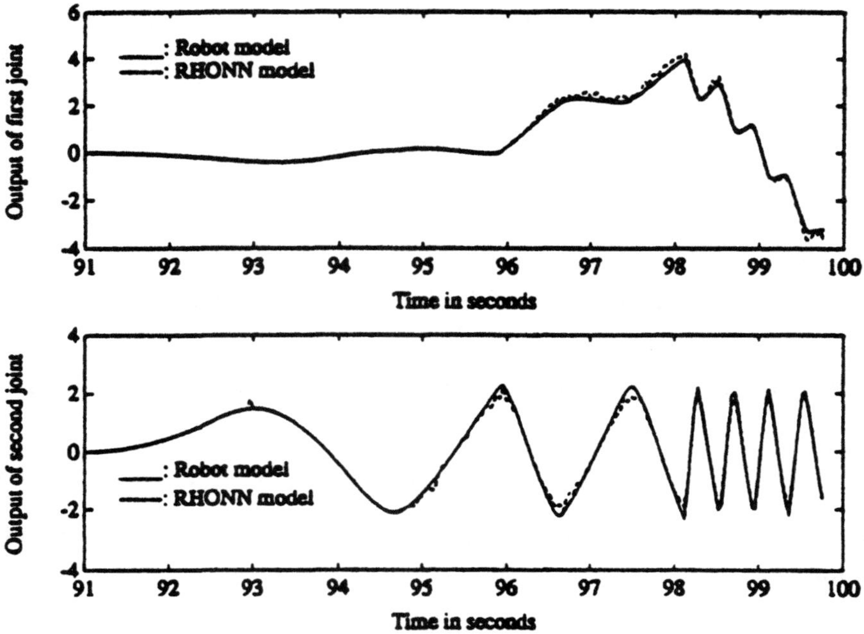

Figure 1: θ_1 and θ_2 trajectories of the robot model and the trained neural network, obtained with the same input patterns.

first 2 seconds of training (i.e. the first 2000 steps) random torques in the interval $[-1,1]$ were used for both joints 1 and 2. For the next 2000 steps the torques $\tau_1 = \sin 0.5t^3$ and $\tau_2 = \cos 0.5t^3$ were used. Finally for the steps 4001 to 6000 the input torques were selected as $\tau_1 = \sin 0.001t^2$ and $\tau_2 = \cos 0.002t^2$. The overall training period was extended up to 96 seconds (i.e. 96000 steps) during which the above three training input patterns were repeatedly applied. After the end of the training period the same input patterns (as for the first 6000 steps) were applied to both the robot model and the neural network.

The resulting trajectories $\theta_1(t)$ and $\theta_2(t)$ are as shown in Fig 2 One can see that after 96 seconds of training the network outputs follow very precisely the outputs of the robot model.

Example 3: System Control Using a Conventional RNN [5]

The 1-link nonlinear robot model of Example 1 will be controlled by an RNN in two stages:

1. Cancellation of plant's nonlinear terms via feedback linearisation

2. Generalised linearised control (GLC) with PI tuning.

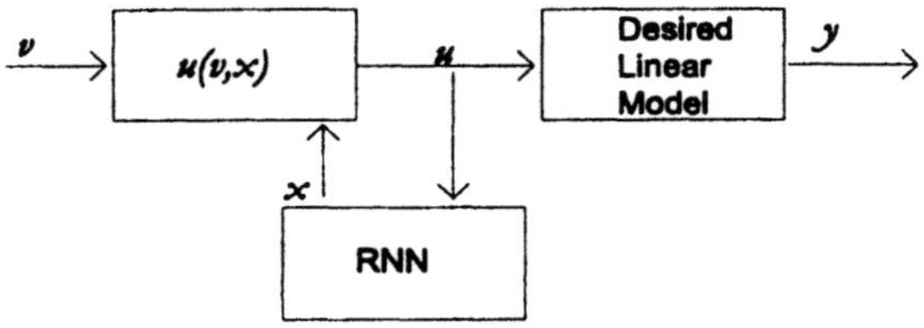

Figure 2: Structure of feedback linearisation via RNN

The desired linear dynamics of the system are selected as $y/v = 1/(s^2 + 5s + 6)$ i.e.

$$\ddot{y}(t) + 5\dot{y}(t) + 6y(t) = v(t) \tag{51}$$

The trained RNN (see Example 1) was used to calculate the linearised state feedback law:

$$u = \frac{v - L_{\hat{f}}^2 \hat{h}(x) - \beta_1 L_{\hat{f}} \hat{h}(x) - \beta_2 \hat{h}(x)}{L_{\hat{g}} L_{\hat{f}} \hat{h}(x)} \triangleq u(v, x)$$

with $\beta_1 = 5$ and $\beta_2 = 6$. The linearisation process has the structure shown in Fig 3

Now, the linearised plant is introduced into the GLC structure (see Fig 3), where the goal is to ensure that the overall closed-loop matches a well damped 2nd-order system:

$$\frac{y_d}{r} = \frac{1}{(s+1)^2} \tag{52}$$

The linear controller is selected of the PI type:

$$v = (K_p + K_i/s)e \tag{53}$$

where $K_p = 1$ and K_i is adjusted.

Figure 5a shows the desired open-loop response (equation (51)) and the linearised plant response for a random noise input. The corresponding value of the error function is $J = 0.0187$.

Figure 5b shows the desired closed-loop response (equation (52)) and the actual response of the GLC structure for the case where $r = 0.5$. The corrsponding value of the error (performance) index $J = (\frac{1}{t_f} \int_0^{t_f} e^2(t)dt)^{1/2}$, $e(t) = y(t) - y_d(t)$ is $J = 0.0121$ with $K_i = 3.8$

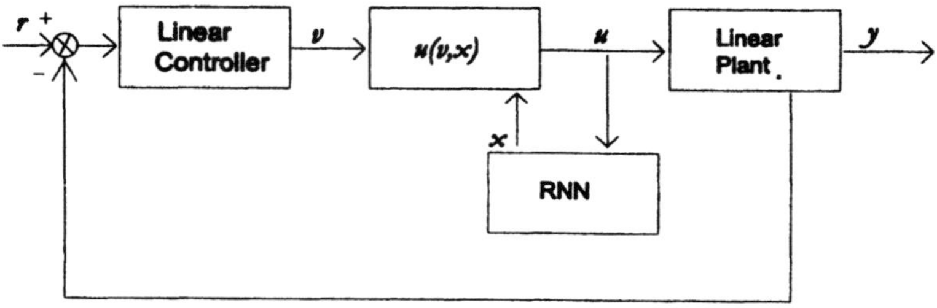

Figure 3: Globally linearising control (GLC) structure

13. Concluding Remarks

Neural networks offer a promising approach to problems of function/mapping approximation, pattern recognition, optimisation and dynamic system identification and control. The nonlinear functional mapping properties of neural networks play a vital role in control. The Hopfield networks belong to the important class of recurrent (feedback) neural networks (RNN) which were first introduced in the context of CAM (Content Addressable Memory) problems for pattern recognition. Several training algorithms have been proposed for the class of recurrent networks which are best studied and designed using Lyapunov's stability theory. However, although the stability (convergence) of the learning scheme(s) is guaranteed, sometimes one can only obtain a local minimum which is not desirable. The main disadvantage of the Hopfield network is that for satisfactory operation it needs a large number of nodes. This paper has reviewed the concepts underlying in the Hopfield neural network model including stability/convergence and estimation of the capacity of the discrete Hopfield network.

Particular attention was given to the problems of identifying (modelling) and controlling complex nonlinear plants (e.g. robotic manipulators). In this respect both the conventional RNNs and the so-called high-order recurrent neural networks (HORNNs) were investigated, their effectiveness been illustrated by simple but representative examples.

The main conclusions in the system identification and control area are:

- the RNN can identify both the relative degree and the dynamics of nonlinear systems.

- after training the RNN is capable of synthesising a linearising state feedback which cancels the plant nonlinearities and produces any desired linear dynamics.

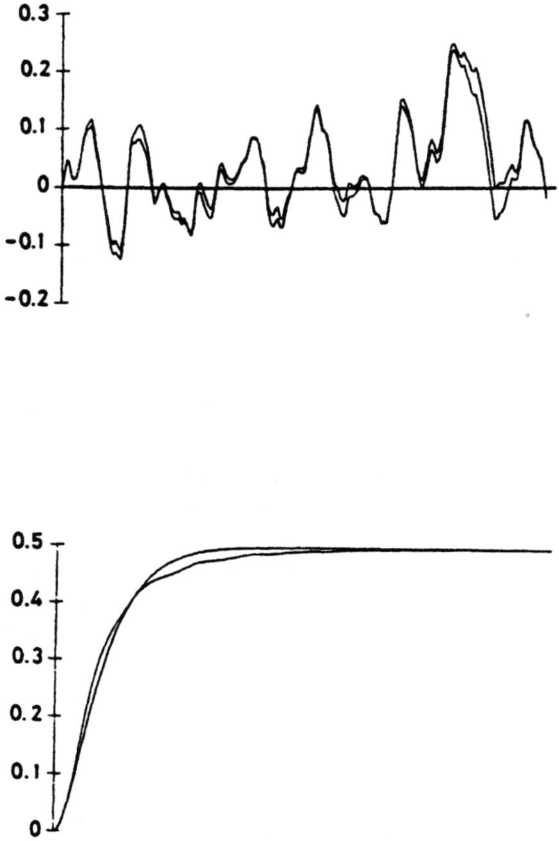

Figure 4: (a) Desired open-loop response versus actual plant response for a random input v
(b) Desired closed-loop response and actual GLC response for a step reference input $r = 0.5$.

- the HORNN can approximate (identify) the input-output dynamics of any nonlineart system with any desired degree of accuracy.
- likewise, the HORNN model can be used for feedback linearisation and control purposes with success.

Due to space limitations many details and proofs were omitted but they can be found in the references and the references cited therein.

References

[1] D. Amit, H. Gutfreund, and H. Sompolinsky. Spin-glass models of neural networks. *Physical Review A*, 32:1007–1018, 1985.

[2] D. Amit, H. Gutfreund, and H. Sompolinsky. Storing infinite numbers of patterns in a spin-glass model of neural networks. *Physical Review Letters*, 55:1530–1533, 1985.

[3] J. A. Anderson and E. Rosenfeld, editors. *Neurocomputing: Foundations of Research*. MIT Press, Cambridge, 1988.

[4] M. A. Cohen and S. Grossberg. Absolute stability of global pattern formation and parallel memory storage by competitive neural networks. *IEEE Transactions on Systems, Man, and Cybernetics*, 13:815–826, 1983.

[5] A. Delgado, C. Kambhampati, and K. Warwick. Dynamic recurrent neural networks for system identification and control. *IEE Proc. - Control Theory Application*, 142(4):307–314, 1995.

[6] T. Fukai, J. Kim, and M. Shiino. Retrieval properties of analog neural networks and the nonmonotonicity of transfer functions. *Neural Networks*, 8(3):391–404, 1995.

[7] CE Garcia and M. Marari. Internal model control. 1. a unifying review and some new results. *Ind. Eng. Chem. Process Des. Dev*, 21:801–806, 1982.

[8] J. A. Hertz, A. Krogh, and R. G. Palmer. *Introduction to the Theory of Neural Computation*. Addison-Wesley Publishing Company, Redwood City, CA, 1991.

[9] J. J. Hopfield. Neural networks and physical systems with emergent collective computational abilities. *Proceedings of the National Academy of Sciences, USA*, 79, 1982. Reprinted in [3].

[10] J. J. Hopfield. Neurons with graded responses have collective computational properties like those of two-state neurons. *Proceedings of the National Academy of Sciences, USA*, 81, 1984. Reprinted in [3].

[11] K.J. Hunt, D. Sbarbaro, R. Zbikowski, and P.J. Gawthrop. Neural networks for control systems: A survey. *Automatica*, 26(6):1083–1112, 1992.

[12] C. Kambhampati, R.J. Craddock, M. Tham, and K. Warwick. Inverse model control using recurrent networks. *Mathematics and Computers in Simulation*, 1999.

[13] E. Kosmatopoulos, M.M. Polycarpou, and M.A.Christodoulou. High-order neural network structures for identification of dynamical systems. *IEEE Transactions on Neural Networks*, 6(2):422–431, 1995.

[14] E.B. Kosmatopoulos, A.G. Chassiakos, and M.A. Christodoulou. Robot identification using dynamical neural networks. In *Proceedings of the 30th IEEE Conference on Decision and Control (CDC'91, Brighton, U.K.*, 1991.

[15] R.J. McEliece, E.C. Psoner, E.R. Rodemich, and S.S. Venkatesh. The capacity of the hopfield associative memory. *IEEE Transactions on Information Theory*, 33:461–482, 1987.

[16] M. Morita and Masahiko. Associative memory with nonmonotonic dynamics. *Neural Networks*, 6(1):115–126, 1992.

[17] J.R. Raol. Parameter estimation of state-space models by recurrent neural networks. *IEE Proc. - Control Theory Applications*, 142(2):114–118, 1995.

[18] S.G. Tzafestas. Neural networks in robot control. In S.G. Tzafestas and H.B. Verbruggen, editors, *Artificial Intelligence in Industrial Decision Making, Control and Automation*, pages 327–387. Kluwer, Dordrecht/Boston, 1995.

[19] C. Weisbuch and F. Fogelamn-Soulie. Scaling laws for the attractors of hopfield networks. *Journal de Physiques Lettres*, 46:623–630, 1985.

[20] S. Yoshizawa, M. Morita, and S. Amari. Capacity of associative memory using a nonmonotonic neuron model. *Neural Networks*, 6:167–176, 1993.

30

COOPERATIVE CONTROL USING MULTIPLE MANIPULATORS WITH DIFFERENT COMMAND INPUTS

K. IZUMI*, K. WATANABE**, T. HIRAO**
and M. MISUMI**

**Department of Mechanical Engineering,*
Faculty of Science and Engineering
Saga University, 1-Honjomachi, Saga 840-8502
Japan

*** Department of Advanced Systems Control Engineering*
Saga University, 1-Honjomachi, Saga 840-8502
Japan

****Department of Mechanical Engineering,*
Graduate School of Science and Engineering
Saga University, 1-Honjomachi, Saga 840-8502
Japan

1. Introduction

A lot of robot manipulators has been applied for the industrial world. An industrial manipulator can achieve a simple task such as welding, painting, pick and place, and grinding, etc. However, multiple manipulators are required to perform more complex tasks than those mentioned in above. There are some advantages due to multiple manipulators are listed below.

1. The handling task of a heavy or big, or flexible object is performed, whereas a single manipulator can not successfully perform such tasks.
2. The complicated task which needs a skill is performed.

S.G. Tzafestas (ed.), Advances in Intelligent Systems, 335–346.

3. The mechanical system constructs a closed-link mechanism, so that the system has a high mechanical stiffness.

Control strategies for multiple manipulators have been proposed up to now. The strategies are classified into the following groups:

1. *Maser/Slave method* [1], [2]: In this method proposed by Nakano [1], each manipulator has a strategy for the trajectory tracking control or for the force control. A manipulator is called the master, which takes charge of the trajectory tracking control. Another manipulator is called the slave, which takes charge of the force control.
2. *Hybrid method* [3]~[7]: We consider to control 6 degree-of-freedom (DOF) object motions and 6-DOF inner forces. The hybrid position and force control is implemented by using $6n$-DOF due to n multiple manipulators.
3. *Compliance/Impedance method* [8]~[11]: In this method, an object is grasped through the compliance or impedance which is realized by a hardware or software. The method is robust against the geometric errors of system.
4. *Dynamic method* [12], [13]: The motion of the manipulators with an object is expressed by the mathematical model. The total system for the manipulator with the object is solved using a dynamic control method.

Conventional cooperative methods have used multiple-manipulators with the same configuration or with the same unit for the control inputs. This paper describes a cooperative system with an industrial manipulator having a force sensor and a pantagraph-type manipulator with 2-DOF. The industrial manipulator is MOVEMASTER EX with 5-DOF made by MITSUBISHI ELECTRIC Co., Ltd. The industrial manipulator has a control box whose command unit is the position. The control box can not accept a next command, unless the manipulator have completed a command. The master/slave method is applied for the cooperative system to handle and carry on an object. The master role is assigned to the pantagraph-type manipulator for tracking the trajectory, whereas the slave role is assigned to the MOVEMASTER for controlling the force. In order to improve the force control, a fuzzy reasoning is used to add the compensative input for the slave manipulator. The effectiveness of the proposed method is illustrated through some experiments.

2. Force Control of Slave Manipulator

The cooperation system and the global coordinate considered here are illustrated by **Fig. 1**, where the slave manipulator is allocated at the left-hand side, and the master manipulator is allocated at the right-hand side. The

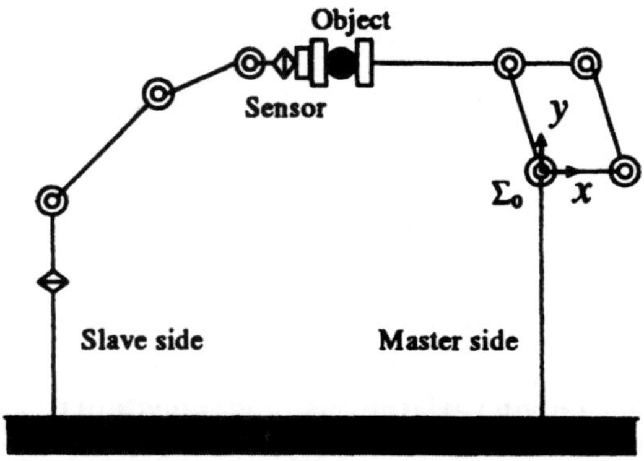

Figure 1. Coordinate of manipulators

control inputs to the slave manipulator are the position vector at x-, y-, and z-coordinates, and the orientation vector consisting of role and pitch angles. The force control is realized by using the position information, so that the control input is the position given by the following relation:

$$\boldsymbol{r}_{sld}(l) = \boldsymbol{r}_{sld}(l-1) + \Delta\boldsymbol{r}_{md}(l) \tag{1}$$

where $l = 1, 2, \cdots$ is the index of discrete time, $\Delta\boldsymbol{r}_{md}(l)$ is the increment of a moving distance during one sampling interval.

The present force control is based on the concept of a stiffness control. It is assumed that the spring constant of the object is given by

$$K = \begin{bmatrix} k_x & 0 & 0 \\ 0 & k_y & 0 \\ 0 & 0 & k_z \end{bmatrix} \tag{2}$$

where k_x, k_y and k_z are the spring constants for each direction. The force is measurable as $\boldsymbol{f}(l) = [f_x(l),\ f_y(l),\ f_z(l)]^T$ by the force sensor, where the reference force is assumed to be constant as $\boldsymbol{f}_d = [f_{xd},\ f_{yd},\ f_{zd}]^T$. Defining $\Delta\boldsymbol{f}(l) = \boldsymbol{f}_d - \boldsymbol{f}(l)$, it results in

$$\Delta\boldsymbol{r}_c(l) = K^{-1}\Delta\boldsymbol{f}(l) \tag{3}$$

where $\Delta\boldsymbol{r}_c(l)$ denotes the compensative input against the force error. Therefore, the final control input is given by

$$\boldsymbol{r}^*_{sld}(l) = \boldsymbol{r}_{sld}(l) + \Delta\boldsymbol{r}_c(l). \tag{4}$$

3. Estimation of Spring Constant Using Fuzzy Reasoning

The spring constant of the object has to be known to produce the compensative input. However, it is difficult to measure the precise spring constant of the object. Therefore, the spring constant of the object is estimated by using a simplified fuzzy reasoning. The spring constant for each direction is assumed to be estimated independently. Consider, for example, the spring constant of x-direction in here. Input variables to the antecedent part are the force error $e_x(l) = f_{xd} - f_x(l)$ and its increment $\Delta e_x(l) = e_x(l) - e_x(l-1)$, and k_x is the consequent variable (usually, constant). Then, the fuzzy rule is expressed by

$$\begin{array}{lll} \mathrm{R}_i: & \mathrm{IF} & e_x(l) = A_{i1} \quad \text{and} \quad \Delta e_x(l) = A_{i2} \\ & \mathrm{THEN} & \Delta k_x(l) = h_i \end{array}$$

where R_i denotes the ith fuzzy rule; A_{ij} is the fuzzy label for j-th input at the i-th rule; h_i is the constant variable at the ith rule. The confidence of the antecedent part at the i-th rule is given by

$$w_i = \mu_{A_{i1}}(e_x(l)) \wedge \mu_{A_{i2}}(\Delta e_x(l)) \tag{5}$$

where $\mu_X(\cdot)$ denotes a confidence of a fuzzy set (or membership function) labeled by X. Therefore, the final fuzzy reasoning output, $\Delta\hat{k}_x$, is given by

$$\Delta\hat{k}_x(l) = \frac{\sum_{i=1}^{n} h_i w_i}{\sum_{i=1}^{n} w_i} \tag{6}$$

where n is the number of fuzzy rules. Finally, the estimated spring constant is obtained by

$$\hat{k}_x(l) = \hat{k}_x(l-1) + \Delta\hat{k}_x(l) \tag{7}$$

The spring constants of other directions, $\hat{k}_y$ and $\hat{k}_z$, are similarly estimated by the same procedure.

Figure 2 illustrates the block-diagram of the control system for the slave manipulator.

4. Experiments

The cooperation system is constructed as **Fig. 3**, where the manipulator allocated at the right-hand side is the master; and the manipulator allocated at the left-hand side is the slave.

4.1. PARAMETER SETTING FOR THE SLAVE MANIPULATOR

The slave manipulator has the restrictions such that the minimum value of position inputs is 0.1 [mm] and the detectable minimum value by the force

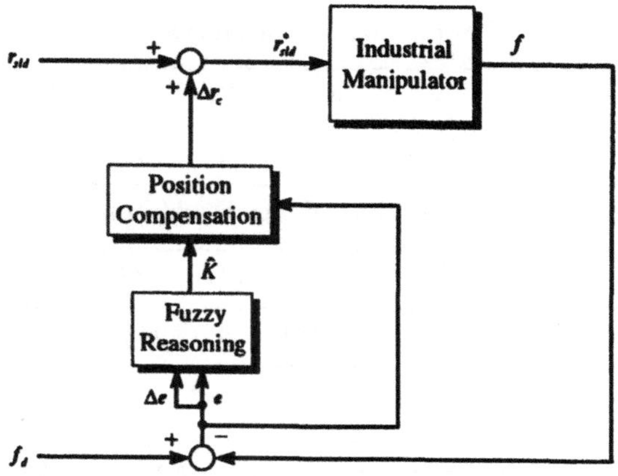

Figure 2. Block diagram of the control system for the slave manipulator

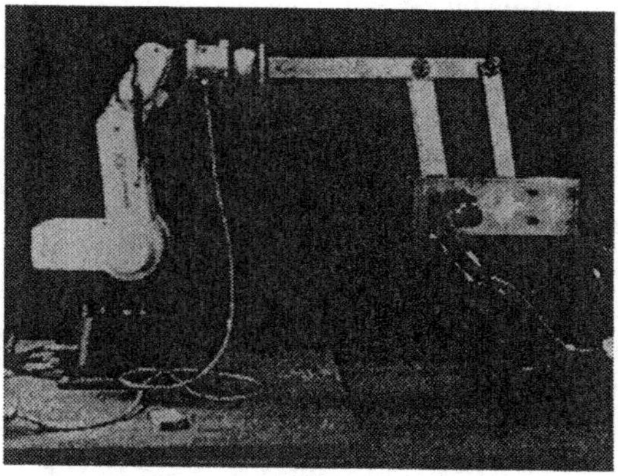

Figure 3. Different kind of manipulators

sensor is 0.098[N], because the measured force is provided by using a unit [kgf].

Experiments are considered for the one-dimensional force control at the x-directional axis, so that the desired force is set to be $f_{xd} = 3.92$ [N] (i.e., 0.4 [kgf]). The initial value is set to be $\hat{k}_x(-1) = 2.0 \times 10^3$ [N/m]. The membership function is the triangle type. The number of membership functions for $e(l)$ is to be 7, and the number of membership functions for $\Delta e(l)$ is to be 7. Such membership functions are allocated as shown in **Table 1** and **Table 2**. The corresponding constant value in the consequent part is allocated as **Table 3**, when the object is considered to be a tennis ball. The sampling interval is to be 0.02 [s]; the total time of an experiment is 120 [s].

TABLE 1. Central values of membership function for $e(l)$

	NB	NM	NS	ZO	PS	PM	PB
α	−0.882	−0.588	−0.294	0	0.294	0.588	0.882

TABLE 2. Central values of membership function for $\Delta e(l)$

	NB	NM	NS	ZO	PS	PM	PB
α	−0.882	−0.588	−0.294	0	0.294	0.588	0.882

4.2. PARAMETER SETTING FOR THE MASTER MANIPULATOR

The strategy of position control for the master manipulator is based on the resolved acceleration control. The equation of motions for the master manipulator is written by

$$\begin{bmatrix} \tau_1 \\ \tau_2 \end{bmatrix} = \begin{bmatrix} J_1 & \beta C_{1-2} \\ \beta C_{1-2} & J_2 \end{bmatrix} \begin{bmatrix} \ddot{\theta}_1 \\ \ddot{\theta}_2 \end{bmatrix} + \begin{bmatrix} 0 & \beta S_{1-2}\dot{\theta}_2 \\ -\beta S_{1-2}\dot{\theta}_1 & 0 \end{bmatrix} \begin{bmatrix} \dot{\theta}_1 \\ \dot{\theta}_2 \end{bmatrix} + \begin{bmatrix} g\,(m_1 S_{c1} + m_3 l_{c3} + m_4 l_1)\, C_1 \\ g\,(m_2 S_{c2} - m_4 l_{c4} + m_3 l_2)\, C_2 \end{bmatrix} \tag{8}$$

where

$$\begin{aligned} J_1 &= m_1 l_{c1}^2 + m_3 l_{c3}^2 + m_4 l_1^2 + I_1 + I_3 \\ J_2 &= m_2 l_{c2}^2 + m_4 l_{c4}^2 + m_3 l_2^2 + I_2 + I_4 \end{aligned}$$

TABLE 3. Constant values in the consequent part

$e_x(l)$	$\Delta e_x(l)$						
	NB	NM	NS	ZO	PS	PM	PB
NB	−0.3	−0.3	−0.3	−0.3	−0.4	−0.5	0
NM	−0.2	−0.2	−0.2	−0.2	−0.3	0	0.2
NS	−0.1	−0.1	−0.1	−0.1	0	0.1	0.1
ZO	0	0	0	0	0	0	0
PS	0.1	0.1	0	−0.1	−0.1	−0.1	−0.1
PM	0.2	0	−0.3	−0.2	−0.2	−0.2	−0.2
PB	0	−0.5	−0.4	−0.3	−0.3	−0.3	−0.3

TABLE 4. Physical parameters for the pantagraph-type manipulator

	Link 1	link 2
mass of link [kg]	5.97×10^{-1}	3.95×10^{-1}
length of link [m]	2.99×10^{-1}	1.45×10^{-1}
length of c.g. [m]	1.41×10^{-1}	7.40×10^{-1}
inertia moment [kgm^2]	8.02×10^{-3}	2.00×10^{-3}

	Link 3	link 4
mass of link [kg]	4.47×10^{-1}	8.78×10^{-1}
length of link [m]	2.99×10^{-1}	5.03×10^{-1}
length of c.g. [m]	1.61×10^{-1}	1.39×10^{-1}
inertia moment [kgm^2]	7.38×10^{-3}	3.18×10^{-2}

$$\begin{aligned}\beta &= m_3 l_{c3} l_2 - m_4 l_{c4} l_1 \\ C_{1-2} &= \cos(\theta_1 - \theta_2) \\ S_{1-2} &= \sin(\theta_1 - \theta_2) \\ C_1 &= \cos(\theta_1)\end{aligned}$$

θ_j, $j = 1, 2$ denote the angles between the link j and x axis; τ_j are the input torques; l_j is the length of the link j; m_i, $i = 1 \sim 4$ is the mass of link i; l_{ci} is the length between the center of gravity of link i and the end of the link i; and I_i is the inertia moment around the center of gravity. Physical parameters of the pantagraph-type manipulator are listed in **Tabele 4**. The proportional gain K_P and differential gain K_D for the resolved acceleration control are defined such as

$$K_P = \begin{bmatrix} 20.25 & 0.0 \\ 0.0 & 20.25 \end{bmatrix}, \quad K_D = \begin{bmatrix} 9.0 & 0.0 \\ 0.0 & 9.0 \end{bmatrix}.$$

The initial position is set to be $[x(0) \quad y(0)] = [-0.358 \quad 0.299]$. The desired values of each direction are generated by

$$x_d(t) = \begin{cases} -0.001t - 0.358 & \text{if } t \leq 60 \text{ [s]} \\ 0.001(t-60) - 0.418 & \text{if } t > 60 \text{ [s]} \end{cases} \tag{9}$$

$$y_d = 0.299 \tag{10}$$

where t denotes the continuous time.

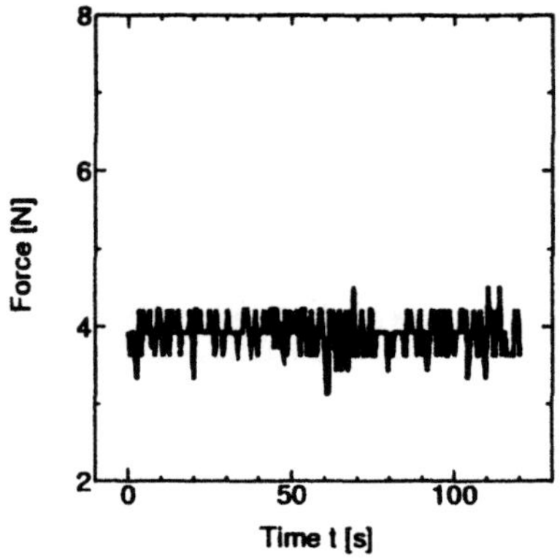

Figure 4. The result of contact force f_x with tennis ball

4.3. RESULTS

Responses of contact force f_x, estimated spring constant $\hat{k}_x$ and the end of link of the slave manipulator are illustrated by **Fig. 4~Fig. 6**. These results are very good because this situation is a learning environment. The contact force has a small oscillation, whose amplitude is 2 or 3 times of the minimum resolution in the force sensor. This result is acceptable, because the manipulators are not rigid completely.

Next, the proposed method is applied to the hard object, which is the hard rubber ball used in a Japanese version of baseball. **Figure 7~Figure 9** illustrate the responses. The contact force has a small oscillation as same as the learning case. It is found in Fig. 8 that the estimated spring constant is higher than one of the learning environment.

Finally, the proposed method is also applied to the soft object which is the ball used in soft-tennis. The resultant responses are illustrated by **Fig. 10~Fig. 12**. The contact force has good results as same as the above cases. The estimated spring constant is smaller than one of the learning environment.

5. Conclusions

The cooperative control with the master/slave method has been described by using the industrial manipulator, whose control input is the position, and the pantagraph-type manipulator, whose control input is the torque. In our system, each manipulator had the different input in unit, and the different structure. The force control of the slave manipulator was realized

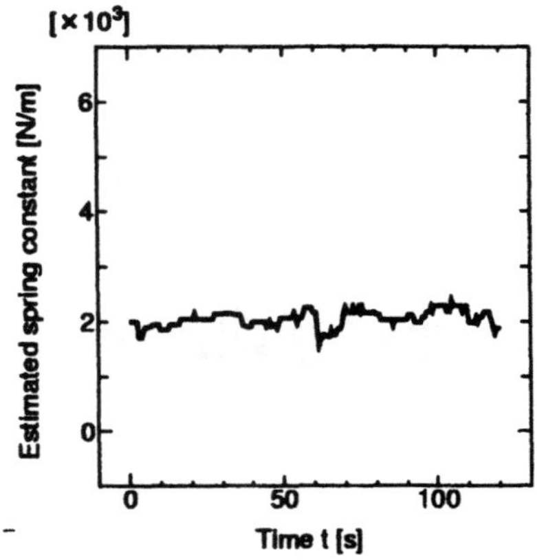

Figure 5. The result of estimated spring constant with tennis ball

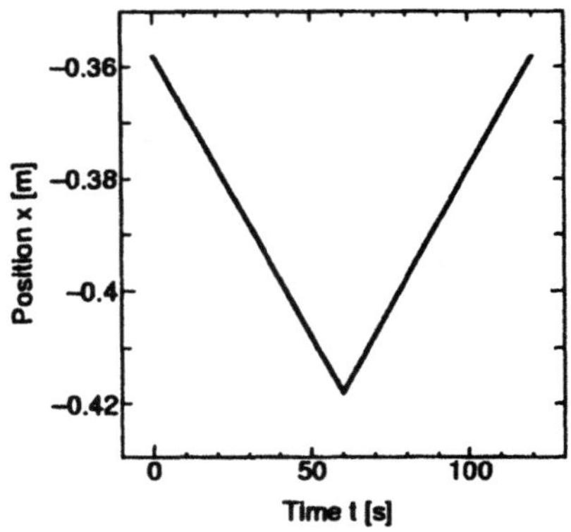

Figure 6. The result of master side position with tennis ball

by the stiffness control using the spring constant estimated by the fuzzy reasoning. The proposed control method was proven to be effective through some experiments.

References

1. E. Nakano, S. Ozaki, T. Ishida, I. Kato, "Cooperational Control of the Anthropomorphous Manipulator 'MELARM'," *Procs. of 4th International Symposium on Industrial Robots*, 1974, pp. 251-260.
2. J.M. Tao, J.Y.S. Luh and Y.F. Zheng, "Compliant Coordination Control of Two

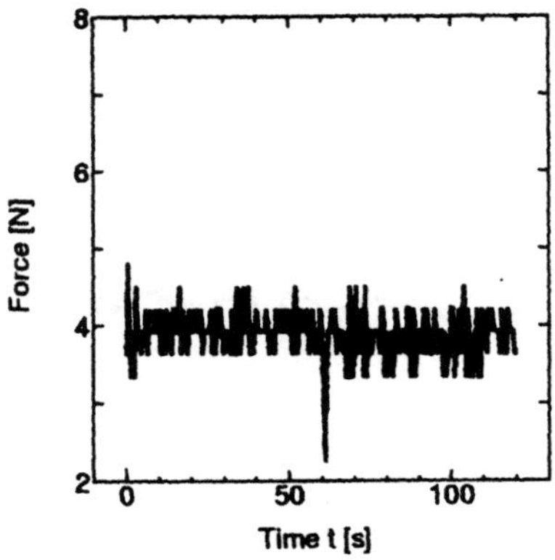

Figure 7. The result of contact force f_x with hard rubber ball

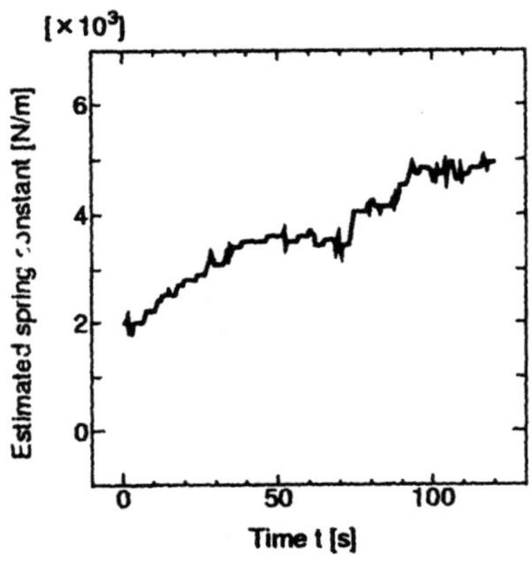

Figure 8. The result of estimated spring constant with hard rubber ball

Moving Industrial Robots," *IEEE Trans. Robotics and Automation*, 1990, Vol. 6, No. 3, pp. 322-330.

3. S. Hayati, "Hybrid Position/Force Control of Multi-Arm Cooperation Robots," *Porcs. of 1986 IEEE Int. Conf. on Robotics and Automation*, 1986, pp. 89-82.
4. T. Yoshikawa and K. Nagai, "Manipulating and Grasping Forces in Manipulation by Multifingered Robot Hands," *IEEE Trans. Robotics and Automation.*, 1991, Vol. 7, No. 1, pp. 67-77.
5. S.A. Cole, P. Hsu and S.S. Sastry, "Dynamic Control of Sliding by Robot Hands for Regrasping," *IEEE Trans. Robotics and Automation*, 1992, Vol. 8, No. 1, pp. 42-52.
6. N. Xi, T.-J. Tarn and A.K. Bejczy, "Intelligent Planning and Control for Multirobot Coordination: An Event-Based Approach," *IEEE Trans. Robotics and Automation*,

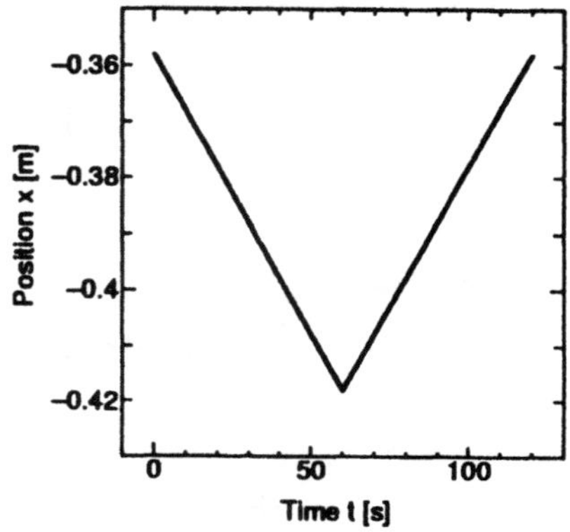

Figure 9. The result of master side position with hard rubber ball

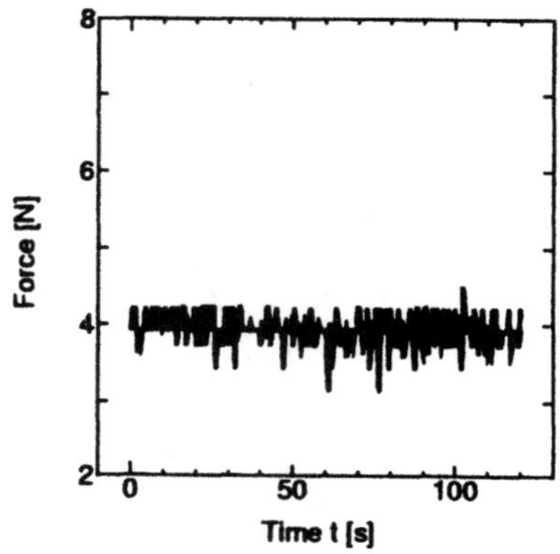

Figure 10. The result of contact force f_x with soft tennis ball

1996, Vol. 12, No. 3, pp. 439-452.
7. M. Zribi, S. Ahmad and S. Luo, "Adaptive Control of Redundant Multiple Robots in Cooperative Motion," *J. of Intelligent and Robotic Systems*, 1996, Vol. 17, No. 2, pp. 169-194.
8. M. Koga, K. Kosuge, K. Furuta and K. Nosaki, "Coordinated Motion Control of Robot Arms Based on the Virtual Internal Model," *IEEE Trans. Robotics and Automation*, 1992, Vol. 8, No. 1, pp. 77-85.
9. S.A. Schneider and R.H. Cannon Jr., "Object Impedance Control for Cooperative Manipulation," *IEEE Trans. Robotics and Automation*, 1992, Vol. 8, No. 3, pp. 383-394.
10. R.G. Bonitz and T.C. Hsia, "Internal Force-Based Impedance Control for Cooper-

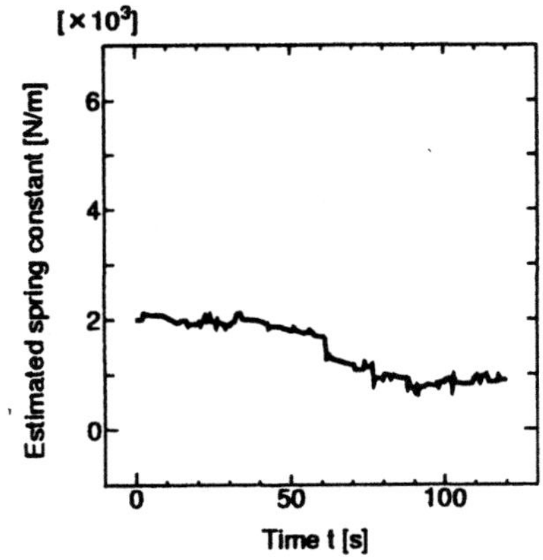

Figure 11. The result of estimated spring constant with soft tennis ball

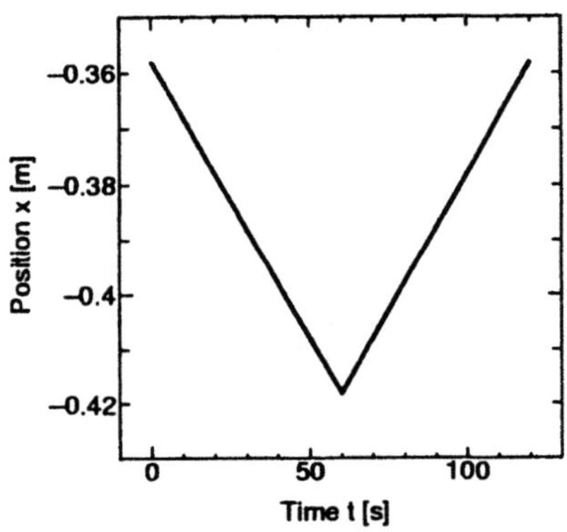

Figure 12. The result of master side position with soft tennis ball

ating Manipulators," *IEEE Trans. Robotics and Automation*, 1996, Vol. 12, No. 1, pp. 78-89.
11. A.-T. Lin and H.-C Tsai, "Impedance Control with On-Line Neural Network Compensator for Dual-Arm Robots," *J. of Intelligent and Robotic Systems*, 1997, Vol. 18, No. 1, pp. 87-104.
12. P. Hsu, "Coordinated Control of Multiple Manipulator," *IEEE Trans. Robotics and Automation*, 1993, Vol. 9, No. 4, pp. 440-409.
13. T. Yoshikawa and X.Z. Zheng, "Coordinated Dynamic Hybrid Position/Force Control for Multiple Robot Manipulators Handling One Constrained Object," *Int. J. of Robotics Research*, 1993, Vol. 12, No. 3, pp. 219-230.

INDEX

International Series on

MICROPROCESSOR-BASED AND INTELLIGENT SYSTEMS ENGINEERING

Editor: Professor S. G. Tzafestas, *National Technical University, Athens, Greece*

1. S.G. Tzafestas (ed.): *Microprocessors in Signal Processing, Measurement and Control.* 1983 ISBN 90-277-1497-5
2. G. Conte and D. Del Corso (eds.): *Multi-Microprocessor Systems for Real-Time Applications.* 1985 ISBN 90-277-2054-1
3. C.J. Georgopoulos: *Interface Fundamentals in Microprocessor-Controlled Systems.* 1985 ISBN 90-277-2127-0
4. N.K. Sinha (ed.): *Microprocessor-Based Control Systems.* 1986 ISBN 90-277-2287-0
5. S.G. Tzafestas and J.K. Pal (eds.): *Real Time Microcomputer Control of Industrial Processes.* 1990 ISBN 0-7923-0779-8
6. S.G. Tzafestas (ed.): *Microprocessors in Robotic and Manufacturing Systems.* 1991 ISBN 0-7923-0780-1
7. N.K. Sinha and G.P. Rao (eds.): *Identification of Continuous-Time Systems.* Methodology and Computer Implementation. 1991 ISBN 0-7923-1336-4
8. G.A. Perdikaris: *Computer Controlled Systems.* Theory and Applications. 1991 ISBN 0-7923-1422-0
9. S.G. Tzafestas (ed.): *Engineering Systems with Intelligence.* Concepts, Tools and Applications. 1991 ISBN 0-7923-1500-6
10. S.G. Tzafestas (ed.): *Robotic Systems.* Advanced Techniques and Applications. 1992 ISBN 0-7923-1749-1
11. S.G. Tzafestas and A.N. Venetsanopoulos (eds.): *Fuzzy Reasoning in Information, Decision and Control Systems.* 1994 ISBN 0-7923-2643-1
12. A.D. Pouliezos and G.S. Stavrakakis: *Real Time Fault Monitoring of Industrial Processes.* 1994 ISBN 0-7923-2737-3
13. S.H. Kim: *Learning and Coordination.* Enhancing Agent Performance through Distributed Decision Making. 1994 ISBN 0-7923-3046-3
14. S.G. Tzafestas and H.B. Verbruggen (eds.): *Artificial Intelligence in Industrial Decision Making, Control and Automation.* 1995 ISBN 0-7923-3320-9
15. Y.-H. Song, A. Johns and R. Aggarwal: *Computational Intelligence Applications to Power Systems.* 1996 ISBN 0-7923-4075-2
16. S.G. Tzafestas (ed.): *Methods and Applications of Intelligent Control.* 1997 ISBN 0-7923-4624-6
17. L.I. Slutski: *Remote Manipulation Systems.* Quality Evaluation and Improvement. 1998 ISBN 0-7932-4822-2
18. S.G. Tzafestas (ed.): *Advances in Intelligent Autonomous Systems.* 1999 ISBN 0-7932-5580-6
19. M. Teshnehlab and K. Watanabe: *Intelligent Control Based on Flexible Neural Networks.* 1999 ISBN 0-7923-5683-7
20. Y.-H. Song (ed.): *Modern Optimisation Techniques in Power Systems.* 1999 ISBN 0-7923-5697-7
21. S.G. Tzafestas (ed.): *Advances in Intelligent Systems.* Concepts, Tools and Applications. 2000 ISBN 0-7923-5966-6
22. S.G. Tzafestas (ed.): *Computational Intelligence in Systems and Control Design and Applications.* 2000 ISBN 0-7923-5993-3

9 781402 003943

GPSR Compliance

The European Union's (EU) General Product Safety Regulation (GPSR) is a set of rules that requires consumer products to be safe and our obligations to ensure this.

If you have any concerns about our products, you can contact us on ProductSafety@springernature.com

In case Publisher is established outside the EU, the EU authorized representative is:

Springer Nature Customer Service Center GmbH
Europaplatz 3
69115 Heidelberg, Germany

Batch number: 10620143

Printed by Printforce, the Netherlands